T.Y. Lin, S. Ohsuga, C.J. Liau, X. Hu, S. Tsumoto (Eds.)

Foundations of Data Mining and Knowledge Discovery

Studies in Computational Intelligence, Volume 6

Editor-in-chief
Prof. Janusz Kacprzyk
Systems Research Institute
Polish Academy of Sciences
ul. Newelska 6
01-447 Warsaw
Poland
E-mail: kacprzyk@ibspan.waw.pl

Further volumes of this series
can be found on our homepage:
springeronline.com

Vol. 1. Tetsuya Hoya
Artificial Mind System – Kernel Memory Approach, 2005
ISBN 3-540-26072-2

Vol. 2. Saman K. Halgamuge, Lipo Wang (Eds.)
Computational Intelligence for Modelling and Prediction, 2005
ISBN 3-540-26071-4

Vol. 3. Bożena Kostek
Perception-Based Data Processing in Acoustics, 2005
ISBN 3-540-25729-2

Vol. 4. Saman Halgamuge, Lipo Wang (Eds.)
Classification and Clustering for Knowledge Discovery, 2005
ISBN 3-540-26073-0

Vol. 5. Da Ruan, Guoqing Chen, Etienne E. Kerre, Geert Wets (Eds.)
Intelligent Data Mining, 2005
ISBN 3-540-26256-3

Vol. 6. Tsau Young Lin, Setsuo Ohsuga, Churn-Jung Liau, Xiaohua Hu, Shusaku Tsumoto (Eds.)
Foundations of Data Mining and Knowledge Discovery, 2005
ISBN 3-540-26257-1

Tsau Young Lin
Setsuo Ohsuga
Churn-Jung Liau
Xiaohua Hu
Shusaku Tsumoto
(Eds.)

Foundations of Data Mining and Knowledge Discovery

Professor Tsau Young Lin
Department of Computer Science
San Jose State University
95192-0103, San Jose, CA
U.S.A.
E-mail: tylin@cs.sjsu.edu

Professor Setsuo Ohsuga
Emeritus Professor of
University of Tokyo
Tokyo
Japan
E-mail: ohsuga@fd.catv.ne.jp

Dr. Churn-Jung Liau
Institute of Information Science
Academia Sinica
128 Academia Road
Sec. II, 115 Taipei
Taiwan
E-mail: liaucj@iis.sinica.edu.tw

Professor Xiaohua Hu
College of Information Science
and Technology
Drexel University
3141 Chestnut Street 19104-2875
Philadelphia
U.S.A.
E-mail: thu@cis.drexel.edu

Professor Shusaku Tsumoto
Department of Medical Informatics
Shimane Medical University
Enyo-cho 89-1, 693-8501
Izumo, Shimane-ken
Japan
E-mail: tsumoto@computer.org

ISSN print edition: 1860-949X
ISSN electronic edition: 1860-9503
ISBN-10 3-642-43228-X Springer Berlin Heidelberg New York
ISBN-13 978-3-642-43228-6 Springer Berlin Heidelberg New York

Springer is a part of Springer Science+Business Media
springeronline.com

Softcover re-print of the Hardcover 1st edition 2005

Typesetting: by the authors and TechBooks using a Springer LaTeX macro package
Printed on acid-free paper SPIN: 11498186 55/TechBooks 5 4 3 2 1 0

Preface

While the notion of knowledge is important in many academic disciplines such as philosophy, psychology, economics, and artificial intelligence, the storage and retrieval of data is the main concern of information science. In modern experimental science, knowledge is usually acquired by observing such data, and the cause-effect or association relationships between attributes of objects are often observable in the data.

However, when the amount of data is large, it is difficult to analyze and extract information or knowledge from it. Data mining is a scientific approach that provides effective tools for extracting knowledge so that, with the aid of computers, the large amount of data stored in databases can be transformed into symbolic knowledge automatically.

Data mining, which is one of the fastest growing fields in computer science, integrates various technologies including database management, statistics, soft computing, and machine learning. We have also seen numerous applications of data mining in medicine, finance, business, information security, and so on. Many data mining techniques, such as association or frequent pattern mining, neural networks, decision trees, inductive logic programming, fuzzy logic, granular computing, and rough sets, have been developed. However, such techniques have been developed, though vigorously, under rather ad hoc and vague concepts. For further development, a close examination of its foundations seems necessary. It is expected that this examination will lead to new directions and novel paradigms.

The study of the foundations of data mining poses a major challenge for the data mining research community. To meet such a challenge, we initiated a preliminary workshop on the foundations of data mining. It was held on May 6, 2002, at the Grand Hotel, Taipei, Taiwan, as part of the 6th Pacific-Asia Conference on Knowledge Discovery and Data Mining (PAKDD-02). This conference is recognized as one of the most important events for KDD researchers in Pacific-Asia area. The proceedings of the workshop were published as a special issue in [1], and the success of the workshop has encouraged us to organize an annual workshop on the foundations of data mining. The workshop, which started in 2002, is held in conjunction with the IEEE International Conference on Data Mining (ICDM). The goal is to bring together individuals interested in the foundational aspects of data mining to foster the exchange of ideas with each other, as well as with more application-oriented researchers.

This volume is a collection of expanded versions of selected papers originally presented at the IEEE ICDM 2002 workshop on the Foundation of Data Mining and Discovery, and represents the state-of-the-art for much of the current research in data mining. Each paper has been carefully peer-reviewed again to ensure journal quality. The following is a brief summary of this volume's contents.

The papers in Part I are concerned with the foundations of data mining and knowledge discovery. There are eight papers in this part.[1] In the paper *Knowledge Discovery as Translation* by S. Ohsuga, discovery is viewed as a translation from non-symbolic to symbolic representation. A quantitative measure is introduced into the syntax of predicate logic, to measure the distance between symbolic and non-symbolic representations quantitatively. This makes translation possible when there is little (or no) difference between some symbolic representation and the given non-symbolic representation. In the paper *Mathematical Foundation of Association Rules-Mining Associations by Solving Integral Linear Inequalities* by T. Y. Lin, the author observes, after examining the foundation, that high frequency expressions of attribute values are the utmost general notion of patterns in association mining. Such patterns, of course, include classical high frequency itemsets (as conjunctions) and high level association rules. Based on this new notion, the author shows that such patterns can be found by solving a finite set of linear inequalities. The results are derived from the key notions of isomorphism and canonical representations of relational tables. In the paper *Comparative Study of Sequential Pattern Mining Models* by H.C. Kum, S. Paulsen, and W. Wang, the problem of mining sequential patterns is examined. In addition, four evaluation criteria are proposed for quantitatively assessing the quality of the mined results from a wide variety of synthetic datasets with varying randomness and noise levels. It is demonstrated that an alternative approximate pattern model based on sequence alignment can better recover the underlying patterns with little confounding information under all examined circumstances, including those where the frequent sequential pattern model fails. The paper *Designing Robust Regression Models* by M. Viswanathan and K. Ramamohanarao presents a study of the preference among competing models from a family of polynomial regressors. It includes an extensive empirical evaluation of five polynomial selection methods. The behavior of these five methods is analyzed with respect to variations in the number of training examples and the level of

[1] There were three keynotes and two plenary talks. S. Smale, S. Ohsuga, L. Xu, H. Tsukimoto and T. Y. Lin. Smale and Tsukimoto's papers are collected in the book Foundation and advances of Data Mining W. Chu and T. Y. Lin (eds).

noise in the data. The paper *A Probabilistic Logic-based Framework for Characterizing Knowledge Discovery in Databases* by Y. Xie and V.V. Raghavan provides a formal logical foundation for data mining based on Bacchus' probability logic. The authors give formal definitions of "pattern" as well as its determiners, which were "previously unknown" and "potentially useful". They also propose a logic induction operator that defines a standard process through which all the potentially useful patterns embedded in the given data can be discovered. The paper *A Careful Look at the Use of Statistical Methodology in Data Mining* by N. Matloff presents a statistical foundation of data mining. The usage of statistics in data mining has typically been vague and informal, or even worse, seriously misleading. This paper seeks to take the first step in remedying this problem by pairing precise mathematical descriptions of some of the concepts in KDD with practical interpretations and implications for specific KDD issues. The paper *Justification and Hypothesis Selection in Data Mining* by T.F. Fan, D.R. Liu, and C.J. Liau presents a precise formulation of Hume's induction problem in rough set-based decision logic and discusses its implications for research in data mining. Because of the justification problem in data mining, a mined rule is nothing more than a hypothesis from a logical viewpoint. Hence, hypothesis selection is of crucial importance for successful data mining applications. In this paper, the hypothesis selection issue is addressed in terms of two data mining contexts. The paper *On Statistical Independence in a Contingency Table* by S. Tsumoto gives a proof showing that statistical independence in a contingency table is a special type of linear independence, where the rank of a given table as a matrix is equal to 1. By relating the result with that in projective geometry, the author suggests that a contingency matrix can be interpreted in a geometrical way.

The papers in Part II are devoted to methods of data mining. There are nine papers in this category. The paper *A Comparative Investigation on Model Selection in Binary Factor Analysis* by Y. An, X. Hu, and L. Xu presents methods of binary factor analysis based on the framework of Bayesian Ying-Yang (BYY) harmony learning. They investigate the BYY criterion and BYY harmony learning with automatic model selection (BYY-AUTO) in comparison with typical existing criteria. Experiments have shown that the methods are either comparable with, or better than, the previous best results. The paper *Extraction of Generalized Rules with Automated Attribute Abstraction* by Y. Shidara, M. Kudo, and A. Nakamura proposes a novel method for mining generalized rules with high support and confidence. Using the method, generalized rules can be obtained in which the abstraction of attribute values is implicitly carried out without the requirement of additional information, such as information on conceptual hierarchies. The paper *Decision Making Based on Hybrid of Multi-knowledge and Naïve Bayes Classifier* by Q. Wu et al. presents a hybrid approach to making decisions for unseen instances, or for instances with missing attribute values. In this approach, uncertain rules are introduced to represent multi-knowledge. The experimental results show that the decision accuracies for unseen instances are higher than those obtained by using other approaches in a single body of knowledge. The paper *First-Order Logic Based Formalism for Temporal Data Mining* by P. Cotofrei and K. Stoffel presents a formalism for a methodology whose purpose is the discovery of knowledge, represented in the form of general Horn clauses, inferred from databases with a temporal dimension. The paper offers the possibility of using statistical approaches in the design of algorithms for inferring higher order temporal rules, denoted as temporal meta-rules. The paper *An Alternative Approach to Mining Association Rules* by J. Rauch and M. Šimůnek presents an approach for mining association rules based on the representation of analyzed data by suitable strings of bits. The procedure, 4ft-Miner, which is the contemporary application of this approach, is described therein. The paper *Direct Mining of Rules from Data with Missing Values* by V. Gorodetsky, O. Karsaev, and V. Samoilov presents an approach to, and technique for, direct mining of binary data with missing values. It aims to extract classification rules whose premises are represented in a conjunctive form. The idea is to first generate two sets of rules serving as the upper and lower bounds for any other sets of rules corresponding to all arbitrary assignments of missing values. Then, based on these upper and lower bounds, as well as a testing procedure and a classification criterion, a subset of rules for classification is selected. The paper *Cluster Identification using Maximum Configuration Entropy* by C.H. Li proposes a normalized graph sampling algorithm for clustering. The important question of how many clusters exist in a dataset and when to terminate the clustering algorithm is solved via computing the ensemble average change in entropy. The paper *Mining Small Objects in Large Images Using Neural Networks* by M. Zhang describes a domain independent approach to the use of neural networks for mining multiple class, small objects in large images. In the approach, the networks are trained by the back propagation algorithm with examples that have been taken from the large images. The trained networks are then applied, in a moving window fashion, over the large images to mine the objects of interest. The paper *Improved Knowledge Mining with the Multimethod Approach* by M. Lenič presents an overview of the multimethod approach to data mining and its concrete integration and possible improvements. This approach combines different induction methods in a unique manner by applying different methods to the same knowledge model in no predefined order. Although each method may contain inherent limitations, there is an expectation that a combination of multiple methods may produce better results.

The papers in Part III deal with issues related to knowledge discovery in a broad sense. This part contains four papers. The paper *Posting Act Tagging Using Transformation-Based Learning* by T. Wu et al. presents the application of transformation-based learning (TBL) to the task of assigning tags to postings in online chat conversations. The authors describe the templates used for posting act tagging in the context of template selection, and extend traditional approaches used in part-of-speech tagging and dialogue act tagging by incorporating regular expressions into the templates. The paper *Identification of Critical Values in Latent Semantic Indexing* by A. Kontostathis, W.M. Pottenger, and B.D. Davison deals with the issue of information retrieval. The authors analyze the values used by Latent Semantic Indexing (LSI) for information retrieval. By manipulating the values in the Singular Value Decomposition (SVD) matrices, it has been found that a significant fraction of the values have little effect on overall performance, and can thus be removed (i.e., changed to zero). This makes it possible to convert a dense term by dimensions and a document by dimension matrices into sparse matrices by identifying and removing such values. The paper *Reporting Data Mining Results in a Natural Language* by P. Strossa, Z. Černý, and J. Rauch represents an attempt to report the results of data mining in automatically generated natural language sentences. An experimental software system, AR2NL, that can convert implicational rules into both English and Czech is presented. The paper *An Algorithm to Calculate the Expected Value of an Ongoing User Session* by S. Millán et al. presents an application of data mining methods to the analysis of information collected from consumer web sessions. An algorithm is given that makes it possible to calculate, at each point of an ongoing navigation, not only the possible paths a viewer may follow, but also the potential value of each possible navigation.

We would like to thank the referees for their efforts in reviewing the papers and providing valuable comments and suggestions to the authors. We are also grateful to all the contributors for their excellent works. We hope that this book will be valuable and fruitful for data mining researchers, no matter whether they would like to uncover the fundamental principles behind data mining, or apply the theories to practical application problems.

San Jose, Tokyo, Taipei, Philadelphia, and Izumo
February, 2005

T.Y. Lin
S. Ohsuga
C.J. Liau
X. Hu
S. Tsumoto

References

1. T.Y. Lin and C.J. Liau (2002) Special Issue on the Foundation of Data Mining, *Communications of Institute of Information and Computing Machinery*, Vol. 5, No. 2, Taipei, Taiwan.

Contents

Part I Foundations of Data Mining

Part II Methods of Data Mining

Part III General Knowledge Discovery

Part I

Foundations of Data Mining

Knowledge Discovery as Translation

Setsuo Ohsuga

Emeritus Professor of University of Tokyo
ohsuga@fd.catv.ne.jp

Abstract. This paper discusses a view to capture discovery as a translation from non-symbolic to symbolic representation. First, a relation between symbolic processing and non-symbolic processing is discussed. An intermediate form was introduced to represent both of them in the same framework and clarify the difference of these two. Characteristic of symbolic representation is to eliminate quantitative measure and also to inhibit mutual dependency between elements. Non-symbolic processing has opposite characteristics. Therefore there is a large gap between them. In this paper a quantitative measure is introduced in the syntax of predicate. It enables to measure the distance between symbolic and non-symbolic representations quantitatively. It means that even though there is no general way of translation from non-symbolic to symbolic representation, it is possible when there is some symbolic representation that has no or small distance from the given non-symbolic representation. It is to discover general rule from data. This paper discussed a way to discover implicative predicate in databases based on the above discussion. Finally the paper discusses some related issues. The one is on the way of generating hypothesis and the other is the relation between data mining and discovery.

1 Introduction

The objective of this paper is to consider knowledge discovery in data from the viewpoint of knowledge acquisition in knowledge-based systems in order for realizing self-growing autonomous problem-solving systems. Currently developed methods of data mining however are not necessarily suited for this purpose because it is to find some local dependency relation between data in a subset of database. What is required is to find one or a finite set of symbolic expression to represent whole database. In order to clarify the point of issue, the scope of applicability of knowledge is discussed first. In order for making knowledge-based systems useful, every rule in a knowledge base should have as wide applicability for solving different problems as possible. It means that knowledge is made and represented in the form free from any specific application but must be adjustable to different applications. In order to generate such knowledge from data a new method is necessary.

S. Ohsuga: *Knowledge Discovery as Translation*, Studies in Computational Intelligence (SCI) **6**, 3–19 (2005)
www.springerlink.com

In this paper discovery is defined as a method to obtain rules in the declarative language from data in non-symbol form. First, applicability of knowledge for problem solving is considered and the scope of knowledge application is discussed from knowledge processing point of view in Sect. 2. It is also discussed that there is a substantial difference between current data mining method and wanted method of knowledge discovery in data.

In Sect. 3 it is shown that discovery is a translation between different styles of representations; one is ovserved data and another is linguistic representation of discovered knowledge. It is pointed out that in general there is a semantic gap between them and because of this gap not necessarily every data but only those meeting special condition can be translated into knowledge. After a general discussion on symbolic and non-symbolic processing in Sect. 4, a mathematical form is introduced to represent both symbolic and non-symbolic processing in the same framework in Sect. 5. With this form the meaning of discovery as translation is made clear. In Sect. 6 the syntax of predicate logic is extended to come closer to non-symbolic system. In Sect. 7 a method of quick test for discovery is discussed. Some related issues such as a framework of hypothesis creation and the relation between discovery and ordinary data mining are discussed in Sect. 8. Section 9 is conclusion.

2 Scope of Knowledge at Application

One of the characteristics of declarative knowledge at problem solving is that rules are mostly independent from specific application and the same rule is used for solving different problems. Hereafter predicate logic is considered as a typical declarative knowledge representation. For the purpose of comprehension a typed logic is used. In this logic every variable is assigned explicitly a domain as a set of instances. For example, instead of writing "man is mortal" like $(\forall x)$ [man $(x) \rightarrow$ mortal (x)] as ordinary first order logic, it is written $(\forall x/\mathrm{MAN})$ mortal (x) where MAN is a domain set of the variable x. $(\forall x)$ denotes "for all". This representation is true for x in this domain, that is, $x \in \mathrm{MAN}$. Each rule includes variables and depending on the problem a value or a set of values is substituted into each variable by inference operation at problem solving. The substitution is possible only when the domain of the variable in the rule includes the value or the set of values included in the problem. For example "is Socrates mortal?" is solved as true because Socrates $\in$ MAN. Not only for single value as in this example, it holds for a set of values. Foe example, "are Japanese mortal" is true because Japanese $\subset$ MAN.

The larger the domain is, the wider class of conclusions can be deduced from the rule. In this case the rule is said to have large scope of applicability. From the knowledge acquisition point of view, the knowledge with the larger scope is the more desirable to be generated because then the narrower scope knowledge can be deduced from it. Let assume a formula $(\forall x/D)$ predicatel (x) and its domain D is divided into a set $\{D1, D2, -, Dn\}$. Then $(\forall x/Di)$

predicatel (x), $(Di \subset D)$ is a formula with narrower domain. $(\forall x/D)$ predicatel (x) implies $(\forall x/Di)$ predicatel (x), $(Di \subset D)$, for all i and can replace all the latter predicates.

Applicability concerns not only the domain of variable but also the way of representing the set of instances. Assume a set of data {(1,2), (2,3), (3,1), (4, 4)}. The data can be represented by a formula to represent, for example, a mathematical function that paths through the points (1,2), (2,3), (3,1), (4, 4) in this order in x-y plane. It also can be represented by a couple of other formulas to represent different functions that path through (1,2), (2,3), (3,1) and (2,3), (3,1), (4, 4) respectively. These functions are different to each other. The first one is more suited for representing the original data set than the last two.

Many data mining method currently developed are not necessarily desirable in this point of view for finding rule because the scopes of rules discovered in these methods are usually very narrow. These methods intend to discover a set of local relations between attributes of observed data that appear more frequently than the others. If a rule to cover wider range of data is discovered, it is more desirable.

3 An Approach to Knowledge Discovery

3.1 Objective of Discovery

The objective of discovery in data is to know clearly about an object in which man has interest. It is assumed that there is a latent structure in the object. For example there can be some causal relations among some aspects of the object. Typically it is to know the causal relation of an object between stimuli from outside (input) and responses to it (output). By observing and analyzing the data representing the aspects of the object one can infer the hidden structure. If, as a typical case, there is no structure at all, observed data appear randomly, and no information on the object can be derived from the data. If there is some inner structure there must be some dependency relation among data to represent different aspects of the object, say stimuli and response. If one could know this structure of object, then he/she can use easily this information to applications.

It is desirable that this information is to represents the object's inner structure totally. It is not always possible to get such information. If there is no such dependency in the object at all it is not possible. Even if such dependency exists, if the dependency relation is complicated but the amount of data is not enough to represent it, it is hardly possible.

Most data mining methods currently developed however do not get such information but attempt to capture some local dependency relations between variables to represent different aspects of the object by a statistical or its equivalent method. Even if the inner structure of an object is complicated and to find it is difficult, it is possible to analyze local dependency of observed data and use the result for applications that need the relations. In this sense the data mining methods have a large applicability. For example, if it is found that there is close relation between amounts of sale of goods A and B in a supermarket, the manager can use the information to make decision on the arrangement of the goods in the store. But this information is useful only for this manager of the super-market in the environment it stands for making this kind of decision. Data mining is often achieved therefore in request to the specific application. The scope of the results of data mining is narrow.

The information to represent object totally has a wide scope such that every local dependency can be derived from it. But even if an object has such a latent structure, it cannot be obtained generally from the results of current data mining method. In order to get it another approach is necessary. In the sequel the objective of discovery is to find a latent structure as a total relation between input and output of an object.

3.2 Discovery as Translation

As has been mentioned, the objective of discovery in data is to know about an object in which a person has interest. What has been discovered must be represented in a symbolic form in whatever the way. Today, various styles of representations are used in information technology such as those based on procedural language, on declarative language, on neural network mechanism, etc. A specific method of processing information is defined to every representation style. Therefore every style has its own object to which the processing style is best suited. Person selects a specific style depending on his/her processing objective. A specific information-processing engine (IPE in the sequel) can be implemented for each style based on the representation scheme and processing method. Computer, for example, is an IPE of the procedural processing style. A neural network has also an IPE. In principle any style of representation can be selected for representing discovered result but ordinary a declarative language is used because it is suited for wider applications. An instance embodying a representation style with its IPE forms an information-processing unit (IPU in the sequel). A computer program is an IPU of procedural processing style. A specific neural network is also an IPU. Each unit has its own scope of processing.

A scope of problems that each IPU can deal with is limited however and often is very narrow. It is desirable from user's point of view that the scope is as wide as possible. Furthermore, it is desirable that the different styles can be integrated easily for solving such complex problems that require the scope over that of any single IPU. In general, however, it is not easy to integrate IPUs with different styles. In many cases it has been done manually in an ad hoc way for each specific pairs of IPUs to be integrated. The author has discussed in [2] a framework of integration of different representation schemes. The capability depends on the flexibility of each IPU to expand its scope of information processing as well as the expressive power of representation scheme of either or both of IPU to be integrated. It depends on representation scheme. Some scheme has a large expandability but the others have not. If one or both of IPUs has such a large expandability, the possibility of integrating these IPUs increases. Among all schemes that are used today, only purely declarative representation scheme meets this condition. A typical example is predicate logic. In the following therefore classic predicate logic is considered as a central scheme. Then discovery is to obtain knowledge through observation on an object and to represent it in the form of predicate logic. Every object has a structure or a behavioral pattern (input-output relation). In many cases however structure/behavioral-pattern is not visible directly but only superficial data is observed. These raw data are in non-symbolic form. Discovery is therefore to transform the data into symbolic expressions, here predicate logic. If a database represents a record of behavior (input-output relation) of an object and is translated into a finite set of predicate formulae, then it is discovery in data.

3.3 Condition to Enable Translation

Translation between systems with the different semantics must be considered. Semantics is considered here as the relation between world of objects (universe of discourse) and a system of information to represent it. The objects being described are entity, entity's attribute, relation between entities, behavior/activity of entity and so on. Translation is to derive a representation of an object in a representation scheme from that of another representation scheme for the same object. Both systems must share the same object in the respective universe of discourse. If one does not have the object in its universe of discourse, then it cannot describe it. Hence the system must expand its universe by adding the objects before translation. Corresponding to this expansion of the universe its information world must also be expanded by being added the description of the object. This is the expandability.

Translation is possible between these two systems if and only if both systems can represent the same objects and also there is one-to-one correspondence between these representations. In other words, discovery is possible if and only if non-symbolic data representation and predicate logic meet this condition. In the next section the relation between non-symbolic processing and symbolic processing is discussed, then the condition for enabling translation from the former to the latter is examined

4 Symbolic and Non-Symbolic Processing

Symbolic representation and non-symbolic representation are the different way of formal representation to refer some objects in the universe of discourse. A non-symbolic expression has a direct and hard-wired relation with the object.

Representation in non-symbolic form is strictly dependent on a devise for measurement that is designed specifically for the object. On the other hand, symbolic representation keeps independence from object itself. The relation between an object and its symbolic representation is made indirectly via a (conceptual) mapping table (dictionary). This mapping table can be changed. Then the same symbolic system can represent the different objects.

The different symbolic systems have been defined in this basic framework. Some is fixed to a specific universe of discourse and accordingly the mapping table is fixed. This example is seen in procedural language for computers. Its universe of discourse is fixed to the computer. Some others have the mapping tables that can be changed. The universe of discourse is not fixed but the same language expression can be used to represent the object in the different universes of discourse by changing the mapping table. This example is seen in most natural language and predicate logic as their mathematical formalization. These languages have modularity in representation. That is, everything is represented in a finite (in general, short) length of words. Thanks to this flexibility of mapping and its modularized representation scheme it gets a capability to accept new additional expressions any time. Therefore when new objects are added in the universe, the scope of predicate logic can expand by adding new representations corresponding to these new objects. This is called expandability of predicate logic in this paper. It gives predicate logic a large potentiality of integrating various IPUs.

For example, let a case of integrating predicate logic system as a symbolic IPU with another IPU of the different style be considered. Let these IPUs have the separate worlds of objects. That is, the description systems of the different IPUs have no common object. Then the universe of discourse of the symbolic IPU is expanded to include the objects of non-symbolic IPU and symbolic representations for the new objects are added to the information world of the symbolic IPU. Then these two IPUs share the common objects and these objects have different representations in the different styles. If an IPU can represent in its representation scheme the other IPU's activity on its objects, then these two IPUs can be integrated. It is possible with predicate logic but not with the other styles of representation.

It is also possible with predicate logic to find unknown relation between objects in the information world by logical inference. These are the characteristics that give predicate logic a large potentiality of integrating various IPUs [2].

The expandability however is mere a necessary condition of an IPU for being translated formally to the other IPU but it is not sufficient. In general there is no one-to-one correspondence between non-symbolic expression and logical expression because of a substantial difference between their syntax as will be discussed below. Furthermore, the granularity of expression by predicate logic is too course comparing to non-symbolic expressions. Some method to expand the framework of predicate logic while preserving its advantages is necessary. In the sequel a quantitative measure is introduced into classical predicate logic and a symbol processing system that represents non-symbol processing approximately is obtained.

There is another approach to merge symbol and non-symbol processing, say a neural network. It is to represent a neural network by means of special intuitive logic. Some attempts have been made so far and some kinds of intuitive logic have been proved equivalent to neural network [3]. But these approaches lose some advantages of classic logic such as expandability and completeness of inference. As the consequence these systems cannot have large usability as the classic logic system. This approach merely shifts the location of the gap from between symbolic processing and non-symbolic processing to between the classic logic and the special intuitive logic. Because of this reason, this paper does not take the latter approach but take an approach to approximate neural network by extended classical logic.

5 Framework To Compare Symbolic and Non-Symbolic Systems

5.1 An Intermediate Form

In order to compare symbolic system and non-symbolic system directly a mathematical form is introduced to represent both systems in the same framework. Predicate logic is considered to represent a symbolic system. For the above purpose a symbolic implicative typed-formula $(\forall x/D)[F(x) \rightarrow G(x)]$ is considered as a form to represent a behavior of an object. Here D is a set of elements, $D = (a, b, c, -z)$, and x/D means $x \in D$. Let the predicates F (and G) be interpreted as a property of x in D, that is, "$F(x)$; an element x in D has a property F". Then the following quantities are defined.

First a state of D is defined as a combination of $F(x)$ for all x/D. For example, "$F(a)$: True", "$F(b)$: False", "$F(c)$: False",–,"$F(z)$: True" forms a state, say SF_I, of D with respect to F. Namely, $SF_I = (F(a), -F(b), -F(c), -, F(z))$. There are $N = 2^n (= 2^{**}n)$ different states.

Let "$F(x)$: True" and "$F(x)$: False" be represented by 1 and 0 respectively. Then SF_I as above is represented (1, 0, 0, –, 1). Let the sequence is identified by a binary number $I = 100-1$ obtained by concatenating 0 or 1 in the order of arrangement. Also let SF_I be I-th state in N states. By arranging all states in the increasing order of I, a state vector $\boldsymbol{Sf}$ is defined. That is, $\boldsymbol{Sf} = (SF_0, SF_1, -, F_{N-1})$. Among them, $\boldsymbol{Sf}_\forall = \{(1, 1, -, 1)\} = (\forall x/D)F(x)$ and $\boldsymbol{Sf}_\exists = \{\boldsymbol{Sf} - (0, 0, -0)\} = (\exists x/D)F(x)$ are the only vectors that the ordinary predicate can represent. $(\exists x)$ denotes "for some x".

If the truth or false of F for one of the elements in D changes, then the state of D changes accordingly. Let this change occurs probabilistically. Then a state probability PF_I is defined to a state SF_I as probability of D being in the state SF_I and a probability vector $\boldsymbol{Pf}$ are also defined as $\boldsymbol{Pf} = (PF_0, PF_1, -, PF_{N-1})$.

	p0	p1	p2	p3	p4	p5	p6	p7	p8	p9	p10	p11	p12	p13	p14	p15
p0	x	x	x	x	x	x	x	x	x	x	x	x	x	x	x	x
p1	0	x	0	x	0	x	0	x	0	x	0	x	0	x	0	x
p2	0	0	x	x	0	0	x	x	0	0	x	x	0	0	x	x
p3	0	0	0	x	0	0	0	x	0	0	0	x	0	0	0	x
p4	0	0	0	0	x	x	x	x	0	0	0	0	x	x	x	x
p5	0	0	0	0	0	x	0	x	0	0	0	0	0	x	0	x
p6	0	0	0	0	0	0	x	x	0	0	0	0	0	0	x	x
p7	0	0	0	0	0	0	0	x	0	0	0	0	0	0	0	x
p8	0	0	0	0	0	0	0	0	x	x	x	x	x	x	x	x
p9	0	0	0	0	0	0	0	0	0	x	0	x	0	x	0	x
p10	0	0	0	0	0	0	0	0	0	0	x	x	0	0	x	x
p11	0	0	0	0	0	0	0	0	0	0	0	x	0	0	0	x
p12	0	0	0	0	0	0	0	0	0	0	0	0	x	x	x	x
p13	0	0	0	0	0	0	0	0	0	0	0	0	0	x	0	x
p14	0	0	0	0	0	0	0	0	0	0	0	0	0	0	x	x
p15	0	0	0	0	0	0	0	0	0	0	0	0	0	0	0	1

x; non-negative value with row sum = 1

Fig. 1. Transition matrix to represent logical expression $(Ax/d)[F(x) \to G(x)]$

Then it is shown that a logical inference $F \wedge [F \to G] \Rightarrow G$ is equivalent to a mathematical form, $\boldsymbol{Pg} = \boldsymbol{Pf} \times T$, if a transition matrix $T = |t_{IJ}|$ satisfies a special condition as is shown in Fig. 1 as an example. This matrix is made as follows. Since $F \to G = -F \vee G$ by definition, if "$F(x)$; True" for some x in D, then $G(x)$ for the x must be true. That is, there is no transition from a state SF_I including "$F(x)$; True" to a state SG_J of D in regard to G including "$G(x)$; False" and t_{IJ} for this pair is put zero. The other elements of the transition matrix can be any positive values less than one. The above form is similar to a stochastic process. Considering the convenience of learning from database as will be shown later, the correspondence of the logical process and stochastic process is kept and a condition of the row sum of the matrix is made equal to one for every row.

It should be noted that many elements in this matrix are zero [1]. In case of non-symbolic system there is no such restriction to the transition matrix. This is the substantial difference between symbolic system and non-symbolic system.

This method is extended to $(\forall x/D)[F1(x) \wedge F2(x) \to G(x)]$, $(\forall x/D)(\forall y/E)$ $[F1(x) \wedge F2(x,y) \to G(y)]$ and to the more general cases. If the premise of an implicative formula includes two predicates with the same variable like $(\forall x/D)[F1(x) \wedge F2(x) \to G(x)]$ as above, then two independent states $\boldsymbol{Sf}1$ and $\boldsymbol{Sf}2$ of D are made corresponding to $F1(x)$ and $F2(x)$ respectively. Then a compound state $\boldsymbol{Sf}$ such as $\boldsymbol{Sf} = \boldsymbol{Sf}1 \times \boldsymbol{Sf}2$ is made as the Cartesian product. From its compound probability vector $\boldsymbol{Pf}$ a probability vector $\boldsymbol{Pg}$ for the state $\boldsymbol{Sg}$ is derived in the same way as above. In this case the number of states in $\boldsymbol{Sf}$ is $2^{**}(2^{**}n)$ and transition matrix T becomes $2^{**}(2^{**}n) \times 2^{**}(2^{**}n)$ matrix. Or it can be represented in a three-dimensional space by a $(2^{**}n) \times (2^{**}n) \times (2^{**}n)$ matrix and is called a Cubic Matrix. Each of three axes represents a predicate in the formula, that is, either $F1$ or $F2$ or G. It is a convenient way for a visual understanding and making the matrix consistent with logical definition. (I, J)-th element in each plane is made in such a way that it represents a consistent relation with the definition of logical imply when the states of D with respect to $F1$ and also to $F2$ are I and J respectively. For example, in a plane of the state vector $\boldsymbol{Sg}$ including $G(a) = 0$, (I, J)-th element corresponding to the states I and J of $\boldsymbol{Sf}1$ and $\boldsymbol{Sf}2$ including $F1(a) = 1$, $F2(a) = 1$ must be zero. It is to prevents a contradictory case of $F1(a) = 1$, $F2(a) = 1$ and $G(a) = 0$ to occur.

There can be cases in which more than two predicates are included in the premise. But in principle, these cases are decomposed to the case of two predicates. For example, $F1(x) \wedge F2(x) \wedge F3(x) \rightarrow G(x)$ can be decomposed into $F1(x) \wedge F2(x) \rightarrow K(x)$ and $K(x) \wedge F3(x) \rightarrow G(x)$ by using an internal predicate $K(x)$.

Further extension is necessary for more than two variables, for example, $(\forall x/D)(\forall y/E)[F1(x) \wedge F2(x, y) \rightarrow G(y)]$. In this case a new variable z defined over the set $D \times E$ is introduced and a cubic matrix can be made. The following treaty is similar to the above case. In this way the set of logical implicative forms with the corresponding transition matrices is extended to include practical expressions. As the matter of course, the more complex is a predicate, the more complex becomes the representation of its matrix.

The computation $\boldsymbol{Pg}_J = \Sigma_I \boldsymbol{Pf}_I \times t_{IJ}$ for $\boldsymbol{Pg} = \boldsymbol{Pf} \times T$ is formally the same as that included in an ordinary non-symbolic operation for transforming inputs to outputs. A typical example is neural network of which the input and output vectors are $\boldsymbol{Pf}_I$ and $\boldsymbol{Pg}_J$ respectively and the weight of an arc between nodes I and J is t_{IJ} . A neural network includes a non-linear transformation after this linear operation. Usually a function called Sigmoid Function is used. At the moment this part is ignored because this is not really an operation between non-symbolic representations but to represent a special way of translation from a non-symbolic expression into a symbolic expression.

A transition matrix for representing predicate logic has many restrictions comparing to a matrix to represent a non-symbolic system. First, since the former represents a probabilistic process, every element in this matrix must be in the interval [0,1] while any weight value is allowed for neural network. But this is to some extent the matter of formalization of representation. By preprocessing the input values the different neural network of which the range of every input value becomes similar to probability may be obtained with substantially the same functionality as original one. Thus the first difference is not substantial one.

Second, in order for a matrix to keep the same relation as logical implication, it has to satisfy a constraint as shown in Fig. 1, while the matrix to represent non-symbolic systems is free from such a restriction. A non-symbolic system can represent an object at the very fine level and in many cases to continuous level. In other words, granularity of representation is very fine in the framework of neural network. But the framework is rigid for the purpose of expanding its scope of representation. For example in order to add a new element to a system a whole framework of representation must be changed. Therefore integration of two or more non-symbolic systems is not easy. Persons must define an ad hoc method for integration for every specific case. The granularity of logical predicate on the other hand is very course. Predicate logic however can expand the scope with the sacrifice of granularity of representation at fine level. Therefore predicate logic cannot represent non-symbolic systems correctly. In general, it is difficult to translate non-symbolic system into symbolic systems. In other words, only such non-symbolic systems that are represented in the same matrix as shown in Fig. 1 are translatable into symbolic systems. Therefore before going into discovery process, it is necessary to examine whether an object is translatable into symbolic system or not.

5.2 Condition of Database being Translated into Predicate

Whether a database can be translated into predicate or not is examined by comparing the matrix generated from the database with that of predicate. Since the latter matrix is different for every predicate formula, a hypothetical predicate is created first that is considered to represent the database. The matrix to represent this formula is compared with that of database. If they do not match to each other the predicate as a hypothesis is changed to the other. Thus this is an exploratory process.

The matrix for the database is created in an ordinary learning process, that is, IJ-th element of transition matrix is created and modified by data in the database. In an ordinary learning, if there is a datum to show the positive evidence the corresponding terms are increased by the small amount while for the negative data these are decreased. In this case an initial value must be decided in advance for every element. If there is no prior knowledge to decide it, the initial values of every element are made the same. In the case being discussed, there are some elements that correspond to every positive data satisfying the formula, i.e. those that are increased by the small amount. In the matrix, these are at the cross points of those rows corresponding to the states SF_I of premise and the columns corresponding to the states SG_J of consequence meeting the condition as hypothetical predicate. The other elements are decreased by some amount such that the row sum keeps one for every row. There are many such cross points corresponding to SF_I and SG_J including the data. For example, in the case of the simplest example $(Ax/D)\{F(x) \rightarrow G(x)\}$, if there are a pair of $F(a)$ and $G(a)$ in the database, all states in $\boldsymbol{Sf}$ and $\boldsymbol{Sg}$ including "$F(a)$; True" and "$G(a)$; True" make up such cross points.

If the matrix made in this way approaches to that as shown in Fig. 1, then it is concluded that the object at the background of the data is represented by the predicate.

Since however some errors can be included in the observation and also not always enough data for letting the learning process to converge are not expected two matrices hardly to match exactly. Therefore an approach to enable an approximate matching is taken by expanding the syntax of an orthodox predicate logic toward to include probabilistic measure in the sequel.

6 Extending Syntax of Logical Expression

The syntax of predicate logic is expanded to include probability of truth of a logical expression while preserving its advantages of expandability.

In the representation of matrix form a probability vector $\boldsymbol{Pf}$ of the state vector $\boldsymbol{Sf}$ represented an occurrence probability of logical states. In the formal syntax of classical first order logic however only two cases of $\boldsymbol{Pf}$ actually can appear. These are (0, 0, 0, –, 1) and (0, *, *, –, *) that correspond to $(\forall x/D)F(x)$ and $(\exists x/D)F(x)$ respectively. Here * denotes any value in [0, 1]. Since a set $D = \{a, b, c, -, z\}$ is assumed finite, $(\forall x/D)F(x) = F(a) \wedge F(b) \wedge - \wedge F(z)$. Even if the probability of "$F(x)$; True" is different for every element, that is, for $x = a$ or $x = b$ or – or $x = z$, ordinary first order logic cannot represent it. In order to improve it a probability measure is introduced there. Let a probability of "$F(x)$: True" be $p(x)$ for $D \ni x$. Then the syntax of logical fact expressions $(\forall x/D)F(x)$ is expanded to $(\forall x/D)\{F(x), p(x)\}$ meaning "for every x of D, $F(x)$ is true with probability $p(x)$".

Since $p(x)$ is a distribution over the set D, it is different from $\boldsymbol{Pf}$ that is a distribution over the set of states $\boldsymbol{Sf}$. It is possible to obtain $\boldsymbol{Pf}$ from $p(x)$ and vice versa. Every state in $\boldsymbol{Sf}$ is defined as the combination of "$F(x)$; True" or "$F(x)$; False" for all elements in D. I-th element of $\boldsymbol{Sf}$ is SF_I . An element in $\boldsymbol{Pf}$ corresponding to SF_I is $\boldsymbol{Pf}_I$. Let "$F(x)$; True" for the element $x; i, j, -$ and "$F(y)$; False" for $y; k, l, -$ in SF_I. Then $\boldsymbol{Pf}_I = p(i) \times p(j) \times - \times (1 - p(k)) \times (1 - p(l)) \times -$.

On the other hand, let an operation to sum up all positive components with respect to i in $\boldsymbol{Pf}$ be $\Sigma_{*i \in I} \boldsymbol{Pf}_I$. Here the "positive component with respect to i" is $\boldsymbol{Pf}_I$ corresponding to SF_I in which "$F(x)$;True" for i-th element x in D. This represents a probability that i-th element x in D is in the state "$F(x)$; True". That is, $\Sigma_{*i \in I} \boldsymbol{Pf}_I = p(x)$.

Implicative formula is also expanded. Let an extension of an implicative formula $(\forall x/D)[F(x) \to G(x)]$ be considered as an example. The detail of the quantitative measure is discussed later. Whatever it may be it is to generate from $(\forall x/D)\{F(x), p(x)\}$ a conclusion in the same form as the premise with its own probability distribution, i.e. $(\forall x/D)\{G(x), r(x)\}$. In general $r(x)$

must be different from $p(x)$ because an implicative formula may also have some probabilistic uncertainty and it affects the probability distribution of consequence.

The matrix introduced in Sect. 5.1 to represent a logical formula to generate a conclusion for a logical premise gives a basis for extension of implicative formula. If one intends to introduce a probabilistic measure in the inference, the restriction imposed to the matrix is released in such a way that any positive value in [0, 1] is allowed to every element under the only constraint that row sum is one for every row. With this matrix and an extended fact representation (non-implicative form) as above, it is possible to get a probability distribution of the conclusion in the extended logical inference as follows.

(1) Generate $\boldsymbol{Pf}$ from $p(x)$ of $(\forall x/D)\{F(x), p(x)\}$.
(2) Obtain $\boldsymbol{Pg}$ as the product of $\boldsymbol{Pf}$ and the expanded transition matrix.
(3) Obtain $r(x)$ of $(\forall x/D)\{G(x), r(x)\}$ from $\boldsymbol{Pg}$.

Thus if a matrix representation is available for predicate logic, it represents an extension of predicate logic because it includes continuous values and allows the same process as non-symbolic operation. But it has drawback in two aspects. First, it needs to hold a large matrix to every implicative representation and second, and the more important, it loses modularity that was the largest advantage of predicate logic for expanding the scope autonomously. Modularity comes from the mutual independence of elements in D in a logical expression. That is, the mutual independence between elements in D is lost in the operation $\boldsymbol{Pg} = \boldsymbol{Pf} \times T$ for the arbitrarily expanded matrix and it causes the loss of modularity. This is an operation to derive $\boldsymbol{Pg}_J$ by $\boldsymbol{Pg}_J = \Sigma_I \boldsymbol{Pf}_I \times t_{IJ} = \boldsymbol{Pf}_1 \times t_{1J} + \boldsymbol{Pf}_2 \times t_{2J} + - + \boldsymbol{Pf}_N \times t_{NJ}$. That is, J-th element of $\boldsymbol{Pg}$ is affected by the other elements of $\boldsymbol{Pf}$ than J-th element. If this occurs, logical value of an element in D is not decided independently but is affected by the other elements. Then there is no modularity any more.

In order to keep the independence of logical value, and therefore the modularity of predicate at inference it is desirable to represent logical implication in the same form as the fact representation like $(\forall x/D)\{[F(x) \to G(x)], q(x)\}$. It is read "for every x of $D, F(x) \to G(x)$ with probability $q(x)$". In this expression $q(x)$ is defined to each element in D independently. Then logical inference is represented as follows.

$(\forall x/D)\{F(x), p(x)\} \wedge (\forall x/D)\{[F(x) \to G(x)], q(x)\}] \Rightarrow (\forall x/D)\{G(x), r(x)\}$,

$$r(x) = f(p(x), q(x)) .$$

If it is possible to represent logical inference in this form, the actual inference operation can be divided into two parts. The first part is the ordinary logical inference such as, $(\forall x/D)F(x) \wedge (\forall x/D)\{F(x) \to G(x)\} \Rightarrow (\forall x/D)G(x)$.

The second part is the probability computation $r(x) = f(p(x),\ q(x))$. This is the operation to obtain $r(x)$ as the function only of $p(x)$ and $q(x)$ with the same variable and is performed in parallel with the first part. Thus logical operation is possible only by adding the second part to the ordinary inference operation.

This is the possible largest extension of predicate logic to include a quantitative evaluation meeting the condition for preserving the modularity. This extension reduces the gap between non-symbolic and symbolic expression to a large extent. But it cannot reduce the gap to zero but leaves a certain distance between them. If this distance can be made small enough, then predicate logic can approximate non-symbolic processing. Here arises a problem of evaluating the distance between arbitrarily expanded matrix and the matrix with the restricted expansion.

Coming back to the matrix operation, the probability of the consequence of inference is obtained for i-th element as

$$r(x_i) = \Sigma_{*i \in I} \boldsymbol{P} g_I = \Sigma_{*i \in I} (\Sigma_I \boldsymbol{P} f_I \times t_{IJ}), (x_i \text{ is } i\text{–th element of } D)$$

This expression is the same as non-symbolic processing. On the other hand an approximation is made that produces an expression like the one shown as above.

First the following quantities are defined.

$q(x_k) = \Sigma_{*k \in I} t_{NJ}$, $r'(x_k) = (\Sigma_{*k \in J} t_{NJ})(\Sigma_{*i \in I} \boldsymbol{P} f_I)$, ($x_k$ is k-th element in D)

$r'(x)$ is obtained by replacing every IJ-th element by NJ-th element, that is, by the replacement, $t_{IJ} \leftarrow t_{NJ}$ in a transition matrix. Since every row is replaced by last row, the result of operations with this matrix is correct only when input vector is $\boldsymbol{P} f = (0, 0, -, 1)$, that is, $(\forall x/D)\ F(x)$ holds true with certainty. If some uncertainty is included in $(\forall x/D)\ F(x)$, then there is some finite difference between the true value $r\ (x_k)$ and its approximation $r'(x_k)$. It is:

$$\begin{aligned} r(x_k) - r'(x_k) &= \Sigma_{*k \in J} \left[(\Sigma_I \boldsymbol{P} f_I \times t_{IJ}) - t_{NJ} \times (\Sigma_I \boldsymbol{P} f_I) \right] \\ &= \Sigma_{*k \in J} (\Sigma_I \boldsymbol{P} f_I \times (t_{IJ} - t_{NJ})) = \Sigma_I (\Sigma_{*k \in J} \boldsymbol{P} f_I \times (t_{IJ} - t_{NJ})) \\ &= \Sigma_I \boldsymbol{P} f_I \times (\Sigma_{*k \in J} (t_{IJ} - t_{NJ})) \end{aligned}$$

This is not always small enough for a general non-symbolic processing. But if this error reduces to zero as data increases for a non-symbolic system, then this system is the one that is translatable into a symbolic system. Thus, by the estimation of this error, whether the database can be translated into predicate formula as a whole or not is decided.

This is a process of hypothesis creation and test. It proceeds as follows.

(1) A hypothetical predicate assumed to represent the given databases is generated.
(2) A transition matrix is generated from the database with respect to the predicate.
(3) Calculate the error.

(4) If the error is below the threshold, then accept the predicate. Otherwise generate the other hypothesis and repeat the above process.

In case of success, the probability distribution $r(x)$ of the conclusion of inference is obtained as an inner product of $p(x)$ and $q(x)$, i.e., $r(x) = p(x) \times q(x)$.

This discussion on a very simple implicative formula holds true for the more general formulas. The importance of this approximation is that, first of all, predicate logic can be expanded without destroying the basic framework of logical inference by simply adding a part to evaluate quantitatively the probability. If it is proved in this way that the predicate formula(s) represents the databases, then this discovered knowledge has the larger applicability for wider class of problems than the database itself.

In general non-symbolic processing assumes mutual dependency between elements in a set and includes computations of their cross terms. On the other hand predicate logic stands on the premise that every element is independent to each other. This is the substantial difference between symbolic and non-symbolic representation/processing. In the above approximation this cross term effects are ignored.

7 Quick Test of Hypothesis

Generation of hypothesis is one of the difficulties included in this method. Large amount of data is necessary for hypothesis testing by learning. It needs lot of computations to come to conclusion. Rough but quick testing is desirable based on small amount of data.

Using an extended inference,

$$(\forall x/D)\ (F(x), p(x)) \wedge (\forall x/D)\{F(x) \rightarrow G(x), q(x)\} \Rightarrow ((\forall x/D)\ (G(x), r(x))\ ,$$

$$r(x) = p(x) \times q(x)\ ,$$

$q(x)$ is obtained directly by learning from the data in a database.

Assuming that every datum is error-free, there can be three cases such as (1) datum to verify the implicative logical formula exists, (2) datum to deny the logical formula exists and (3) some datum necessary for testing hypothesis does not exist.

The way of coping with the data is different by a view to the database. There are two views. In one view, it is assumed that a database represents every object in the universe of discourse exhaustively or, in other words, a closed world assumption holds to this database. In this case if some data to prove the hypothesis does not exists in the database the database denies the hypothesis. On the other hand, it is possible to assume that a database is always incomplete but is open. In this case, even if data to prove a predicate do not exist, it does not necessarily mean that the hypothesis should be rejected.

The latter view is more natural in case of discovery and is taken in this paper. Different from business databases in which every individual datum has its own meaning, the scope of data to be used for knowledge discovery cannot be defined beforehand but is augmented by adding new data. A way of obtaining the probability distribution for a hypothesis is shown by an example.

Example: A couple of databases, $FG(D, E)$ and $H\ (D, E)$, be given.

$$FG(D,E) = (-, (a1, b1),\ (a1, b2),\ (a1, b4),\ (a2, b2), -)\ ,$$
$$H(D,E) = (-, (a1, b1),\ (a1, b2),\ (a1, b3), (a2, b2), -)\ ,$$

Where $D = (a1, a2, -, am)$ and $E = (b1, b2, -, bn)$.

Assume that a logical implicative formula $(\forall x/D)(\forall y/E)\{[F(x) \wedge H(x, y) \to G(y)]q(x)\}$ is made as a hypothesis. At the starting time, every initial value in the probability distribution $q(x)$ is made equal to 0.5. Then since $F(a1)$ holds true for an element $a1$ and $H(a1, b1)$, $H(a1, b2)$, $H(a1, b3)$ hold true in the database, $G(b1), G(b2), G(b3)$ must hold true with this hypothesis. But there is no datum to prove $G(b3)$ in the databases. Thus for 2 cases out of 3 required cases the hypothesis is proved true actuary by data. The probability distribution $q(x)$ of the logical formula is obtained as a posterior probability starting from the prior probability 0.5 and evaluating the contribution of the existing data to modify the effective probability like $q(a1) = 0.5 + 0.5 \times 2/3 = 5/6$

By calculating the probability for every data in this way, a probability distribution, $q(x)$, is obtained approximately. If for every element of D the probability is over the pre-specified threshold value, this hypothetical formula is accepted. When database is very large, small amount of data is selected from there and hypotheses are generated by this rough method and then precise test is achieved.

8 Related Issues

8.1 Creating Hypothesis

There still remains a problem of constructing hypothesis. There is no definite rule for constructing it except a fact that it is in a scope of variables included in the database. Let a set of variables (colums) in the database be $\mathbf{X} = (X1, X2, -, XN)$ and the objective of discovery be to discover a predicate formula such as $Pi1 \wedge Pi2 \wedge - \wedge Pir \to G(XN)$ for XN of which the attribute is G.

For any subset of it a predicate is assumed. Let i-th subset of $\mathbf{X}$ be $(Xi1, Xi2, -, Xik)$. Then, a predicate $Pi(Xi1, Xi2, -, Xik)$ is defined.

Let a set of all predicates thus generated be $\boldsymbol{P}$. For any subset $\boldsymbol{Pj} = (Pj1, Pj2, -, Pjm)$ in $\boldsymbol{P}$, i.e. $\boldsymbol{P} \supset \boldsymbol{Pj}$, $Pj1 \wedge - \wedge Pj2 \wedge - \wedge Pjm \to G(XN)$ can be a candidate of discovery that may satisfy the condition discussed so far. That is, this formula can be a hypothesis.

In more general case, the form $Pi1\#Pi2\# - Pir \rightarrow G(XN)$ is considered instead of $Pi1 \wedge Pi2 \wedge - \wedge Pir \rightarrow G(XN)$ where the symbol # denotes either $\wedge$ or $\vee$.

Since there can be so many hypotheses this is not to give a practical method of hypothesis creation but shows a framework of hypotheses to be created. Human insight is required to select a few possible candidates out of them.

8.2 Data Mining and Knowledge Discovery

The relation of ordinary data mining and discovery is discussed. Coming back to the simplest case of Fig. 1, assume that non-symbolic representation does not meet the condition of translatability into predicate formula, that is, some finite non-zero values appear to the positions that must be zero for the set of instances D as the variable domain. More generally referring to the extended representation, assume that the error exceeds the pre-defined value. However some reduced set Di of D may meet the condition where Di is a subset of D, $Di \subset D$. Unless all elements are distributed evenly in the matrix, probability of such a subset to occur is large. Data mining is to find such subsets and to represent the relation among the elements. In this sense the data mining method is applicable to any object.

Assume that an object has a characteristic that enables the discovery as discussed in this paper. In parallel with this it is possible to apply ordinary data mining methods to the same object. In general however it is difficult to deduce the predicate formula to represent the database as a whole, i.e. discovery as discussed so far, from the result of data mining. In this sense these approaches are different.

9 Conclusion

This paper stands on an idea that discovery is a translation from non-symbolic raw data to symbolic representation. It has discussed first a relation between symbolic processing and non-symbolic processing. Predicate logic was selected as the typical symbolic representation. A mathematical form was introduced to represent both of them in the same framework. By using it the characteristics of these two methods of representation and processing are analyzed and compared. Predicate logic has capability to expand its scope. This expandability brings the predicate logic a large potential capability to integrate different information processing schemes. This characteristic is brought into predicate logic by elimination of quantitative measure and also of mutual dependency between elements in the representation. Non-symbolic processing has opposite characteristics. Therefore there is a gap between them and it is difficult to reduce it to null. In this paper the syntax of predicate logic was extended so that some quantitative representation became possible. It reduces the gap to a large extent. Even though this gap cannot be eliminated completely, this extension is useful for some application including knowledge discovery from database because it was made clear that translation from the non-symbolic to symbolic representation, that is discovery, is possible only for the data of which this gap is small. Then this paper discussed a way to discover one or more implicative predicate in databases using the above results.

Finally the paper discussed some related issues. One is the framework of hypothesis creation and the second is the relation between data mining and discovery.

References

1. S. Ohsuga; Symbol Processing by Non-Symbol Processor, Proc. PRICAI'96
2. S. Ohsuga; The Gap between Symbol and Non-Symbol Processing – An Attempt to Represent a Database by Predicate Formulae, Proc. PRICAI'200
3. S. Ohsuga; Integration of Different Information Processing Methods, (to appear in) DeepFusion of Computational and Symbolic Processing, (eds. F. Furuhashi, S. Tano, and H.A. Jacobsen), Springer, 2000
4. H. Tsukimoto: Symbol pattern integration using multi-linear functions, (to appear in) Deep Fusion of Computational and Symbolic Processing, T. Furuhashi, S. Tano, and H.A. Jacobsen), Springer, 2000

Mathematical Foundation of Association Rules – Mining Associations by Solving Integral Linear Inequalities

T.Y. Lin

Department of Computer Science, San Jose State University, San Jose, California 95192-0103
tylin@cs.berkeley.edu

Summary. Informally, data mining is derivation of patterns from data. The mathematical mechanics of association mining (AM) is carefully examined from this point. The data is table of symbols, and a pattern is any algebraic/logic expressions derived from this table that have high supports. Based on this view, we have the following theorem: A pattern (generalized associations) of a relational table can be found by solving a finite set of linear inequalities within a polynomial time of the table size. The main results are derived from few key notions that observed previously: (1) Isomorphism: Isomorphic relations have isomorphic patterns. (2) Canonical Representations: In each isomorphic class, there is a unique bitmap based model, called granular data model

Key words: attributes, feature, data mining, granular, data model

1 Introduction

What is data mining? There are many popular citations. To be specific, [6] defines data mining as the non-trivial process of identifying valid, novel, potentially useful, and ultimately understandable patterns from data. Clearly it serves more as a guideline than a scientific definition. "Novel," "useful," and "understandable," involve subjective judgments; they cannot be used for scientific criteria. In essence, it says data mining is

- Drawing useful patterns (high level information and etc.) from data.

This view spells out few key ingredients:

1. What are the *data*?
2. What are the *pattern*?
3. What is the logic system for *drawing* patterns from data?
4. How the pattern is related to real world? (usefulness)

T.Y. Lin: *Mathematical Foundation of Association Rules – Mining Associations by Solving Integral Linear Inequalities*, Studies in Computational Intelligence (SCI) **6**, 21–42 (2005)
www.springerlink.com

This paper was motivated from the research on the foundation of data mining (FDM). We note that

- The goal of FDM is not looking for new data mining methods, but is to understand how and why the algorithms work.

For this purpose, we adopt the axiomatic method:

1. Any assumption or fact (data and background knowledge) that are to be used during data mining process are required to be explicitly stated at the beginning of the process.
2. Mathematical deductions are the only accepted reasoning modes.

The main effort of this paper is to provide the formal answers to the questions. As there is no formal model of real world, last question cannot be in the scope of this paper. The axiomatic method fixes the answer of question three. So the first two question will be our focusing. To have a more precise result, we will focus on a very specific, but very popular special techniques, namely, the association (rule) mining.

1.1 Some Basics Terms in Association Mining (AM)

A relational table (we allow repeated rows) can be regarded as a knowledge representation $K : V \longrightarrow C$ that represents the universe (of entities) by attribute domains, where V is the set of entities, and C is the "total" attribute domain. Let us write a relational table by $K = (V, A)$, where K is the table, V is the universe of entities, and $A = \{A^1, A^2, \ldots A^n\}$ is the set of attributes.

In AM, two measures, support and confidence, are the criteria. It is well-known among researchers, support is the essential one. In other words, high frequency is more important than the implications. We call them high frequency patterns, undirected association rules or simply associations.

Association mining originated from the market basket data [1]. However, in many software systems, the data mining tools are applied to *relational tables*. For definitive, we have the following translations and will use interchangeably:

1. An item is an attribute value,
2. A q-itemset is a subtuple of length q, in short, q-subtuple
3. A q-subtuple is a high frequency q-itemset or an q-association, if its occurrences are greater than or equal to a given threshold.
4. A q-association or frequent q-itemset is a pattern, but a pattern may have other forms.
5. All attributes of a relational are assumed to be distinct (non-isomorphic); there is no loss in generality for such an assumption; see [12].

2 Information Flows in AM

In order to fully understand the mathematical mechanics of AM, we need to understand how the data is created and transformed into patterns. First we need a convention:

- A symbol is a string of "bit and bytes;" it has no real world meaning. A symbol is termed a word, if the intended real world meaning *participates* in the formal reasoning.

We would like to caution the mathematicians, in group theory, the term "word" is our "symbol."

Phase One: A slice of the real world → a relational table of words.
The first step is to examine how the data are created. The data are results of a knowledge representation. Each word (an attribute name or attribute value) in the table represents some real world facts. Note that the semantics of each word are not implemented and rely on human support (by traditional data processing professionals). Using AI's terminology [3], those attribute names and values (column names and elements in the tables) are the semantic primitives. They are primitives, because they are undefined terms inside the system, yet the symbols do represent (*unimplemented*) human-perceived semantics.

Phase Two: A table of words → A table of symbols.
The second step is to examine how the data are processed by data mining algorithms. In AM, a table of words is used as a table of symbols because data mining algorithms do not consult humans for the semantics of symbols and the semantics are not implemented. Words are treated as "bits and bytes" in AM algorithm.

Phase Three: A table of symbols → high frequency subtuples of symbols.
Briefly, the table of symbols is the only available information in AM. No background knowledge is assumed and used. From a axiomatic point of view, this is where AM is marked very differently from clustering techniques (both are core techniques in data mining [5]); in latter techniques, background knowledge are utilized. Briefly in AM the data are the only "axioms," while in clustering, besides the data, there is the geometry of the ambient space.

Phase Four: Expressions of symbols → expressions of words.
Patterns are discovered as expressions of symbols in the previous phase. In this phase, those individual symbols are interpreted as words again by human experts using the meaning acquired in the representation phase. The key question is: Can such interpreted expressions be realized by some real world phenomena?

3 What are the Data? – Table of Symbols

3.1 Traditional Data Processing View of Data

First, we will re-examine how the data are created and utilized by data processing professionals: Basically, a set of attributes, called relational schema, are selected. Then, a set of real world entities are represented by a table of words. These words, called attribute values, are meaningful words to humans, but their meanings are *not* implemented in the system. In a traditional data processing (**TDP**) environment, DBMS, under *human commands*, processes these data based on human-perceived semantics. However, in the system, for example, COLOR, yellow, blue, and etc are "bits and bytes" without any meaning; they are pure symbols. Using AI's terminology [3], those attribute names and values (column names, and elements in the tables) are the semantic primitives. They are primitives, because they are undefined terms inside the system, yet the symbols do represent (*unimplemented*) human-perceived semantics.

3.2 Syntactic Nature of AM – Isomorphic Tables and Patterns

Let us start this section with an obvious, but a somewhat surprising and important observation. Intuitively, data is a table of symbols, so if we change some or all of the symbols, the mathematical structure of the table will not be changed. So its patterns, e.g., association rules, will be preserved. Formally, we have the following theorem [10, 12]:

Theorem 3.1. *Isomorphic relations have isomorphic patterns.*

Isomorphism classifies the relation tables into isomorphic classes. So we have the following theorem, which implies the syntactic nature of AM. They are patterns of the whole isomorphic class, even though many of isomorphic relations may have very different semantics; see next Sect. 3.3.

Corollary 3.2. *Patterns are property of isomorphic class.*

3.3 Isomorphic but Distinct Semantics

The two relations, Table 1, 2, are isomorphic, but their semantics are completely different, one table is about (hardware) parts, the other is about suppliers (sales persons). These two relations have isomorphic associations;

1. Length one: TEN, TWENTY, MAR, SJ, LA in Table 1 and 10, 20, SCREW, BRASS, ALLOY in Table 2
2. Length two: (TWENTY, MAR), (MAR, SJ), (TWENTY, SJ) in Table 1, (20, SCREW), (SCREW, BRASS), (20, BRASS), Table 2

Table 1. A Table K

V	K	($S\#$	Business Amount (in m.)	Birth Day	CITY)
v_1	$\longrightarrow$	(S_1	TWENTY	MAR	NY
v_2	$\longrightarrow$	(S_2	TEN	MAR	SJ
v_3	$\longrightarrow$	(S_3	TEN	FEB	NY
v_4	$\longrightarrow$	(S_4	TEN	FEB	LA
v_5	$\longrightarrow$	(S_5	TWENTY	MAR	SJ
v_6	$\longrightarrow$	(S_6	TWENTY	MAR	SJ
v_7	$\longrightarrow$	(S_7	TWENTY	APR	SJ
v_8	$\longrightarrow$	(S_8	THIRTY	JAN	LA
v_9	$\longrightarrow$	(S_9	THIRTY	JAN	LA

Table 2. An Table K'

V	K	($S\#$	Weight	Part Name	Material
v_1	$\longrightarrow$	(P_1	20	SCREW	STEEL
v_2	$\longrightarrow$	(P_2	10	SCREW	BRASS
v_3	$\longrightarrow$	(P_3	10	NAIL	STEEL
v_4	$\longrightarrow$	(P_4	10	NAIL	ALLOY
v_5	$\longrightarrow$	(P_5	20	SCREW	BRASS
v_6	$\longrightarrow$	(P_6	20	SCREW	BRASS
v_7	$\longrightarrow$	(P_7	20	PIN	BRASS
v_8	$\longrightarrow$	(P_8	30	HAMMER	ALLOY
v_9	$\longrightarrow$	(P_9	30	HAMMER	ALLOY

However, they have non-isomorphic interesting rules:
We have assumed: Support $\geq$ 3

1 In Table 1, (TWENTY, SJ) is interesting rules; it means the business amount at San Jose is likely 20 millions.

1' However, it is isomorphic to (20, BRASS), which is not interesting at all, because 20 is referred to PIN, not BRASS.

2 In Table 2, (SCREW, BRASS) is interesting; it means the screw is most likely made from BRASS.

2' However, it is isomorphic to (MAR, SJ), which is not interesting, because MAR is referred to a supplier, not to a city.

4 Canonical Models of Isomorphic Class

From *Corollary* 3.2 of Sect. 3.2, we see that we only need to conduct AM in one of the relations in an isomorphic class. The natural question is: Is there a canonical model in each isomorphic class, so that we can do efficient AM in this canonical model. The answer is "yes;" see [10, 12]. Actually, the canonical model has been used in traditional data processing, called bitmap indexes [7].

4.1 Tables of Bitmaps and Granules

In Table 3, the first attributes, F, would have three bit-vectors. The first, for value 30, is 11000110, because the first, second, sixth, and seventh tuple have $F = 30$. The second, for value 40, is 00101001, because the third, fifth, and eighth tuple have $F = 40$; see Table 4 for the full details.

Table 3. A Relational Table K

V	$\rightarrow$	F	G
e_1	$\rightarrow$	30	*foo*
e_2	$\rightarrow$	30	*bar*
e_3	$\rightarrow$	40	*baz*
e_4	$\rightarrow$	50	*foo*
e_5	$\rightarrow$	40	*bar*
e_6	$\rightarrow$	30	*foo*
e_7	$\rightarrow$	30	*bar*
e_8	$\rightarrow$	40	*baz*

Table 4. The bit-vectors and granules of K

F-Value	=Bit-Vectors	=Granules
30	= 11000110	$=(\{e_1, e_2, e_6, e_7\})$
40	= 00101001	$=(\{e_3, e_5, e_8\})$
50	= 00010000	$=(\{e_4\})$
G-Value	=Bit-Vectors	=Granules
Foo	= 10010100	$=(\{e_1, e_4, e_6\})$
Bar	= 01001010	$=(\{e_2, e_7\})$
Baz	= 00100001	$=(\{e_3, e_5, e_8\})$

Using Table 4 as the translation table, the two tables, K and K' in Table 3 are transformed into table of bitmaps, TOB(K) (Table 5. It should be obvious that we will have the exact same bitmap table for K', that is, TOB(K) = TOB(K').

Next, we note that a bit-vector can be interpreted as a subset, called granule, of V. For example, the bit vector, 11000110, of $F = 30$ represents the subset $\{e_1, e_2, e_6, e_7\}$, similarly, 00101001, of $F = 40$ represents the subset $\{e_3, e_5, e_8\}$. As in the bitmap case, Table 3 is transformed into table of granules (TOG), Table 6. Again, it should be obvious that TOG(K) = TOG(K').

Table 5. Table of Symbols K and Table of Bitmaps $TOB(K)$)

	Table K		TOB(K)	
V	F	G	F-bit	G-bit
e_1	30	*foo*	11000110	10010100
e_2	30	*bar*	11000110	01000010
e_3	40	*baz*	00101001	00100101
e_4	50	*foo*	00010000	10010100
e_5	40	*baz*	00101001	01001001
e_6	30	*foo*	11000110	10010100
e_7	30	*bar*	11000110	01000010
e_8	40	*baz*	00101001	00101001

Table 6. Table of Symbols K, and Table of Granules TOG(K)

	Table K		TOG(K)	
V	F	G	F-granule	G-granule
e_1	30	*foo*	$\{e_1, e_2, e_6, e_7\}$	$\{e_1, e_4, e_6\}$
e_2	30	*bar*	$\{e_1, e_2, e_6, e_7\}$	$\{e_2, e_7\}$
e_3	40	*baz*	$\{e_3, e_5, e_8\}$	$\{e_3, e_5, e_8\}$
e_4	50	*foo*	$\{e_4\}$	$\{e_1, e_4, e_6\}$
e_5	40	*baz*	$\{e_3, e_5, e_8\}$	$\{e_3, e_5, e_8\}$
e_6	30	*foo*	$\{e_1, e_2, e_6, e_7\}$	$\{e_1, e_4, e_6\}$
e_7	30	*bar*	$\{e_1, e_2, e_6, e_7\}$	$\{e_2, e_7\}$
e_8	40	*baz*	$\{e_3, e_5, e_8\}$	$\{e_3, e_5, e_8\}$

Proposition 4.1. *Isomorphic tables have the same TOB and TOG.*

4.2 Granular Data Model (GDM) and Association Mining

We will continue our discussions on the canonical model, focusing on the granular data model and its impact on association mining. Note that the collection of F-granules forms a partition, and hence induces an equivalence relation, Q^F; for the same reason, we have Q^G. In fact, this is a fact that has been observed by Tony Lee (1983) and Pawalk (1982) independently [8, 21].

Proposition 4.2. *A subset B of attributes of a relational table K, in particular a single attribute, induces an equivalence relation Q^B on V.*

Pawlak called the pair $(V, \{Q^F, Q^G\})$ a knowledge base. Since knowledge base often means something else, instead, we have called it a granular structure or a granular data model (GDM) in previous occasions. Pawlak stated casually that $(V, \{Q^F, Q^G\})$ and K determines each other; this is slightly inaccurate. The correct form of what he observed should be the following:

Proposition 4.3.

1. *A relational table K determines TOB(K), TOG(K) and GDM(K).*
2. *GDM(K), TOB(K) and TOG(K) determine each other.*
3. *By naming the partitions (giving names to the equivalence relations and respective equivalence classes), GDM(K), TOG(K) and TOB(K) can be converted into a "regular" relational table K′, which is isomorphic to the given table K; there are no mathematical restrictions (except, distinct entities should have distinct names) on how they should be named.*

We will use examples to illustrate this proposition. We have explained how K, and hence TOG(K), determines the GDM(K). We will illustrate the reverse, constructing TOG from GDM. For simplicity, from here on, we will drop the argument K from those canonical models, when the context is clear. Assume we are given a GDM, say a set $V = \{e_1, e_2, \ldots, e_8\}$ and two partitions:

1. Q^1=Q^F=$\{\{e_1, e_2, e_6, e_7\}$, $\{e_4\}$, $\{e_3, e_5, e_8\}\}$,
2. $Q^2 = Q^G = \{\{e_1, e_4, e_6\}, \{e_2, e_7\}, \{e_3, e_5, e_8\}\}$.

The equivalence classes of Q^1 and Q^2 are called elementary granules (or simply granules); and their intersections are called derived granules. We will show next how TOG can be constructed: We place (1) the granule, $gra_1 = \{e_1, e_2, e_6, e_7\}$ on Q^1-column at 1st, 2nd, 6th and 7th rows (because the granule consists of entities e_1, e_2, e_6, and e_7) indexed with ordinals 1st, 2nd, 6th and 7th;

(2) $gra_2 = \{e_4\}$ on Q^1-column at 4th row; and (3) $gra_3 = \{e_3, e_5, e_8\}$ on Q^1-column at 3rd, 5th and 8th rows; these granules fill up Q^1-column.

We can do the same for Q^2-column. Now we have the first part of the proposition; see the right-handed side of the Table 6 and Table 4. To see the second part, we note that by using F and G to name the partitions $Q^j, j = 1, 2$, we will convert TOG and TOB back to K; see the left-handed side of the Table 6 and Table 4.

Previous analysis allows us to term TOB, TOG, and GDM the canonical model. We regard them as different representations of the canonical model: TOB is a bit table representation, TOG is a granular table representation, and GDM is a non-table representation. To be definite, we will focus on GDM; the reasons for such a choice will be clear later. *Proposition* 4.1 and 2, and (*Theorem* 3.1.) allow us to summarize the followings:

Theorem 4.4.

1. *Isomorphic tables have the same canonical model.*
2. *It is adequate to do association mining (AM) in granular data model (GDM).*

In [14], we have shown the efficiency of association mining in such representations.

5 What are the Patterns?

What are the patterns in data mining? We will not tackle the problem with utmost generality, instead we will analyze what has been accepted by the AM community: Associations (high frequent itemsets) and generalized associations (associations in a table with AOG (Attribute Oriented Generalization [4]) are well accepted notion of patterns. Loosely speaking, an association is a conjunction of attribute values, which has high support. Observe that AOG is giving some disjunctions of attribute values new names. So generalized association is a conjunction of these new names, which has high support. If we ignoring the new names generalized association is merely a conjunction of disjunctions, that has high supports. We will generalize the "conjunction of disjunctions" to a general formula. So a high frequency pattern is any formula that has high support. In next section, we will show that such high frequency pattern can be found by solving linear inequalities.

5.1 Horizontal Generalization – Relation Lattice

Let $K = (V, A)$ be a given relational table and its GDM be (V, Q), where $Q = \{Q^1, Q^2, \ldots Q^n\}$ is a set of equivalence relations induced by $A = \{A^1, A^2, \ldots A^n\}$. Tony Lee considers the table $(V, 2^A)$ with the power set 2^A as attributes; see Table 7.

Table 7. Table $(V, 2^A)$ with power set 2^A as attributes

V	F	G	$(F,G) = F \wedge G$
e_1	30	foo	$(30, foo) = 30 \wedge foo$
e_2	30	bar	$(30, bar) = 30 \wedge bar$
e_3	40	baz	$(40, baz) = 40 \wedge baz$
e_4	50	foo	$(50, foo) = 50 \wedge foo$
e_5	40	baz	$(40, baz) = 40 \wedge baz$
e_6	30	foo	$(30, foo) = 30 \wedge foo$
e_7	30	bar	$(30, bar) = 30 \wedge bar$
e_8	40	baz	$(40, baz) = 40 \wedge baz$

Let $\Delta(V)$, called the partition lattice, be the lattice of all partitions (equivalence relations) on V; MEET ($\wedge$) is the standard intersection of partitions, while JOIN($\vee$) is the smallest partition that is coarser than its components. Lee has observed that [8]:

Proposition 5.1. *There is a map $\theta : 2^A \longrightarrow \Delta(V)$ that respects the meet, but not the join.*

Lee called the image, $Im\theta$, the relation lattice and note that

1. The join in $Im\theta$ is different from that of $\Delta(V)$.
2. So $Im\theta$ is a subset, but not a sublattice, of $\Delta(V)$.

Lee established many connections between database concepts and lattice theory. However, such an embedding θ is an unnatural one, so we have, instead, considered the natural embedding

Definition 5.2. *The smallest lattice generated by $Im\theta$, by abuse of language, has been called the (Lin's) relation lattice, denoted by $L(Q)$ [12].*

This definition will not cause confusing, since we will not use Lee's notion at all. The difference between $L(Q)$ and $Im\theta$ is that former contains all the joins of distinct attributes.

The pair $(V, L(Q))$ is called the GDM of relation lattice. It should be obvious that θ maps a q-columns of (V, Q) to 1-column of $(V, L(Q))$. For example,

1. $(Q^1, Q^2) \in$ Table 6$\longrightarrow Q^1 \wedge Q^2 \in$ Table 8.
2. $(\{e_1, e_2, e_6, e_7\}, \{e_2, e_7\}) \longrightarrow \{e_1, e_2, e_6, e_7\} \wedge \{e_2, e_7\} = \{e_2, e_7\}$ in the attribute $Q^1 \wedge Q^2$ of $(V, L(Q))$.

Previous discussions allow us to conclude:

Proposition 5.3. *A q-subtuple in (V, Q) corresponds an 1-subtuple in (V, L(Q)), which is an elementary granule. Moreover, the q-subtuple is an association, if the corresponding elementary granule is large (has adequate supports).*

Table 8. A TOG = $(V, L(Q)$ with a relation lattice $L(Q)$ as attributes

V	Q^1 F-granule	Q^2 G-granule	$Q^1 \wedge Q^2$ (Meet) $F \wedge G$-granule	$Q^1 \vee Q^2$ (Join) $F \vee G$-granule
e_1	$\{e_1, e_2, e_6, e_7\}$ $=30$	$\{e_1, e_4, e_6\}$ $=foo$	$\{e_1, e_2, e_6, e_7\} \wedge \{e_1, e_4, e_6\}$ $= \{e_1, e_6\} = 30 \wedge foo$	$\{e_1, e_2, e_3, e_4, e_5, e_6, e_7, e_8\}$
e_2	$\{e_1, e_2, e_6, e_7\}$ $=30$	$\{e_2, e_7\}$ $=bar$	$\{e_1, e_2, e_6, e_7\} \wedge \{e_2, e_7\}$ $= \{e_2, e_7\} = 30 \wedge bar$	$\{e_1, e_2, e_3, e_4, e_5, e_6, e_7, e_8\}$
e_3	$\{e_3, e_5, e_8\}$ $=40$	$\{e_3, e_5, e_8\}$ $=baz$	$\{e_3, e_5, e_8\} \wedge \{e_3, e_5, e_8\}$ $= \{e_3, e_5, e_8\} = 40 \wedge baz$	$\{e_1, e_2, e_3, e_4, e_5, e_6, e_7, e_8\}$
e_4	$\{e_4\}$ $=50$	$\{e_1, e_4, e_6\}$ $=foo$	$\{e_4\} \wedge \$\{e_1, e_2, e_4, e_6, e_7\}$ $= \{e_4\} = 50 \wedge foo$	$\{e_1, e_2, e_3, e_4, e_5, e_6, e_7, e_8\}$
e_5	$\{e_3, e_5, e_8\}$ $=40$	$\{e_3, e_5, e_8\}$ $=baz$	$\{e_3, e_5, e_8\} \wedge \{e_3, e_5, e_8\}$ $= \{e_3, e_5, e_8\} = 40 \wedge baz$	$\{e_1, e_2, e_3, e_4, e_5, e_6, e_7, e_8\}$
e_6	$\{e_1, e_2, e_6, e_7\}$ $=30$	$\{e_1, e_4, e_6\}$ $=foo$	$\{e_1, e_2, e_6, e_7\} \wedge \{e_1, e_2, e_6, e_7\}$ $= \{e_1, e_6\} = 30 \wedge foo$	$\{e_1, e_2, e_3, e_4, e_5, e_6, e_7, e_8\}$
e_7	$\{e_1, e_2, e_6, e_7\}$ $=30$	$\{e_2, e_7\}$ $=bar$	$\{e_1, e_2, e_6, e_7\} \wedge \{e_2, e_7\}$ $= \{e_2, e_7\} = 30 \wedge bar$	$\{e_1, e_2, e_3, e_4, e_5, e_6, e_7, e_8\}$
e_8	$\{e_3, e_5, e_8\}$ $=40$	$\{e_3, e_5, e_8\}$ $=baz$	$\{e_3, e_5, e_8\} \wedge \{e_3, e_5, e_8\}$ $= \{e_3, e_5, e_8\} = 40 \wedge baz$	$\{e_1, e_2, e_3, e_4, e_5, e_6, e_7, e_8\}$

Note that this proposition only makes assertions about elementary granules that correspond to some q-subtuples. There are elementary granules in $(V, L(Q))$ that do not correspond to any q-subtuples. Here is the first generalization of associations.

Definition 5.4. *An elementary granule(1-subtuple) in $(V,\ L(Q))$ is a generalized association, if and only if the granule is large (has adequate support).*

To illustrate the proposition, we use K as well as GDM forms of Table 8 and Table 7: (Threshold is 2 tuples)

1. Q^1-association:
   a) $Q^1_1 = \{e_1, e_2, e_6, e_7\}$ is large $\longleftrightarrow$ 30 is an association; cardinality = support = 5,
   b) $Q^1_2 = \{e_3, e_5, e_8\}$ is large $\longleftrightarrow$ 40 is an association; cardinality = support = 3.
2. Q^2-association:
   a) $Q^2_1 = \{e_1, e_4, e_6\}$ is large $\longleftrightarrow$ *foo* is an association; cardinality = support = 3,
   b) $Q^2_2 = \{e_2, e_7\}$ is large $\longleftrightarrow$ *bar* is an association; cardinality = support = 2,
   c) $Q^2_3 = \{e_3, e_5, e_8\}$ is large $\longleftrightarrow$ *baz* is an association; cardinality = support = 3.
3. $Q^1 \ \wedge\ Q^2$-association:
   a) $c_1 = \{e_1, e_2, e_6, e_7\} \wedge \{e_1, e_4, e_6\} = \{e_1, e_6\}$ is large $\longleftrightarrow$ $30 \wedge foo$ is an association; cardinality = support = 2,
   b) $c_2 = \{e_1, e_2, e_6, e_7\} \wedge \{e_2, e_7\} = \{e_2, e_7\}$ is large $\longleftrightarrow$ ($30 \wedge bar$ is an association; cardinality = support = 2,
   c) $c_3 = \{e_3, e_5, e_8\} \wedge \{e_3, e_5, e_8\} = \{e_3, e_5, e_8\}$ is large $\longleftrightarrow$ $40 \wedge baz$ is association; cardinality = support = 2.
   d) $c_4 = \{e_4\} \wedge \{e_1, e_4, e_6\} = \{e_4\}$ is not large $\longleftrightarrow$ $40 \wedge foo$ is not an association (cardinality = support = 1)
4. $Q^1 \ \vee\ Q^2$-association: This information can not be considered in K. It is normally considered in AOG on K, in which generalization (the partition) is the whole domain. The partition often named "all." $\{e_1, e_2, e_3, e_4, e_5, e_6, e_7, e_8\}$ is an association; cardinality = support = 8.

Let us write a relational table by (K, A), where K is the table and A is the attributes. Using this notation, $(V, L(Q))$ is almost equivalent to consider a table $(K, 2^A)$; an attribute of latter table is a subset $A^{j_1}A^{j_2}\ldots A^{j_q}$ of original table; see any Database text book, such as [7]. A typical attribute value take the form of a $\wedge$-conjunction format.

5.2 Vertical Generalizations = Attribute Oriented Generalizations

We will recall the most well known generalizations, called Attribute Oriented Generalizations (AOG) or knowledge base [4, 19]. As AOG is more well noted, we will use it. Traditional AOG is based on a given concept hierarchy, which is a nested sequence of named partitions on an (active) attribute domain. However, the essential idea is how to add a named partition as a new attribute into the given relational table. By a named partition we mean that the partition and its equivalence classes all have names; these names have no mathematical consequence. They (for the benefit to humans) can be viewed as summaries of the informational content or interpretations of the real world phenomena.

Again, we will illustrate by examples: The attribute domain that we will look at is the set of F-attribute values in Table 3. Suppose we are given a new named partition, $\{\{30, 50\}, \{40\}\}$, together with the name *Odd* for the first equivalence class, *Even* the second equivalence class, and GF the partition. Such a named partition can be regarded as a new attribute GF (column) to the given table. Similarly a new attribute GG, based on G, is also exhibited in Table 9.

Table 9. A Generalized Table GK with new attributes named GF and GG

V	$\to$	F	G	GF	GG
e_1	$\to$	30	foo	$(30 \vee 50) = Odd$	$(foo) = FO$
e_2	$\to$	30	bar	$(30 \vee 50) = Odd$	$(bar \vee baz) = BA$
e_3	$\to$	40	baz	$(40) = Even$	$(bar \vee baz) = BA$
e_4	$\to$	50	foo	$(30 \vee 50) = Odd$	$(foo) = FO$
e_5	$\to$	40	baz	$(40) = Even$	$(bar \vee baz) = BA$
e_6	$\to$	30	foo	$(30) = Even$	$(foo) = FO$
e_7	$\to$	30	bar	$(30 \vee 50) = Odd$	$(bar \vee baz) = BA$
e_8	$\to$	40	baz	$(40) = Even$	$(bar \vee baz) = BA$

A typical new attribute value in the AOGed Table GK is in the form of a new name or a $\vee$-disjunction. Since GK is an extension, they are exactly the same on original attributes.

As our primary interest is in applying AOG to canonical models, we should ignore the names and focus on those algebraic expressions in Table 9. We will call the associations involving new names symbolic associations. If all the new names in a symbolic association are replaced by the algebraic expression ($\vee$-expressions), we will call such "associations" algebraic associations; see the following examples (support *ge* 2).

1. $Odd \wedge FO = (30 \vee 50) \wedge foo$; supported by $\{e_1, e_4\}$
2. $Odd \wedge BA = (30 \vee 50) \wedge (bar \vee baz)$; supported by $\{e_2, e_7\}$
3. $Even \wedge BA = 40 \wedge (bar \vee baz)$; supported by $\{e_3, e_5, e_8\}$

The purpose of AOG is to get more general patterns. So let us state the results.

Proposition 5.5. *A symbolic association, a $\wedge$-conjunction involving new names in a new AOGed table is an algebraic association, a $\wedge$-conjunction of $\vee$-disjunctions, of symbols from the original table.*

Next, we will summarize the previous discussions in GDM theory: Table 9 induces the following GDM: Note that those symbolic names are not part of the formal mathematical structures; they, hopefully, are used to help readers to identify the sets.

$GDM_{given} = (V, Q^1, Q^2)$

1. $V = \{e_1, e_2, \ldots, e_8\}$
2. $Q^1 = Q^F = \{\{e_1, e_2, e_6, e_7\}(=30), \{e_4\}(=60), \{e_3, e_5, e_8\}(=40)\}$,
3. $Q^2 = Q^G = \{\{e_1, e_4, e_6\}(=foo), \{e_2, e_7\}(=bar), \{e_3, e_5, e_8\}(=baz)\}$.

Note that that the partition $\{\{30, 50\}, \{40\}\}$ of F-attribute induces a new partition on V: From Table 4, we can see 30 defines the granule $\{e_1, e_2, e_6, e_7\}$, 50 defines $\{e_4\}$, and 40 defines $\{e_3, e_5, e_8\}$. *Odd* defines the granule $\{e_1, e_2, e_6, e_7\} \cup \{e_4\} = \{e_1, e_2, e_4, e_6, e_7\}$, and *Even* defines a new granule $\{e_3, e_5, e_8\}$. These two new granules define a new partition (equivalence relation). Similarly BA and FO define a new partition on V. So AOG transforms GDM_{given} to: Again the symbols are not part of mathematics.

$GDM_{extend} = (V, Q^1, Q^2, Q^3, Q^4)$, where the two new partitions are:

1. $Q^3 = \{\{e_1, e_2, e_4, e_6, e_7\}(= odd = 30 \vee 50), \{e_3, e_5, e_8\}(= even = 40)\}$ is a coarsening of Q^1
2. $Q^4 = \{\{e_1, e_4, e_6\}(= FO = foo), \{e_2, e_3, e_5, e_7, e_8\}(= BA = bar \vee baz)\}$ is a coarsening of Q^3

5.3 Full Generalization

The essential idea in the final generalizations is to do AOG on GDM $(V, L(Q))$. In previous expositions we have used rather intuitive terms to review various notions. Now we are ready to provide a precise mathematical formulation, so we will turn to mathematical terminology, the "attribute" is a mathematical "partition," a "generalization" of an attribute is a "coarser" partition. Let us set up some notations: $Q^B = \{Q^{B^1}, Q^{B^2} \ldots Q^{B^q}\}$ be subset of Q, where Q is a set of partitions on V induced by attributes. A subset Q^B of Q defines a new partition in terms of $\wedge$.

$$\bigwedge(Q^B) = Q^{B^1} \wedge Q^{B^2} \ldots \wedge Q^{B^q}.$$

We write $G(Q^B)$ for the set of all generalizations, that is, the set of all partitions that are coarser than $\bigwedge(Q^B)$:

$$G(Q^B) = \{P \mid P \succ \bigwedge(Q^B)\},$$

where $\succ$ means "coarser than". Let Q_1^B and Q_2^B be two subsets. Then $(Q_1^B \cap Q_2^B)$ and $(Q_1^B \cup Q_2^B)$ are the usual set theoretical intersection and union. Let Q_j^B be a typical subset of Q; when the index j varies the typical subset varies through all non-empty subsets of Q. For each Q_j^B, we set

> $G(Q_j^B)$ to be the set of all possible generalization (=partitions that are coarser than $\bigwedge(Q^B)$).

$AG(Q) = \bigcup_j G(Q_j^B)$ is the set of all possible partitions that are coarser than $\bigwedge(Q^B)$ for all possible subsets Q^B.

Recall that $L(Q)$ is the smallest sublattice of $\Delta(V)$ that contains Q. Now we will apply AOG to $(V, L(Q))$. Formally, we define

$$L^*(Q) = \{P \mid \exists X \in L(Q) \text{ such that } P \succ X\}$$

We will show that $L^*(Q)$ is the smallest sublattice of $\Delta(V)$ that contains all coarsening of Q. Formally

Proposition 5.6.

1. $L^*(Q)$ is a lattice.
2. $L^*(Q) = \{P \mid P \succ Min(L(Q))\}$.

where $Min(L(Q))$ is the smallest element in $L(Q)$.

Proof. Let G_1 and G_2 be two elements of $L^*(Q)$. By definition, there are two subsets (Q_j^B) of Q such that

$$G_j \succ \bigwedge(Q_j^B), j = 1,\ 2.$$
$$G_1 \wedge G_2 \succ (\bigwedge(Q_1^B)) \wedge G_2 \succ (\bigwedge(Q_1^B)) \wedge (\bigwedge(Q_2^B)) = \bigwedge(Q_1^B \cap Q_2^B)\ .$$

Obviously, $(Q_1^B \cap Q_2^B)$ is a subset of Q, So $G_1 \wedge G_2 \succ \bigwedge(Q_1^B \cap Q_2^B)$ implies $G_1 \wedge G_2 \in L^*(Q)$; This prove that $L^*(Q)$ is closed under $\wedge$. For $\vee$, it is similar:

$$G_1 \vee G_2 \succ (\bigwedge(Q_1^B)) \vee G_2 \succ (\bigwedge(Q_1^B)) \vee (\bigwedge(Q_2^B)) = \bigwedge(Q_1^B \cup Q_2^B).$$

So $L^*(Q)$ is closed under $\vee$; this proves $L^*(Q)$ is a lattice, the first item.

For the second item, note that $Min(L(Q))$ is the smallest element in $L(Q)$. By definition, any P that is coarser than coarser than $Min(L(Q))$, an attribute in $L(Q)$, is an element in $L^*(Q)$; so the second item is proved.

Definition 5.7. *$(V, L^*(Q))$ is called the Universal Model.*

In [12], the universal means all possible attributes (partitions). Here we will include more, basically, it means all possible information is represented by some elementary granule in the model.

6 Universal Model

The universal model $(V, L^*(Q))$ is universal in two sense: Its attributes (partitions) contains all possible generalized attributes (features) of the given table and all possible information (outputs of all possible SQL-statements).

6.1 Complete Information in a Relational Table

In association mining, vertical generalizations are done implicitly; association of length q (= *wedge* of some attribute values) is horizontal generalizations. The traditional generalized association is constructed as follows: First, do the vertical generalization, namely, AOG, then do the vertical generalization implicitly (by looking at all q-subtuples). The consequence is that the search of patterns is incomplete. The patterns found in traditional approach are $\wedge$-conjunctions (horizontal generalization implicitly) of *some*$\vee$-disjunctions (vertical generalizations, AOG) (*Proposition* 5.5.) Our approach reverses the order, it does horizontal generalization first, then vertical generalization. The consequence is that we $L^*(Q)$ produces $\vee$-disjunctions (vertical generalization) of $\wedge$-conjunctions (horizontal generalization). Such formulas represent all possible information in a relational table. They consist of all possible outputs of SQL statements. Or equivalently, it contains outputs of all relational algebra.

Complete Information Theorem 6.1. Let $(V, L^*(Q))$ be the universal model. Its granule represents $\vee$-disjunctions of $\wedge$-conjunctions of all attributes values in the original table K.

6.2 The Complete Set of Attributes/Features

Feature and attribute have been used interchangeably. In the classical data model, an attribute or a feature is a representation of property, characteristic, and etc.; see e.g., [20]. It represents a human perception about the data; each perception is represented by a symbol, and has been called attribute and the set of attributes a schema. Though the symbols are stored in the system, but not their meanings. So each symbol is a notation or name of a column or a partition. These names play no roles in the processing of columns or partitions. The processing can be carried out without these names. Human interpretations of the data structures are unimportant in AM; of course the data structures themselves are very important in classical data processing.

We often hear such an informal statement "a new feature (attribute) F is selected, extracted, or constructed from a subset $B = \{B^1, B^2, \ldots B^q\}$ of attributes in the table K." What does such a statement mean? Let us assume the new feature F is so obtained. We insert this new feature into the table K, and denote the new table by $(K + F)$. The informal statement probably means in the new table, F should be functionally depends on B. As extraction and construction are informal words, we can use the functional dependency as formal definition of feature selection, extraction and constructions. Formally we define

Definition 6.1. *F is a feature derived (selected, extracted and constructed feature) from B, if F is functional dependent on B in the new table $(K + F)$.*

Let us consider the canonical model $G_{(K+F)}$ in the form of TOG. Then by Lemma, Q_F is coarser than Q_B. In the language of last section, F is a generalization over the set B. So the notion of generalizations (AOG on a set of attributes) include the notion of new features that are selected, extracted, and constructed from given features.

Proposition 6.2. *Derived features (selected, extracted and constructed features) are attribute generalizations (AOG on a set of attributes).*

Feature Completion Theorem 6.2. *Universal model $(V, L^*(Q))$ consist of all derived features.*

Proof. From last proposition, derived features are subset of generalization. On the other hand a generalization should meet the requirement of *Definition* 6.1., hence the set of derived features is the same as the set of generalizations.

This theorem is rather anti-intuitive. Taking human's view there are infinitely many features. But the theorem says there are only finitely many features (as $L^*(Q)$ is a finite set). How one can reconcile the contradiction? Where did the finite-ness slip in? Our analysis says it comes in at the representation phase. We represent the universe by finite words; as these words are interpreted by human, infinite variation may be still there. However, in phase two, suddenly these words are reduced to symbols. Thus the infinite world now is encoded by a finite data set. A common confusing often comes from confusing the data mining as "facts" mining.

7 Generalized Associations in GDM

Let the granules (equivalence classes) of $Min(L(Q))$ be $c_1, c_2, \dots c_m$. Let g be a typical granule in a partition $P \in L^*(Q)$. Since P is a coarsening of $Min(L(Q))$,

$$g = c_{j_1} \cup c_{j_2} \cup \dots c_{j_k} \ ,$$

where $c_{j_1} \cup c_{j_2} \cup \dots c_{j_k}$, is a subset of $c_1, c_2, \dots c_m$.

As we will consider many g, each is a union of some c_j. So we will develop a universal representation. Let us define some notations: We define an operation, denoted by X*x, of a set X and binary number x by the following two equations:

$$X * x = X, \text{ if } x = 1 \text{ and } X \neq \emptyset$$
$$X * x = \emptyset, \text{ if } x = 0 \text{ or } X = \emptyset \ .$$

Then, the expression of g given above can be re-written as

$$g = c_1 * x_1 \cup c_2 * x_2 \cup \dots c_m * x_m$$
$$= \bigcup_i (c_i * x_i), \text{ where } x_i = 1 \text{ if } i \in \{j_1, j_2, \dots j_k\}, \text{ otherwise } x_i = 0.$$

By varying $x_i = 1, \ldots m$, we enumerate all possible granules of $L^*(Q)$. Next, we write $\|g\|$ to represent the cardinality of the set g. Since $c_1, c_2, \ldots c_m$ are disjoints (they are equivalence classes), so the $\cup$ equation induces sumation.

$$\begin{aligned}\|g\| &= \|c_1\| * x_1 + \|c_2\| * x_2 + \ldots \|c_m\| * x_m \\ &= \sum_i (c_i * x_i) .\end{aligned}$$

Now we can state our main theorem

Main Theorem 7.1. *Let $(V, L^*(Q))$ be the universal model,*

1. *(Information Generator) Then every elementary granule, $g \in P$ for some $P \in L^*(Q)$, can be represented by:*
   $g = c_1 * x_1 \cup c_2 * x_2 \cup \ldots c_m * x_m = \bigcup_i (c_i * x_i)$,
   or equivalently
   $g = (c_1, c_2, \ldots c_m) \circ (x_1, x_2, \ldots x_m) = \bigcup_i (c_i * x_i)$, *where $\circ$ is dot product of vectors.*
2. *(Pattern Generator) The granule g is a generalized association if*
   $$** \qquad \sum \|c_i\| * x_i \geq th,$$
   where the notation $\| \bullet \|$ means the cardinality of the set $\bullet$ and th is the given threshold.

Complexity Remarks 7.2.

1. The cardinal number $\|L^*(Q)\|$ is bounded by the Bell number [2] of $m = \|Min(L(Q))\|$. In other words, the total number of generalizations (all partitions $\succ Min(L(O)))$ is bounded by B_m
2. The equation (**) concerns with granules (distinct attribute values), not the partitions (attributes).
3. A solution of the equation (**) is said to be a minimal solution, if no granules in the solution can be dropped. A minimal solution together with any c_j is another solution. So to find the minimal solution is the essential complexity problem of the equation.
4. The length (cardinality) of minimal solution is bounded by the threshold th. So the total number of minimal solution is bounded by$\binom{m}{th}$ of m. The worst case occurs when $n = m$. Therefore the complexity of minimal solution is bounded by $\binom{n}{th}$.

Some Computation: Suppose the table has 97250 rows, and Min(L(Q)) has 28 granules and their cardinalities are: 9500, 9000, 8500, 8000, 7500, 7000, 6500, 6000, 5500, 5000, 4500, 4000, 3500, 3000, 2500, 2000, 1500, 1000, 500, 450, 400, 350, 300, 250, 200, 150, 100, 50.

1. the threshold is 70,000: There are 11,555,001 generalized associations, and 20,406 minimal solution.
2. the threshold is 70,000/2: There are 23,0962,642 generalized associations, and 494,852 minimal solution.
3. the threshold is 70,000/5: There are 267,968,385 generalized associations, and 15,868 minimal solution.

4. the threshold is 70,000/10: There are 268,405,290 generalized associations, and 1,586 minimal solution.
5. the threshold is 70,000/40: There are 268,434,578 generalized associations, and 126 minimal solution.

It is clear if the size of each equivalence class in $Min(L(Q))$ is small then this equation cannot be very useful. But if the size of equivalence classes are large, then the theorem may be very useful. Roughly, if we are looking for short associations, then this theorem may be useful. If we look at the last item, even though there are almost 270 millions associations, the minimal solutions are 126 only.

8 Find Generalized Association Rule by Linear Inequalities – An Example

We will apply this theorem to the relational table given previously: To give a more concise representation, we will suppress the e from the entities. The universe $V = \{e_1, e_2, \ldots, e_8\}$ will be represented by the string 12345678, the three components of the GDM of Table 3 are compressed as follows:

1. $V = \{e_1, e_2, \ldots, e_8\} = 12345678$,
2. $Q^1 = Q^F = \{\{e_1, e_2, e_6, e_7\}, \{e_4\}, \{e_3, e_5, e_8\}\} = 1267, 4, 358$,
3. $Q^2 = Q^G = \{\{e_1, e_4, e_6\}, \{e_2, e_7\}, \{e_3, e_5, e_8\}\} = 146, 27, 358$.

So the given GDM is (V is suppressed).

1. $Q = (Q^1, Q^2) = ((1267, 4, 358), (146, 27, 358))$;

Its relation lattice is:

2. $L = (Q^1, Q^2, Q^1 \wedge Q^2, Q^1 \vee Q^2) = ((1267, 4, 358), (146, 27, 358), (16, 27, 358, 4), (12345678))$

Its complete relation lattice L^* is:

3. L^* = all coarser partitions of $Q^1 \wedge Q^2 = (358, 16, 27, 4)$.
   1. Combine two granules in $Q^1 \wedge Q^2$:
      a) Combine of 358 and the rest: (35816,27,4), (35827,16,4), (3584,16,27)
      b) Combine of 16 and the rest: (358,1627,4), (358,164,27)
      c) Combine of 27 and the rest: (358,16,274)
   2. Grouping granules into groups of 3, 1 and 2, 2 members: (3581627,4), (358164,27), (358274, 16), (15274, 358). (35816,274), (35827,164), (3584, 2716)
   3. Combine all granule into one: (35816274)
   4. Total coarsenings is 15, including the trivial coarsening.

4. The list of all distinct granules $g \in P$ for every $P \in L^*(Q)$: $g_1 = 358, g_2 = 16, g_3 = 27, g_4 = 4, g_5 = 35816, g_6 = 35827, g_7 = 3584, g_8 = 1627, g_9 = 164, g_{10} = 274, g_{11} = 3581627, g_{12} = 358164, g_{13} = 358274, g_{14} = 16274, g_{15} = 35816274$
5. Among the list of all distinct granules: The only non large granule is g_4, so we have 14 generalized associations. In the sequel, we will verify if this solution agrees with the one obtained from the main theorem.

Next we will use *Main Theorem* 7.1. to generate all granules, that is

$$\begin{aligned} g &= (358, 16, 27, 4) \circ (x_1, x_2, x_3, x_4) \\ \|g\| &= (\|358\|, \|16\|, \|27\|, \|4\|) \circ (x_1, x_2, x_3, x_4) \end{aligned}$$

where integer $\circ$ binary is the ordinary product (treat binary number as integer)

1. $(358, 16, 27, 4) \circ (0, 0, 1, X)$
   a) $X = 0 : g_3 = \|27\| * x_3 = 2 \geq 2$
   b) $X = 1 : g_{10} = \|27\| * x_3 + \|4\| * x_4 = 2 + 1 \geq 2$
2. $(358, 16, 27, 4) \circ (0, 1, Y, X)$
   a) $(Y, X) = (0, 0) : g_2 = \|16\| * x_2 = 2 \geq 2$
   b) $(Y, X) = (0, 1) : g_9 = \|16\| * x_2 + \|4\| * x_4 = 2 + 1 \geq 2$
   c) $(Y, X) = (1, 0) : g_8 = \|16\| * x_2 + \|27\| * x_3 = 2 + 2 \geq 2$
   d) $(Y, X) = (1, 1) : g_{14} = \|16\| * x_2 + \|27\| * x_3 + \|4\| * x_4 = 2 + 2 + 1 \geq 2$
3. $(358, 16, 27, 4) \circ (1, Z, Y, X)$
   a) $(Z, Y, X) = (0, 0, 0) : g_1 = \|358\| * x_1 = 3 \geq 2$
   b) $(Z, Y, X) = (0, 0, 1) : g_7 = \|358\| * x_1 + \|4\| * x_4 = 3 + 1 \geq 2$
   c) $(Z, Y, X) = (0, 1, 0) : g_6 = \|358\| * x_1 + \|27\| * x_3 = 3 + 2 \geq 2$
   d) $(Z, Y, X) = (0, 1, 1) : g_{13} = \|358\| * x_1 + \|27\| * x_3 + \|4\| * x_4 = 3 + 2 + 1 \geq 2$
   e) $(Z, Y, X) = (1, 0, 0) : g_5 = \|358\| * x_1 + \|16\| * x_2 = 3 + 2 \geq 2$
   f) $(Z, Y, X) = (1, 0, 1) : g_{12} = \|358\| * x_1 + \|16\| * x_2 + \|4\| * x_4 = 3 + 2 + 1 \geq 2$
   g) $(Z, Y, X) = (1, 1, 0) : g_{11} = \|358\| * x_1 + \|16\| * x_2 + \|27\| * x_3 = 3 + 2 + 2 \geq 2$
   h) $(Z, Y, X) = (1, 1, 1) : g_{15} = \|358\| * x_1 + \|16\| * x_2 + \|27\| * x_3 + \|4\| * x_4 = 3 + 2 + 2 + 1 \geq 2$

In this analysis, we are able to produces all granule by main theorem. The three solutions (first level numbered items) are the minimal solution; they are the "generators" and can be found in $\binom{m}{th}$ $(m = 4, th = 2)$. For this example, Bell number is $B_4 = 15$, that is the number of partitions. While large granules are 14. We will conclude this section with symbolic expressions of each granule:

1. Granules generated by horizontal generalizations
   $g_1 = 358 = 40 \wedge baz$
   $g_2 = 16 = 30 \wedge foo$
   $g_3 = 27 = 30 \wedge bar$
   $g_4 = 4 = 50 \wedge foo$ – non large granule

2. Granules generated by vertical generalizations (AOG)
   $g_5 = 35816 = 40 \wedge baz + 30 \wedge foo$
   $g_6 = 35827 = 40 \wedge baz + 30 \wedge bar$
   $g_7 = 3584 = 40 \wedge baz + 50 \wedge foo$
   $g_8 = 1627 = 30 \wedge foo + 30 \wedge foo$
   $g_9 = 164 = 30 \wedge foo + 50 \wedge foo$
   $g_{10} = 274 = 30 \wedge bar + 50 \wedge foo$
   $g_{11} = 358 + 1627 = 40 \wedge baz + foo \wedge 30 + 30 \wedge bar$
   $g_{12} = 358164 = 40 \wedge baz + foo \wedge +50 \wedge foo$
   $g_{13} = 358274 = 40 \wedge baz + 30 \wedge bar + 50 \wedge foo$
   $g_{14} = 16274 = f00 + 30 \wedge bar + 50 \wedge foo$
   $g_{15} = 35816274 = 40 \wedge baz + foo \wedge 30 + 30 \wedge bar + 50 \wedge foo$

9 Conclusions

Data, patterns, method of derivations, and useful-ness are key ingredients in data mining. In this paper, we formalize the current state of Association Ming: Data are a table of symbols. The patterns are the formulas of input symbols that has high repeatitions. The method of derivations is the most conservative and reliable one, namely, mathematical deductions. The results are somewhat surprising:

1. Patterns are properties of the isomorphic class, not individual relations
2. Un-interpreted attributes (features) are partitions; they can be enumerated.
3. Generalized associations can be found by solving integral linear inequalities. Unfortunately, the number of generalized associations is enormous. This, together with first item, signifies that the current notion of data and patterns (implied by the algorithms) may be too primitive, not mature yet. This also explains why there are so many associations.
4. Real world modeling may be needed to create a much more meaningful notion of patterns. In the current state of Association Mining a pattern is simply repeated data that may have no real world meaning. So we may need to introduce some semantics into the data model; see some initial works in [13, 15, 16].

References

1. R. Agrawal, T. Imielinski, and A. Swami, "Mining Association Rules Between Sets of Items in Large Databases," in Proceeding of ACM-SIGMOD international Conference on Management of Data, pp. 207–216, Washington, DC, June, 1993.
2. Richard A. Brualdi, Introductory Combinatorics, Prentice Hall, 1992.

3. A. Barr and E.A. Feigenbaum, The handbook of Artificial Intelligence, Willam Kaufmann 1981.
4. Y.D. Cai, N. Cercone, and J. Han. "Attribute-oriented induction in relational databases," in Knowledge Discovery in Databases, pp. 213–228. AAAI/MIT Press, Cambridge, MA, 1991.
5. Margaret H. Dunham, Data Mining Introduction and Advanced Topics Prentice Hall, 2003, ISBN 0-13-088892-3.
6. Fayyad U.M., Piatetsky-Sjapiro, G. Smyth, P. (1996) From Data Mining to Knowledge Discovery: An overview. In Fayyard, Piatetsky-Sjapiro, Smyth, and Uthurusamy eds., Knowledge Discovery in Databases, AAAI/MIT Press, 1996.
7. H Gracia-Molina, J. Ullman. & J. Windin, J, Database Systems The Complete Book, Prentice Hall, 2002.
8. T.T. Lee, "Algebraic Theory of Relational Databases," The Bell System Technical Journal Vol 62, No 10, December, 1983, pp. 3159–3204.
9. M. Kryszkiewicz. Concise Representation of Frequent Patterns based on Disjunction-free Generators. In Proceedings of The 2001 IEEE International Conference on Data Mining, San Jose, California, November 29 – December 2, 2001, pp. 305–312.
10. T.Y. Lin, "A Feature/Attribute Theory for Association Mining and Constructing the Complete Feature Set," Chu and Lin (eds), Physica-Verlag, 2005, to appear; see also "Feature Completion," Communication of the Institute of Information and Computing Machinery, Taiwan, Vol 5, No. 2, May 2002, pp. 57–62. (Proceedings of The Workshop on Toward the Foundation of Data Mining in PAKDD2002, May 6–9, Taipei, Taiwan).
11. T.Y. Lin, "Deductive Data Mining: Mathematical Foundation of Database Mining," in: the Proceedings of 9th International Conference, RSFDGrC 2003, Chongqing, China, May 2003, Lecture Notes on Artificial Intelligence LNAI 2639, Springer-Verlag, 403–405.
12. T.Y. Lin, "Attribute (Feature) Completion – The Theory of Attributes from Data Mining Prospect," in: Proceeding of IEEE international Conference on Data Mining, Maebashi, Japan, Dec 9–12, 2002, pp. 282–289.
13. T.Y. Lin, "Data Mining and Machine Oriented Modeling: A Granular Computing Approach," Journal of Applied Intelligence, Kluwer, Vol. 13, No 2, September/October,2000, pp. 113–124.
14. Eric Louie and T.Y. Lin, "Finding Association Rules using Fast Bit Computation: Machine-Oriented Modeling," in: Foundations of Intelligent Systems, Z. Ras and S. Ohsuga (eds), Lecture Notes in Artificial Intelligence #1932, Springer-Verlag, 2000, pp. 486–494. (12th International symposium on methodologies for Intelligent Systems, Charlotte, NC, Oct 11–14, 2000).
15. T.Y. Lin, N. Zhong, J. Duong, S. Ohsuga, "Frameworks for Mining Binary Relations in Data." In: Rough sets and Current Trends in Computing, Lecture Notes on Artificial Intelligence 1424, A. Skoworn and L. Polkowski (eds), Springer-Verlag, 1998, 387–393.
16. E. Louie, T.Y. Lin, "Semantics Oriented Association Rules," In: 2002 World Congress of Computational Intelligence, Honolulu, Hawaii, May 12–17, 2002, 956–961 (paper # 5702).
17. "The Power and Limit of Neural Networks," Proceedings of the 1996 Engineering Systems Design and Analysis Conference, Montpellier, France, July 1–4, 1996, Vol. 7, 49–53.

18. Morel, Jean-Michel and Sergio Solimini, Variational methods in image segmentation: with seven image processing experiments Boston: Birkhuser, 1995.
19. Lin, T.Y. (1989). Neighborhood systems and approximation in database and knowledge base systems. In *Proceedings of the Fourth International Symposium on Methodologies of Intelligent Systems (Poster Session)*, Charlotte, North Carolina, Oct 12, 1989, 75–86.
20. H. Liu and H. Motoda, "Feature Transformation and Subset Selection," IEEE Intelligent Systems, Vol. 13, No. 2, MarchApril, pp. 26–28 (1998).
21. Z. Pawlak, Rough sets. Theoretical Aspects of Reasoning about Data, Kluwer Academic Publishers, 1991.

Comparative Study of Sequential Pattern Mining Models

Hye-Chung (Monica) Kum, Susan Paulsen, and Wei Wang

Computer Science, University of North Carolina at Chapel Hill
(kum, paulsen, weiwang)@cs.unc.edu

Summary. The process of finding interesting, novel, and useful patterns from data is now commonly known as Knowledge Discovery and Data mining (KDD). In this paper, we examine closely the problem of mining sequential patterns and propose a general evaluation method to assess the quality of the mined results. We propose four evaluation criteria, namely (1) recoverability, (2) the number of spurious patterns (3) the number of redundant patterns, and (4) the degree of extraneous items in the patterns, to quantitatively assess the quality of the mined results from a wide variety of synthetic datasets with varying randomness and noise levels. *Recoverability*, a new metric, measures how much of the underlying trend has been detected. Such an evaluation method provides a basis for comparing different models for sequential pattern mining. Furthermore, such evaluation is essential in understanding the performance of approximate solutions. In this paper, the method is employed to conduct a detailed comparison of the traditional frequent sequential pattern model with an alternative approximate pattern model based on sequence alignment. We demonstrate that the alternative approach is able to better recover the underlying patterns with little confounding information under all circumstances we examined, including those where the frequent sequential pattern model fails.

1 Introduction

The process of finding interesting, novel, and useful patterns from data is now commonly known as KDD (Knowledge Discovery and Data mining). In any given KDD problem, the foremost task is to define patterns operationally. The chosen pattern definition will have two impacts: First, it will determine the nature of the algorithms designed to detect the defined patterns, and, second, it will determine how useful the detected patterns are to the data analyst. Thus, the pattern definition must lend itself to the development of efficient algorithms as well as produce patterns of real interest to the end user. The computational efficiency of most KDD definitions/algorithms are usually well documented, however, the efficacy of various definitions in producing useful results has received less attention. Our purpose here is to introduce a general

H.-C. (Monica) Kum et al.: *Comparative Study of Sequential Pattern Mining Models*, Studies in Computational Intelligence (SCI) **6**, 43–70 (2005)
www.springerlink.com

evaluation method to assess the quality of the mined results from sequential pattern mining models, and then to compare the results.

In particular we will focus on the problem of mining sequential patterns. For example, supermarkets often collect customer purchase records in sequence databases in which a sequential pattern would indicate a customer's buying habit. Mining sequential patterns has become an important data mining task with broad applications in business analysis, career analysis, policy analysis, and security.

One common pattern definition in sequential pattern mining is provided by the *support model* [1]: the task is to find the complete set of frequent subsequences in a set of sequences. Much research has been done to efficiently find the patterns defined by the support model, but to the best of our knowledge, no research has examined in detail the usefulness of the patterns actually generated. In this paper, we examined closely the results of the support model to evaluate whether it in fact generates interesting patterns.

To this end, we propose a general evaluation method that can quantitatively assess how well the models can find known patterns in the data. Integral to the evaluation method is a synthetic dataset with known embedded patterns against which we can compare the mined results. For this purpose we have used the well known IBM synthetic data generator [1], which has become a benchmark for evaluating performance in association rule mining and sequential pattern mining. We propose to extend this benchmark to evaluate the quality of the mined results. By mapping the mined patterns back to the base patterns that generated the data, we are able to measure how well the mining models find the real underlying patterns. We can also determine whether or not a model (particular pattern definition) generates any spurious or confounding patterns. All of these aspects of quality can be evaluated under a variety of situations with varying randomness and noise levels.

In addition to the support model, we evaluated an entirely different model for analyzing sequential data, the *multiple alignment sequential pattern mining* model. This model was developed in response to several perceived weaknesses in the support model. For example, the support model fails to efficiently detect trends in sequences if the pattern, as expressed in real data, is subject to noise. As a simple extension from association rule mining (mining patterns in sets), the support model does not efficiently detect trends in sequences.

However, detecting common underlying patterns (called consensus strings or motifs) in simple sequences (strings) has been well studied in computational biology. Current research employs the multiple alignment model to detect consensus strings [6]. Multiple alignment sequential pattern mining extends the work done in string multiple alignment to find the major groups of similar sequences in the database and then uncovers the underlying trend in each group.

Multiple alignment is extended from the simple edit distance problem, where one tries to find an alignment of two sequences such that the edit distance is minimal. In string multiple alignment, the purpose is to find an

Table 1. Multiple alignment of 5 typos of the word *pattern*

seq_1	P A T T T E R N
seq_2	P A {} {} T E R M
seq_3	P {} {} T T {} R N
seq_4	O A {} T T E R B
seq_5	P {} S Y Y R T N
Underlying pattern	P A {} T T E R N

alignment over N strings such that the total pairwise edit distance for all N strings is minimal. A good alignment is one in which similar characters are lined up in the same column. In such an alignment, the concatenation of the common characters in each column would represent the underlying pattern. Table 1 demonstrates how the original word "pattern" was recovered from five typos.

Although our main focus is on developing criteria for evaluating the quality of various sequential pattern mining models, some space will be devoted to describing the two competing models – the support model and the multiple alignment model. We will review how the less well known multiple alignment model is extended to sequences of sets. In summary, we make the following contributions.

- We design an effective evaluation method to assess the quality of the mined results in sequential data.
- We propose a novel model for sequential pattern mining, *multiple alignment sequential pattern mining.*
- We employ the evaluation method for a comparative study of the support model and the alignment model. This paper is the first to examine systematically the results from the support model.
- We derive the expected support for a sequence under the null hypothesis of no patterns in the data. Such analysis allows us to better understand the behavior and limitations of a parameter, *min_sup*, which is integral to the support model.

The remainder of the paper is organized as follows. Sections 2 and 3 provide an overview of the support model and the alignment model respectively. Section 2 includes the analysis of the expected support in random data. Section 4 demonstrates both models through an example. Section 5 details the evaluation method and Sect. 6 reports the results of the comparative study. Section 7 provides an overview of related works. Section 8 concludes with a summary and a discussion of the future works.

Table 2. Notations : a Sequence, an itemset, and the set $\mathcal{I}$ of all items

Items $\mathcal{I}$	Itemset s_{22}	Sequence seq_2
{ A, B, C, $\cdots$, X, Y, Z }	(BCX)	⟨(A)(BCX)(D)⟩

2 Support Model

2.1 Definitions

Let items be a finite set of literals $\mathcal{I} = \{I_1, \ldots, I_p\}$ $(1 \leq p)$. Then an itemset is a set of items from $\mathcal{I}$ and a sequence is an ordered list of itemsets. A database $\mathcal{D}$ is a set of such sequences. We denote a sequence seq_i as $\langle s_{i1} \ldots s_{in} \rangle$ (concatenation of itemsets) and an itemset s_{ij} as $\{x_1, \ldots, x_k\}$ where items x_k are from $\mathcal{I}$. Conventionally, the itemset $s_{ij} = \{x_1, \ldots, x_k\}$ in sequence seq_i is also written as $s_{ij} = (x_{j1} \cdots x_{jk})$. The subscript i refers to the ith sequence, the subscript j refers to the jth itemset in seq_i, and the subscript k refers to the kth item in the jth itemset of seq_i (Table 2).

A sequence $seq_i = \langle X_1 \cdots X_n \rangle$ is called a *subsequence* of another sequence $seq_j = \langle Y_1 \cdots Y_m \rangle$, and seq_j a *supersequence* of seq_i, if $n \leq m$ and there exist integers $1 \leq a_1 < \cdots < a_n \leq m$ such that $X_b \subseteq Y_{a_b}$ $(1 \leq b \leq n)$. That is, seq_j is a supersequence of seq_i and seq_i is a subsequence of seq_j, if and only if seq_i is derived by deleting some items or whole itemsets from seq_j.

Given a sequence database $\mathcal{D}$, the *support* of a sequence seq_p, denoted as $sup(seq_p)$, is the number of sequences in $\mathcal{D}$ that are supersequences of seq_p. Sometimes, $sup(seq_p)$ is expressed as the percentage of sequences in $\mathcal{D}$ that are supersequences of seq_p.

2.2 Problem Statement

Given N sequences of sets and a support threshold, min_sup, find all patterns P such that $sup(P) \geq min\_sup$.

2.3 Methods

The bottleneck in applying the support model occurs when counting the support of all possible frequent subsequences in $\mathcal{D}$. Two classes of algorithms have been developed to efficiently count the support of potential patterns. The apriori based breadth-first algorithms [1, 15] pursue level-by-level candidate-generation-and-test pruning by following the Apriori property: any super-pattern of an infrequent pattern cannot be frequent. In contrast, the projection based depth-first algorithms [2, 13, 19] avoid the costly candidate-generation-and-test process by growing long patterns from short ones. The depth first methods generally do better than the breadth first methods when the data can fit in memory. The advantage becomes more evident when the patterns are long [18].

2.4 Limitations

Although the support model based sequential pattern mining has been extensively studied and many methods have been proposed (e.g., GSP [15], SPAM [2], PrefixSpan [13], and SPADE [19]), three difficulties inherent to the conventional problem definition remain.

First, support alone cannot distinguish between statistically significant patterns and random occurrences. Application of the evaluation method (discussed in Sect. 5) to the support model reveals that for long sequences the model generates a huge number of redundant and spurious patterns. These confounding patterns bury the true patterns in the mined result. Our theoretical analysis of the expected support for short patterns in long sequences confirms that many short patterns can occur frequently simply by chance. To combat a similar problem in association rule mining, Brin has used correlation to find statistically significant associations [4]. Unfortunately, the concept of correlation does not extend easily to sequences of sets.

Second, these methods mine sequential patterns with exact matching. The exact match based model is vulnerable to noise in the data. A sequence in the database supports a pattern if, and only if, the pattern is fully contained in the sequence. However, the exact matching approach may miss general trends in the sequence database. Many customers share similar buying habits, but few follow exactly the same buying pattern. Consider a baby product retail store. Expecting parents who will need a certain number of products over a year constitute a considerable group of customers. Most of them will have similar sequential patterns, but almost none will be exactly the same patterns. Understanding the general trend in the sequential data for expecting parents would be much more useful than finding all frequent subsequences in the group. Thus, to find non-trivial interesting long patterns, we must consider mining *approximate* sequential patterns.

Third, most methods mine the complete set of sequential patterns. When long patterns exist, mining the complete set of patterns is ineffective and inefficient because every sub-pattern of a long pattern is also a pattern [17]. For example, if $\langle a_1 \cdots a_{20} \rangle$ is a sequential pattern, then each of its subsequence is also a sequential pattern. There are $(2^{20} - 1)$ patterns in total! Not only is it very hard for users to understand and manage a huge number of patterns, but computing and storing a huge number of patterns is very expensive – or even computationally prohibitive. In many situations, a user may just want the *long patterns* that cover many short ones.

Mining compact expressions for frequent patterns, such as max-patterns [3] and frequent closed patterns [12], has been recently proposed and studied in the context of frequent itemset mining. However, mining max-sequential patterns or closed sequential patterns is far from trivial. Most of the techniques developed for frequent itemset mining "cannot work for frequent subsequence mining because subsequence testing requires ordered matching which is more difficult than simple subset testing" [17].

Recently, Yan published the first method for mining frequent closed subsequences using several efficient search space pruning methods [17]. Much more work is needed in this area. Nonetheless, in a noisy sequence database, the number of max- or closed sequential patterns can still be huge, and many of them are trivial for users.

2.5 Analysis of Expected Support on Random Data

Implicit in the use of a minimum support for identifying frequent sequential patterns is the desire to distinguish true patterns from those that appear by chance. Yet many subsequences will occur frequently by chance, particularly if (1) the subsequence is short, (2) the data sequence is long, and (3) the items appear frequently in the database. When *min_sup* is set low, the support of those sequences that are frequent by chance can exceed *min_sup* simply by chance. Conversely, an arbitrarily large *min_sup* may miss rare but statistically significant patterns.

We can derive the expected support, $E\{sup(seq)\}$, for a subsequence under the null hypothesis that the probability of an item appearing at a particular position in the sequence is independent of both its position in the sequence and the appearance of other items in the sequence. We will first consider simple sequences of items rather than sequences of itemsets. The analysis is then extended to sequences of itemsets.

The expected support (measured as a percentage) for a subsequence under the null hypothesis is equivalent to the probability that the subsequence appears at least once in an observed sequence. We use the probability of the sequence appearing at least once because an observed sequence in the database can increment the support of a subsequence only once, regardless of the number of times the subsequence appears in it.

$$E\{sup(seq_i)\} = Pr\{seq_i \text{ appears at least once}\} = 1 - Pr\{seq_i \text{ never appears}\} \tag{1}$$

The appendix details the derivation of the expected support for a sequence of length two:

$$\begin{aligned} E\{sup(\mathsf{AB})\} &= 1 - Pr\{\sim (\mathsf{A..B})\} \\ &= 1 - \left(p(\mathsf{A}) \sum_{j=1}^{L} \{[1-p(\mathsf{B})]^{L-j}[1-p(\mathsf{A})]^{j-1}\} + [1-p(\mathsf{A})]^{L} \right) \end{aligned} \tag{2}$$

where A and B represent arbitrary items, not necessarily distinct from each other, $Pr\{(\mathsf{A..B})\}$ denotes the probability that item A will be followed by item B at least one time in a sequence, and $p(\mathsf{A})$ is the probability of item A appearing in the sequence at any particular position.

As illustrated in the appendix, the probability of observing longer subsequences is calculated recursively. To determine $Pr\{\mathsf{A..B..C}\}$ we again calculate $Pr\{\sim (\mathsf{A..B..C})\}$ whose calculation is facilitated by knowing the probability of the shorter sequence B..C:

$$Pr\{\sim (\mathsf{A..B..C})\} = [1 - p(\mathsf{A})]^L + p(\mathsf{A}) \cdot [1 - p(\mathsf{A})]^{j-1} \\ \cdot \sum_{j=1}^{L} \left(p(\mathsf{B}) \sum_{k=1}^{L-j} \{[1 - p(\mathsf{C})]^{L-j-k} [1 - p(\mathsf{B})]^{k-1}\} + [1 - p(\mathsf{B})]^{L-j} \right) \quad (3)$$

from which $Pr\{\mathsf{A..B..C}\}$ can be found by subtraction:

$$E\{sup(\mathsf{ABC})\} = Pr\{(\mathsf{A..B..C}) \text{ appears at least once}\} = 1 - Pr\{\sim (\mathsf{A..B..C})\} \quad (4)$$

The analytical expression for the expected support quickly becomes cumbersome as the length of the subsequence increases, but may be computed for a specific subsequence through recursion.

Now consider a sequence of itemsets. Under some models for how an itemset is constructed the probability of seeing a set of n arbitrary items in the same itemset is simply

$$p(i_1) \cdot p(i_2) \cdots p(i_n) .$$

Then, by substituting the probability of an arbitrary element, $p(i_k)$, with $p(i_1) \cdot p(i_2) \cdots p(i_n)$, in any of the equations above we can derive the expected support of sequence of itemsets. For example, the expected support of a simple two itemset pattern, $\langle (i_1..i_n)(i_1..i_m) \rangle$, can easily be derived from (2) by replacing $p(\mathsf{A})$ with $p(i_1) \cdot p(i_2) \cdots p(i_n)$ and $p(\mathsf{B})$ with $p(i_1) \cdot p(i_2) \cdots p(i_m)$.

The quantities $p(\mathsf{A}), p(\mathsf{B})$, etc., are not known, but must be estimated from the observed database and will necessarily exhibit some sampling error. Consequently, the expression for the expected support of a subsequence, $E\{sup(seq_i)\}$, will also exhibit sampling error. It is, however, beyond the scope of this current work to evaluate its distribution. We merely propose as a first step that both the choice of *min_sup* and the interpretation of the results be guided by an estimate of $E\{sup(seq_i)\}$. In Table 3, we calculated the expected support of $E\{sup(seq = \langle (\mathsf{A})(\mathsf{B}) \rangle)\}$ with respect to L for two items with varying probabilities and for different lengths. As shown in Fig. 1, the expected support grows linearly with respect to L.

3 Multiple Alignment Model

In this section, we present a new model for finding useful patterns in sequences of sets. The power of the model hinges on the following insight: The probability

Table 3. Example expected support

P(A)	P(B)	L=10	L=25	L=50	L=100	L=150
0.01	0.01	0.4266%	2.58%	8.94%	26.42%	44.30%
0.001	0.01	0.0437%	0.28%	1.03%	3.54%	6.83%
0.01	0.001	0.0437%	0.28%	1.03%	3.54%	6.83%
0.001	0.001	0.0045%	0.03%	0.12%	0.46%	1.01%

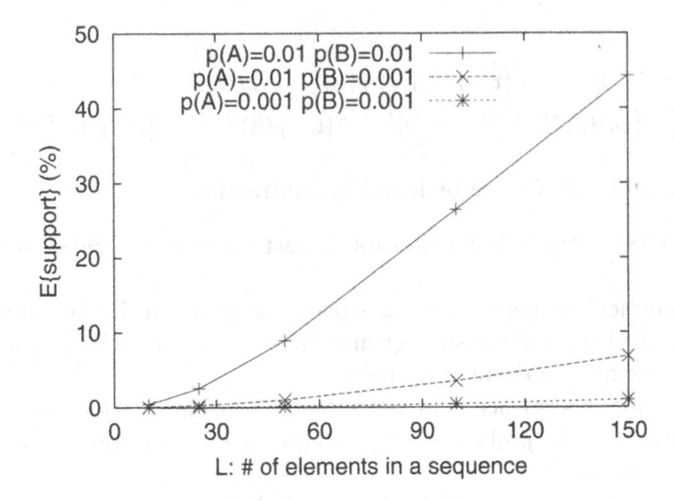

Fig. 1. $E\{sup\}$ w.r.t. L

that any two long data sequences are the same purely by chance is very low. Thus, if a number of long sequences can be aligned with respect to particular items that occur frequently in certain positions, we will have implicitly found sequential patterns that are statistically significant.

3.1 Definitions

This section uses many forward references to the example in Sect. 4, which is used to demonstrate the two sequential mining models. For those unfamiliar with the multiple alignment model, we strongly recommend reading the example section in conjunction with this definition section. Appropriate tables from the example are forward referenced after each definition is given.

The database, $\mathcal{D}$, is defined in the same way as in the support framework.

The *global multiple alignment* of a set of sequences is obtained by inserting a null itemset, (), either into or at the ends of the sequences such that each itemset in a sequence is lined up against a unique itemset or () in all other sequences. In the rest of the paper, alignment will always refer to a global multiple alignment. The upper sections of Tables 7 and 8 show the global multiple alignment of two different group of sequences.

Given two aligned sequences and a distance function for itemsets, the *pairwise score (PS)* between the two sequences is the sum over all positions of the distance between an itemset in one sequence and the corresponding itemset in the other sequence. Given a multiple alignment of N sequences, the *multiple alignment score (MS)* is the sum of all pairwise scores. Then the *optimum multiple alignment* is one in which the multiple alignment score is minimal.

$$PS(seq_i, seq_j) = \Sigma distance(s_{ik_i}(x), s_{jk_j}(x)) \quad \text{(for all aligned positions } x\text{)}$$
$$MS(N) = \Sigma PS(seq_i, seq_j) \quad \text{(over all } 1 \leq i \leq N \wedge 1 \leq j \leq N\text{)} \quad (5)$$

where $s_{ik_i}(x)$ refers to the k_i^{th} itemset for sequence i which has been aligned to position x.

Weighted sequences are an effective method to compress a set of aligned sequences into one sequence. A *weighted itemset*, denoted in (6), is defined as an itemset that has a weight associated with each item in the itemset as well as the itemset itself. Then a weighted sequence, denoted in (6), is a sequence of weighted itemsets paired with a separate weight for the whole sequence. The weight associated with the weighted sequence, n, is the total number of sequences in the set. The weight associated with the itemset ws_j, v_j, represents how many sequences have a nonempty itemset in position j. And the weight associated with each item i_{jk} in itemset ws_j, w_{jk}, represents the total number of the item i_{jk} present in all itemsets in the aligned position j. Tables 7 and 8 give the weighted sequence for each group of aligned sequences.

$$\begin{aligned} &\text{weighted itemset } ws_j = (i_{j1} : w_{j1}, \cdots, i_{jm} : w_{jm}) : v_j \\ &\text{weighted seq } wseq_i \quad = \\ &\langle (i_{11} : w_{11}, \cdots, i_{1s} : w_{1s}) : v_1 \cdots (i_{l1} : w_{l1}, \cdots, i_{lt} : w_{lt}) : v_l \rangle : n \end{aligned} \quad (6)$$

The *strength* of an item, i_{jk}, in an alignment is defined as the percentage of sequences in the alignment that have item i_{jk} present in the aligned position j (7). Clearly, larger strength value indicates that more sequences share the item in the same aligned position.

$$strength(i_{jk}) = w_{jk}/n * 100\% \quad (7)$$

Given a threshold, *min_strength*, and a multiple alignment of N sequences, the *consensus itemset* for position j in the alignment is an itemset of all items that occur in at least *min_strength* sequences in position j. Then a *consensus sequence* is simply a concatenation of the consensus itemsets for all positions excluding any null consensus itemsets. When weights are included it is called a *weighted consensus sequence.*

$$\text{Consensus itemset } (j) = \{x_{jk} | \forall x_{jk} \in ws_j \wedge strength(x_{jk}) \geq min\_strength\} \quad (8)$$

Based on item strengths, items in an alignment are divided into three groups: rare items, common items, and frequent items. The rare items may represent noise and are in most cases not of any interest to the user. The frequent items occur in enough of the sequences to constitute the underlying pattern in the group. The common items do not occur frequently enough to be part of the underlying pattern but occur in enough sequences to be of interest. The common items constitute variations on the general pattern. That is, they are the items most likely to occur regularly in a subgroup of the sequences. Using this categorization we make the final results more understandable by defining two types of consensus sequences corresponding to two thresholds: (1) The *pattern consensus sequence*, which is composed solely of frequent items and (2) the *variation consensus sequence*, an expansion of the pattern consensus sequence to include common items (Table 9). This method presents both the frequent underlying patterns and their variations while ignoring the noise. It is an effective method to summarize the alignment because the user can clearly understand what information is being dropped. Furthermore, the user can control the level of summarization by defining the two thresholds for frequent (θ) and rare (δ) items as desired.

3.2 Problem Statement

Given (1) N sequences, (2) a distance function for itemsets, and (3) strength cutoff points for frequent and rare items (users can specify different cutoff points for each partition), the problem of *multiple alignment sequential pattern mining* is (1) to partition the N sequences into K sets of sequences such that the sum of the K multiple alignment scores is minimum, (2) to find the optimal multiple alignment for each partition, and (3) to find the pattern consensus sequence and the variation consensus sequence for each partition.

3.3 Methods

The exact solution to multiple alignment pattern mining is NP-hard, and therefore too expensive to be practical. An efficient approximation algorithm, ApproxMAP (for APPROXimate Multiple Alignment Pattern mining), was introduced in [7]. ApproxMAP has three steps. First, k nearest neighbor clustering is used to approximately partition the database. Second, for each partition, the optimal multiple alignment is approximated by the following greedy approach: in each partition, two sequences are aligned first, and then a sequence is added incrementally to the current alignment of $p-1$ sequences until all sequences have been aligned. At each step, the goal is to find the best alignment of the added sequence to the existing alignment of $p-1$ sequences. Third, based on user defined thresholds the weighted sequence of each partition is used to generate two consensus sequences per partition, the pattern and the variation consensus sequences. To further reduce the data presented to the user, a simple gamma-corrected color coding scheme is used to represent the item strengths in the patterns. Table 9 illustrates the color-coding scheme.

The definition of the distance function for itemsets, $REPL$, is given in Table 4. $REPL$ is mathematically equivalent to the Sorensen coefficient, an index similar to the Jaccard coefficient except that it gives more weight to the common elements [10]. Thus, it is more appropriate if the commonalities are more important than the differences.

Table 4. Inter itemset distance metric

$REPL(X,Y) = \frac{\|(X-Y)\cup(Y-X)\|}{\|X\|+\|Y\|} = \frac{\|X\|+\|Y\|-2\|X\cap Y\|}{\|X\|+\|Y\|}$	$(0 \leq REPL() \leq 1)$
$INDEL(X) = REPL(X,()) = 1$	

3.4 Discussion

Multiple alignment pattern mining has many practical applications. It is a versatile exploratory data analysis tool for sequential data, because it organizes and summarizes the high dimensional data into something that can be viewed by people. Multiple alignment pattern mining summarizes interesting aspects of the data as follows: (1) The partitioning through k nearest neighbor clustering and subsequent multiple alignment within a cluster organizes the sequences, (2) the weighted sequences provide a compressed expression of the full database, and (3) the weighted consensus sequences provides a summary of each cluster's pattern at a user specified level. In addition, given the appropriate threshold, consensus sequences are patterns that are approximately similar to many sequences in the database. Note that once users have found interesting patterns, they can use the more efficient pattern search methods to do confirmatory data analysis.

Multiple alignment pattern mining is also an effective method for clustering similar sequences. This has the following benefits. First, it enables multiple alignment on a group of necessarily similar sequences to find the underlying pattern (consensus sequence). Second, once sequences are grouped the user can specify the threshold, *min_strength*, specific to each group. This is in contrast to the support model in which *min_sup* is specified against the whole database. Frequent uninteresting patterns, which are usually grouped into large partitions, will have a higher threshold than infrequent interesting patterns, which tend to be grouped into small partitions. Thus, frequent uninteresting patterns will not flood the results as they do in the support model.

4 Example

Table 5 is a sequence database $\mathcal{D}$. Although the data is lexically sorted it is difficult to gather much information from the raw data even in this tiny example.

Table 6 displays the results from the support model. For readability, we manually found the the maximal sequential patterns and show them in bold. Note that finding the maximal patterns automatically in sequential data is a non-trivial task.

The ability to view Table 5 is immensely improved by using the alignment model – grouping similar sequences, then lining them up (multiple alignment), and coloring the consensus sequences as in Tables 7 through 9. Note that the

Table 5. Sequence database $\mathcal{D}$ lexically sorted

ID	Sequences
seq_4	⟨(A) (B) (DE)⟩
seq_2	⟨(A) (BCX) (D)⟩
seq_3	⟨(AE) (B) (BC) (D)⟩
seq_7	⟨(AJ) (P) (K) (LM)⟩
seq_5	⟨(AX) (B) (BC) (Z) (AE)⟩
seq_6	⟨(AY) (BD) (B) (EY)⟩
seq_1	⟨(BC) (DE)⟩
seq_9	⟨(I) (LM)⟩
seq_8	⟨(IJ) (KQ) (M)⟩
seq_{10}	⟨(V) (PW) (E)⟩

Table 6. Results from the support model given $\mathcal{D}$ ($min\_sup = 20\% = 2$ seq)

id	Pattern	Support	id	Pattern	Support	id	Pattern	Support
1	⟨(A)(B)⟩	5	11	**⟨(B)(DE)⟩**	2	21	⟨(A)(B)(C)⟩	2
2	⟨(A)(C)⟩	3	12	⟨(C)(D)⟩	3	22	⟨(A)(B)(D)⟩	2
3	⟨(A)(D)⟩	4	13	⟨(C)(E)⟩	2	23	⟨(A)(B)(E)⟩	2
4	⟨(A)(E)⟩	3	14	⟨(I)(M)⟩	2	24	**⟨(A)(B)(BC)⟩**	2
5	⟨(A)(BC)⟩	3	15	⟨(J)(K)⟩	2	25	⟨(A)(C)(D)⟩	2
6	⟨(B)(B)⟩	3	16	⟨(J)(M)⟩	2	26	**⟨(A)(BC)(D)⟩**	2
7	⟨(B)(C)⟩	2	17	⟨(K)(M)⟩	2	27	⟨(B)(B)(E)⟩	2
8	⟨(B)(D)⟩	4	18	⟨(BC)(D)⟩	3	28	**⟨(J)(K)(M)⟩**	2
9	⟨(B)(E)⟩	4	19	**⟨(BC)(E)⟩**	2	29	**⟨(A)(B)(B)(E)⟩**	2
10	⟨(B)(BC)⟩	2	20	⟨(A)(B)(B)⟩	2			

patterns ⟨(A)(BC)(DE)⟩ and ⟨(IJ)(K)(LM)⟩ do not match any sequence exactly. Therefore, these patterns are not detected in the support model.

Given the input data shown in Table 5 ($N = 10$ sequences), ApproxMAP (1) calculates the $N * N$ sequence to sequence distance matrix from the data, (2) partitions the data into two clusters ($k = 2$), (3) aligns the sequences in each cluster (Tables 7 and 8) – the alignment compresses all the sequences in each cluster into one weighted sequence per cluster, and (4) summarizes the weighted sequences (Tables 7 and 8) into weighted consensus sequences (Table 9).

5 Evaluation Method

In this section, we present a method that objectively evaluates the quality of the results produced by any sequential pattern mining method when applied to the output of a well-known IBM synthetic data generator [1]. The evaluation method is a matrix of four experiments – (1) random data, (2) pattern data, and pattern data with (3) varying degrees of noise, and (4) varying numbers of random sequences – assessed on four criteria: (1) recoverability, (2) the

Table 7. Cluster 1 ($\theta = 50\% \wedge w \geq 4$)

seq_2	⟨(A)	()	(BCX)	()	(D)⟩	
seq_3	⟨(AE)	(B)	(BC)	()	(D)⟩	
seq_4	⟨(A)	()	(B)	()	(DE)⟩	
seq_1	⟨()	()	(BC)	()	(DE)⟩	
seq_5	⟨(AX)	(B)	(BC)	(Z)	(AE)⟩	
seq_6	⟨(AY)	(BD)	(B)	()	(EY)⟩	
seq_{10}	⟨(V)	()	()	(PW)	(E)⟩	
Weighted Sequence	(A:5, E:1, X:1,Y:1):5	(B:3, D:1):3	(B:6, C:4, V:1, X:1):7	(P:1,W:1, Z:1):2	(A:1,D:4, E:5,Y:1):7	7
Pattern Consensus Seq	⟨(A)		(BC)		(DE)⟩	
Weighted PatConSeq	⟨(A:5)		(B:6, C:4)		(D:4, E:5)⟩	7
Variation Consensus Seq	⟨(A)	(B)	(BC)		(DE)⟩	
Weighted VarConSeq	⟨(A:5)	(B:3)	(B:6, C:4)		(D:4, E:5)⟩	7

Table 8. Cluster 2 ($\theta = 50\% \wedge w \geq 2$)

seq_8	⟨(IJ)	()	(KQ)	(M)⟩	
seq_7	⟨(AJ)	(P)	(K)	(LM)⟩	
seq_9	⟨(I)	()	()	(LM)⟩	
Weighted Sequence	⟨(A:1,I:2,J:2):3	(P:1):1	(K:2,Q:1):2	(L:2,M:3):3	3
Pattern Consensus Seq	⟨(IJ)		(K)	(LM)⟩	
Weighted Pattern ConSeq	⟨(I:2, J:2):3		(K:2):2	(L:2, M:3):3⟩	3

Table 9. Results from the alignment model: Consensus Seqs ($\theta = 50\%, \delta = 20\%$)

100%: 85%: 70%: 50%: 35%: 20%		
Pattern Consensus Seq$_1$	strength = 50% = 3.5 > 3 seqs	**(A)(BC)(DE)**
Variation Consensus Seq$_1$	strength = 20% = 1.4 > 1 seqs	**(A)(B)(BC)(DE)**
Pattern Consensus Seq$_2$	strength = 50% = 1.5 > 1 seqs	**(IJ)(K)(LM)**
Variation Consensus Seq$_2$	Not appropriate in this small set	

number of spurious patterns, (3) the number of redundant patterns, and (4) the level of extraneous items in the result. *Recoverability*, defined in Sect. 5.2, provides a good estimate of how well the underlying trends in the data are detected. This evaluation method will enable researchers not only to use the data generator to benchmark performance but also to quantify the quality of the results. Quality benchmarking will become increasingly important as more data mining methods focus on approximate solutions.

Using this method, we compare the quality of the results from the support model and the alignment model. Note that all methods generate the same results for the support model. In contrast, the exact solution to the multiple alignment pattern mining is NP-hard, and in practice all solutions are approximate. Consequently the results of the alignment model are method dependent. We use the results of ApproxMAP to represent the alignment model.

5.1 Synthetic Data

Given several parameters (Table 10), the IBM data generator given in [1] produces the database and reports the base patterns used to generate it. The data is generated in two steps. First, it generates N_{pat} potentially frequent sequential patterns, called base patterns, according to L_{pat} and I_{pat}. Second, each sequence in the database is built by combining base patterns until the size required, determined by L_{seq} and I_{seq}, is met. Along with each base pattern, the data generator reports the expected frequency, $E(F_B)$, and the expected length (total number of items), $E(L_B)$, of the base pattern in the database. The $E(F_B)$ is given as a percentage of the size of the database and the $E(L_B)$ is given as a percentage of the number of items in the base pattern.

Random data is generated by assuming independence between items both within and across itemsets.

Table 10. Parameters for the synthetic database

Notation	Meaning	Default value
N_{seq}	# of data sequences	1000
L_{seq}	Avg. # of itemsets per data sequence	10
I_{seq}	Avg. # of items per itemset in the database	2.5
N_{pat}	# of base pattern sequences	10
L_{pat}	Avg. # of itemsets per base pattern	7
I_{pat}	Avg. # of items per itemset in base patterns	2
$\|\|I\|\|$	# of items	100

5.2 Evaluation Criteria

The effectiveness of a model can be evaluated in terms of how well it finds the real underlying patterns in the database (the base patterns), and whether or not it generates any confounding information. To the best of our knowledge, no previous study has measured how well the various methods recover the known base patterns in the data generator in [1]. In this section, we propose a new measure, *recoverability*, to evaluate the match between the base patterns and the result patterns. Expanding it, we propose a vector of four criteria – (1) recoverability, (2) the number of spurious patterns, (3) the number of redundant patterns, and (4) the level of extraneous items in the result patterns – which can comprehensively depict the quality of the result patterns.

The quality of the results from a pattern mining algorithm is determined by examining each of the result patterns output from a method closely. We only consider result patterns with more than one itemset in the sequence as a valid pattern because one-itemset patterns are obviously not meaningful sequential patterns and can be dismissed easily. Furthermore, in the multiple alignment model, only the pattern consensus sequences are considered.

Table 11. Evaluation Criteria

Notation	Definition
Recoverability	How much of the base patterns has been detected by the result
N_{item}	Total number of items in all the result patterns
N_{extraI}	Total number of extraneous items in all the result patterns
$\frac{N_{extraI}}{N_{item}}$	Level of extraneous items: % of incorrect predictions in the result
N_{total}	Total number of result patterns = N_{max}+N_{spur}+N_{redun}
N_{max}	Total number of max patterns = number base patterns detected
N_{spur}	Total number of spurious patterns
N_{redun}	Total number of redundant patterns

In order to measure these four criteria, we match the resulting patterns from each method to the most similar base pattern. That is, the result pattern, P_j, is matched with the base pattern, B_i, if the longest common subsequence between P_j and B_i, denoted as $B_i \otimes P_j$, is the maximum over all base patterns. The items shared by P_j and B_i, $B_i \otimes P_j$, are defined as *pattern items.* The remaining items in P_j are defined as *extraneous items.* The overall level of extraneous items in the set of result patterns is the total number of extraneous items over the total number of all items in the full set of result patterns.

Depending on the mix of pattern and extraneous items, each result pattern is categorized as either a spurious, max, or redundant pattern. *Spurious patterns* are result patterns with more extraneous items than pattern items. Of the remaining, *max patterns* are result patterns whose $\|B_i \otimes P_j\|$[1] is the longest over all result patterns mapped to a given base pattern B_i. Note that the number of max patterns is equal to the number of base patterns detected. The rest of the result patterns, P_a, are *redundant patterns* as there exists a max pattern, P_{max}, that matches with the same base pattern but better.

Next we measure how well a model detects the underlying base pattern by evaluating the quality of the max patterns. The number of base patterns found or missed is not alone an accurate measure of how well the base patterns were detected because it can not take into account which items in the base pattern were detected or how strong (frequent) the patterns are in the data. The base patterns are only potentially frequent sequential patterns. The actual occurrence of a base pattern in the data, which is controlled by $E(F_B)$ and $E(L_B)$, varies widely. $E(F_B)$ is exponentially distributed then normalized to sum to 1. Thus, some base patterns with tiny $E(F_B)$ will not exist in the data or occur very rarely. Recovering these patterns are not as crucial as recovering the more frequent base patterns. $E(L_B)$ controls how many items in the base patterns, on average, are injected into one occurrence of the base pattern in a sequence. $E(L_B)$ is normally distributed.

For example, for each base pattern given in Table 13 the $E(F_B)$ and $E(L_B)$ is given in parenthesis. The table is sorted by $E(F_B)$ of the base patterns. Here

[1] $\|seq_i\|$=length of seq_i denotes the total number of items in seq_i

we see that for 10 base patterns the most frequent pattern occurs in 21% of the data ($E(F_B)$=21% for $BaseP_1$). On the other hand the weakest pattern in the data occurs in only 0.8% of the data ($E(F_B)$=0.8% for $BaseP_{10}$).

We designed a weighted measure, *recoverability*, which can more accurately evaluate how well the base patterns have been recovered. Specifically, given a set of base patterns, B_i, the expected frequency $E(F_B)$ and the expected length $E(L_B)$ of each base pattern in the data, and a set of result patterns, P_j, we define the recoverability as,

$$\text{Recoverability } \mathcal{R} = \sum_{\text{base pat } \{B_i\}} E(F_{B_i}) \cdot \min \left\{ \begin{array}{c} 1 \\ \left(\frac{\max_{\text{rslt pat } \{P_j(i)\}} \|B_i \otimes P_j\|}{E(L_{B_i}) \cdot \|B_i\|} \right) \end{array} \right. \tag{9}$$

$\max\|B_i \otimes P_j\|$ is the number of items in the max pattern, P_j, shared with the base pattern B_i. Since $E(L_B)$ is an expected value, sometimes the actual observed value, $\max\|B_i \otimes P_j\|$ is greater than $E(L_B) \cdot \|B_i\|$. In such cases, we truncate the value of $\frac{\max \|B_i \otimes P_j\|}{E(L_{B_i}) \cdot \|B_i\|}$ to 1 so that recoverability stays between 0 and 1. When recoverability of the mining is high, major portions of the base patterns have been found.

In summary, a good model would produce high recoverability with small numbers of spurious and redundant patterns and a low level of extraneous items. Table 11 summarizes the evaluation criteria and the notations.

6 Results

6.1 Spurious Patterns in Random Datasets

We generated random datasets with parameters N_{seq}, I_{seq}, and $\|I\|$ as given in Table 10 and varied L_{seq} to test empirically how many spurious patterns are generated from random data.

The support model has no mechanism to eliminate patterns that occur simply by chance. As seen in the theoretical analysis (Sect. 2.5), when sequences are long, short patterns can occur frequently simply by chance. Consequently, support alone cannot distinguish between statistically significant patterns and random sequences.

Consistent with the theoretical analysis, the support model generates huge number of spurious patterns given random data (Fig. 2(a)). The number of spurious patterns increase exponentially with respect to L_{seq}. This follows naturally from Fig. 1 which show that $E(sup)$ of sequences with two itemsets increase linearly with respect to L_{seq}. Thus, the total of all patterns, $1 \leq L_{pat} \leq max_L$, should increase exponentially. When min_sup=5% and $L_{seq} = 30$, there are already 53,471 spurious patterns. In real applications, since $min\_sup << 5\%$, and $N_{seq} > 30$, there will be many spurious patterns mixed in with true patterns.

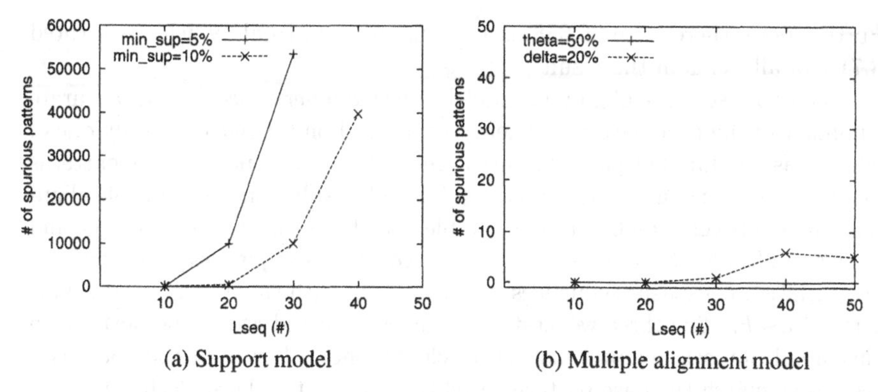

(a) Support model (b) Multiple alignment model

Fig. 2. Number of spurious patterns in random data

In contrast, the probability of a group of random sequences aligning well enough to generate a consensus sequence is negligible. Thus, using default settings ($k = 5$ and $\theta = 50\%$), the multiple alignment model found no spurious patterns in any datasets $L_{seq} = 10..50$ (Fig. 2(b)). In Fig. 2(b), we also report the number of spurious patterns that occur at the default cutoff point for variation consensus sequences, $\delta = 20\%$. We see that when sequences are longer, there are a few negligible number of spurious patterns (1, 6, and 5 when L_{seq}=30, 40, and 50 respectively) generated at this low cutoff point.

6.2 Baseline Study of Patterned Database

This experiment serves several purposes. First, it evaluates how well the models detect the underlying patterns in a simple patterned dataset. Second, it illustrates how readily the results may be understood. Third, it establishes a baseline for the remaining experiments. We generate 1000 sequences from 10 base patterns as specified in Table 10. We tuned both models to the optimal parameters.

The recoverability for both models was good at over 90% (Table 12). However, in the support model it is difficult to extract the 10 base patterns from results that include 253,714 redundant patterns and 58 spurious patterns.

Table 12. Results from patterned data

	Multiple Alignment Model $k = 6$ & $\theta = 30\%$	Support Model $min\_sup = 5\%$
Recoverability	91.16%	91.59%
$\frac{N_{extraI}}{N_{item}}$	$\frac{3}{106} = 2.83\%$	$\frac{66,058}{1,782,583} = 3.71\%$
N_{total}	8	253,782
N_{max}	7	10
N_{spur}	0	58
N_{redun}	1	253,714

Furthermore, there were 66,058 extraneous items in total, which constituted 3.71% of all items in the result patterns.

In comparison, the alignment model returned a very succinct but accurate summary of the base patterns. There was in total only 8 patterns, only one of which was a redundant pattern. Furthermore, there were no spurious patterns and only 3 extraneous items (Table 12). The full result from the multiple alignment model is very small and manageable. The full result fits on one page and is given in Table 13. It shows the pattern consensus sequences, $PatConSeq_i$, the variation consensus sequences, $VarConSeq_i$, with the matching base patterns, $BaseP_i$. The three weakest base patterns missed are at the bottom. In this small dataset, manual inspection clearly shows how well the consensus sequences match the base patterns used to generate the data. Each consensus pattern found was a subsequence of considerable length of a base pattern. The 17 consensus sequences together provide a good overview of the 1000 data sequences. Note that we have 9 clusters but only 8 pattern consensus sequence because one cluster had no items with strength greater than $\theta = 30\%$ – i.e. the pattern consensus sequence $PatConSeq_9$ was a null sequence.

Many of the redundant patterns in the support model are either subsequences of a longer pattern or a small variation on it. They hold little additional information and instead make it difficult to see the real base patterns. Although, finding max or closed patterns has been researched in the context of frequent itemset mining [3, 9], extending these methods to sequential patterns is not trivial.

In contrast, the one redundant pattern in the alignment model has useful information. Sequences in the partition that generated $PatConSeq_3$ (Table 13) were separated out from the partition that generated $PatConSeq_2$ because they were missing most of the seven items at the end of $PatConSeq_3$. Thus, when the redundant patterns in the alignment model are subsequences of a longer pattern, it alerts the user that there is a significant group of sequences that do not contain the items in the longer pattern.

6.3 Robustness with Respect to Noise

We evaluate the robustness of the models with respect to varying degrees of noise in the data. Noise is introduced into the patterned data (as discussed in Sect. 5.1) using a corruption probability α. Items in the database are randomly changed into another item or deleted with probability α. This implies that $1 - \alpha$ is the probability of any item remaining constant. Hence, when $\alpha = 0$ no items are changed, and higher values of α imply a higher level of noise [18].

The results show that the support model is vulnerable to random noise injected into the data. As seen in Table 14 and Fig. 3, as the corruption factor, α, increases, the support model detects less of the base patterns and incorporates more extraneous items in the patterns. When the corruption factor is 20%, the recoverability degrades significantly to 54.94%. Such results are expected since the model is based on exact match. Note that even when

Table 13. Consensus seqs & the matching base patterns ($k = 6$, $\theta = 30\%$, $\delta = 20\%$)

$BaseP_i$ ($E(F_B)$:$E(L_B)$)	‖ P ‖	**Pattern <100: 85: 70: 50: 35: 20>**
$BaseP_1$ (0.21:0.66)	14	<(15 16 17 66) (15) (58 99) (2 74) (31 76) (66) (62) (93) >
$PatConSeq_1$	13	**<(15 16 17 66) (15) (58 99) (2 74) (31 76) (66) (62)>**
$VarConSeq_1$	18	**<(15 16 17 66) (15 22) (58 99) (2 74) (24 31 76) (24 66) (50 62) (93)>**
$BaseP_2$ (0.161:0.83)	22	<(22 50 66) (16) (29 99) (94) (45 67) (12 28 36) (50) (96) (51) (66) (2 22 58) (63 74 99) >
$PatConSeq_2$	19	**<(22 50 66) (16) (29 99) (94) (45 67) (12 28 36) (50) (96) (51) (66) (2 22 58)>**
$VarConSeq_2$	25	**<(22 50 66) (16) (29 99) (22 58 94) (2 45 58 67) (12 28 36) (2 50) (24 96) (51) (66) (2 22 58)>**
$PatConSeq_3$	15	**<(22 50 66) (16) (29 99) (94) (45 67) (12 28 36) (50) (96) (51)>**
$VarConSeq_3$	15	**<(22 50 66) (16) (29 99) (94) (45 67) (12 28 36) (50) (96) (51)>**
$BaseP_3$ (0.141:0.82)	14	< (22) (22) (58) (2 16 24 63) (24 65 93) (6) (11 15 74) >
$PatConSeq_4$	11	**<(22) (22) (58) (2 16 24 63) (24 65 93) (6)>**
$VarConSeq_4$	13	**<(22) (22) (22) (58) (2 16 24 63) (2 24 65 93) (6 58)>**
$BaseP_4$ (0.131:0.90)	15	<(31 76) (58 66) (16 22 30) (16) (50 62 66) (2 16 24 63) >
$PatConSeq_5$	11	**<(31 76) (58 66) (16 22 30) (16) (50 62 66)>**
$VarConSeq_5$	11	**<(31 76) (58 66) (16 22 30) (16) (50 62 66) (16 24)>**
$BaseP_5$ (0.123:0.81)	14	<(43) (2 28 73) (96) (95) (2 74) (5) (2) (24 63) (20) (93) >
$PatConSeq_6$	13	**<(43) (2 28 73) (96) (95) (2 74) (5) (2) (24 63) (20)>**
$VarConSeq_6$	16	**<(22 43) (2 28 73) (58 96) (95) (2 74) (5) (2 66) (24 63) (20)>**
$BaseP_6$ (0.121:0.77)	9	<(63) (16) (2 22) (24) (22 50 66) (50) >
$PatConSeq_7$	8	**<(63) (16) (2 22) (24) (22 50 66)>**
$VarConSeq_7$	9	**<(63) (16) (2 22) (24) (22 50 66)>**
$BaseP_7$ (0.054:0.60)	13	<(70) (58 66) (22) (74) (22 41) (2 74) (31 76) (2 74) >
$PatConSeq_8$	16	**<(70) (58) (22 58 66) (22 58) (74) (22 41) (2 74) (31 76) (2 74)>**
$VarConSeq_8$	18	**<(70) (58 66) (22 58 66) (22 58) (74) (22 41) (2 22 66 74) (31 76) (2 74)>**
$PatConSeq_9$	0	cluster size was only 5 sequences so no pattern consensus sequence was produced
$VarConSeq_9$	8	**<(70) (58 66) (74) (74) (22 41) (74)>**
$BaseP_8$ (0.014:0.91)	17	< (20 22 23 96) (50) (51 63) (58) (16) (2 22) (50) (23 26 36) (10 74) >
$BaseP_9$ (0.038:0.78)	7	< (88) (24 58 78) (22) (58) (96) >
$BaseP_{10}$ (0.008:0.66)	17	< (16) (2 23 74 88) (24 63) (20 96) (91) (40 62) (15) (40) (29 40 99) >

Table 14. Effects of noise ($min\_sup = 5\%$)

$1-\alpha$	Recoverability	$\frac{N_{extral}}{N_{item}}$	N_{total}	N_{max}	N_{spur}	N_{redun}
0%	91.59%	$\frac{66,058}{1,782,583} = 3.71\%$	253,782	10	58	253,714
10%	76.06%	$\frac{14,936}{235,405} = 6.43\%$	46,278	10	9	46,259
20%	54.94%	$\frac{3,926}{39,986} = 9.82\%$	10,670	9	5	10,656
30%	41.83%	$\frac{1,113}{11,366} = 9.79\%$	3,646	10	1	3,635
40%	32.25%	$\frac{360}{4,021} = 8.95\%$	1,477	8	0	1,469
50%	28.35%	$\frac{143}{1,694} = 8.44\%$	701	9	0	692

Table 15. Effects of noise ($k = 6, \theta = 30\%$)

$1-\alpha$	Recoverability	$\frac{N_{extral}}{N_{item}}$	N_{total}	N_{max}	N_{spur}	N_{redun}
0%	91.16%	$\frac{3}{106} = 2.83\%$	8	7	0	1
10%	91.16%	$\frac{1}{104} = 0.96\%$	9	7	0	2
20%	91.16%	$\frac{1}{134} = 0.75\%$	12	7	0	5
30%	90.95%	$\frac{0}{107} = 0\%$	9	7	0	2
40%	88.21%	$\frac{0}{95} = 0\%$	9	7	0	2
50%	64.03%	$\frac{0}{68} = 0\%$	8	5	0	3

recoverability is only 54.94%, the model returns 10,670 patterns of which 10,656 are redundant patterns and 9.82% of the items are extraneous items.

In comparison, the alignment model is robust to noise in the data. Despite the presence of noise, it is still able to detect a considerable number of the base patterns. The results show that recoverability is still 88.21% with no extraneous items when corruption factor is 40% (Table 15 and Fig. 3).

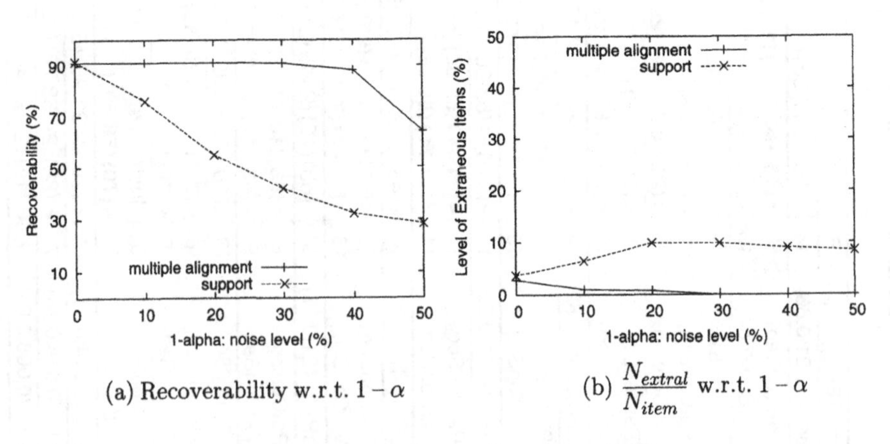

(a) Recoverability w.r.t. $1-\alpha$

(b) $\frac{N_{extral}}{N_{item}}$ w.r.t. $1-\alpha$

Fig. 3. Effects of noise ($k = 6, \theta = 30\%$)

6.4 Robustness w.r.t. Random Sequences

This experiment is designed to test the effect of random sequences added to patterned data. We generated random data as in the first experiment with the same parameters as the patterned data in the second experiment and added them together. The main effect of the random sequences are the weakening of the patterns as a percentage of the database.

Consequently, in the support model all the criteria (recoverability, the number of spurious and redundant patterns, and the level of extraneous items) are reduced when min_sup is maintained at 5% (Table 16). On the other hand, if we maintain $min\_sup = 50$ sequences, obviously the recoverability can be maintained at 91.59%. The tradeoff is that all other evaluation criteria (number of spurious and redundant patterns and the level of extraneous items) increase somewhat (Table 17).

In ApproxMAP, the parameter that controls which patterns are detected is k in the kNN clustering step. In this experiment, when k is maintained at 6, the added random sequences had no effect on the clusters formed or how the patterned sequences were aligned. Each cluster just picks up various amounts of the random sequences which are aligned after all the patterned sequences are aligned. In effect, the random sequences are ignored when it cannot be aligned with the patterned sequences in the cluster. The rest of the random sequences formed separate clusters that generated no patterns, as in the first experiment on random data. Nonetheless, when θ is maintained at 30%, the consensus sequences were shorter, reducing recoverability, because the random sequences increased the cluster size, and thus weakened the signature in the cluster as a percentage of the cluster size (Table 18).

However, as seen in Table 19, we can easily find the longer underlying patterns by adjusting θ to compensate for the random sequences in the data. θ can be lowered to pick up more patterned items. The tradeoff would be that more extraneous items could be picked up as well. Yet, in this experiment we found that ApproxMAP was robust to random sequences in the data. When θ was lowered to the point at which recoverability is maintained at 91.16%, the number of spurious and redundant patterns remained the same even when all 800 random sequences were added to 1000 patterned sequences. The number of extraneous items increase slightly from 3 to 11 giving slightly higher levels of extraneous items in the results.

6.5 Scalability

Methods based on the support model has to build patterns, either in depth first or breath first manner, one at a time. Thus, the lower bound on the time complexity with respect to the patterns is exponential. In contrast, the methods based on the alignment model find the patterns directly through multiple alignment. Thus, the time complexity does not depend on the number of patterns. Instead, computation time is dominated by the clustering step,

Table 16. Effects of Random Sequences ($min\_sup = 5\%$)

$N_{RandSeq}$	N_{PatSeq}	Recoverability	$\frac{N_{extraI}}{N_{item}}$	N_{total}	N_{spur}	N_{redun}
0	1000	91.59%	$\frac{66,058}{1,782,583} = 3.71\%$	253,782	58	253,714
200	1000	91.52%	$\frac{26,726}{882,063} = 3.03\%$	129,061	16	129,035
400	1000	87.67%	$\frac{12,372}{544,343} = 2.27\%$	82,845	7	82,828
600	1000	83.76%	$\frac{6,404}{286,980} = 2.23\%$	47,730	4	47,716
800	1000	80.38%	$\frac{3,611}{158,671} = 2.28\%$	28,559	4	28,546

Table 17. Effects of Random Sequences ($min\_sup = 50$ sequences)

$N_{RandSeq}$	N_{PatSeq}	min_sup	Recov.	$\frac{N_{extraI}}{N_{item}}$	N_{total}	N_{spur}	N_{redun}
0	1000	$\frac{50}{1000} = 5\%$	91.59%	$\frac{66,058}{1,782,583} = 3.71\%$	253,782	58	253,714
200	1000	$\frac{50}{1200} = 4.2\%$	91.59%	$\frac{66,195}{1,783,040} = 3.71\%$	253,939	59	253,870
400	1000	$\frac{50}{1400} = 3.6\%$	91.59%	$\frac{66,442}{1,783,945} = 3.72\%$	254,200	59	254,131
600	1000	$\frac{50}{1600} = 3.1\%$	91.59%	$\frac{66,571}{1,784,363} = 3.73\%$	254,341	59	254,272
800	1000	$\frac{50}{1800} = 2.8\%$	91.59%	$\frac{66,716}{1,784,828} = 3.74\%$	254,505	61	254,434

Table 18. Effect of Random Sequences ($k = 6, \theta = 30\%$)

$N_{RandSeq}$	N_{PatSeq}	Recoverability	$\frac{N_{extraI}}{N_{item}}$	N_{total}	N_{spur}	N_{redun}
0	1000	91.16%	$\frac{3}{106} = 2.83\%$	8	0	1
200	1000	91.16%	$\frac{2}{102} = 1.96\%$	8	0	1
400	1000	88.33%	$\frac{0}{97} = 0\%$	8	0	1
600	1000	82.87%	$\frac{0}{92} = 0\%$	8	0	1
800	1000	72.88%	$\frac{0}{81} = 0\%$	7	0	1

Table 19. Effect of Random Sequences ($k = 6$)

$N_{RandSeq}$	N_{PatSeq}	θ	Recoverability	$\frac{N_{extraI}}{N_{item}}$	N_{total}	N_{spur}	N_{redun}
0	1000	30%	91.16%	$\frac{3}{106} = 2.83\%$	8	0	1
200	1000	25%	91.16%	$\frac{4}{107} = 3.74\%$	8	0	1
400	1000	22%	91.16%	$\frac{6}{109} = 5.50\%$	8	0	1
600	1000	18%	91.16%	$\frac{10}{113} = 8.85\%$	8	0	1
800	1000	17%	91.16%	$\frac{11}{114} = 9.65\%$	8	0	1

which has to calculate the distance matrix and build the k nearest neighbor list. This inherently makes the time complexity $O(N_{seq}^2 L_{seq}^2 I_{seq})$ where N_{seq} is the total number of sequences, L_{seq} is the average number of itemsets in the sequence, and I_{seq} is the average number of items per itemset. However, unlike the support model there is potential for improving on the time complexity by using a sampling based iterative clustering algorithm and stopping the sequence to sequence distance calculation as soon as it is clear that the two sequences are not neighbors. Thus, not surprisingly when the patterns are relatively long, the multiple alignment algorithms tend to be faster than support based algorithms that are exponential in nature to the size of the pattern.

7 Related Work

Many papers have proposed methods for finding frequent subsequences to uncover patterns in sequential data [1, 3, 13, 15, 19]. There are three works in particular, that examine the support model and extend the problem statement to find more interesting and useful patterns. Spiliopoulou extend the model to find rules of the form "if A then B" using the confidence model [14]. Recently, Liang at el. examine ways to detect infrequent dependent event patterns for temporal data [8, 11]. Both these works try to find a particular type of sequential pattern in the data which are different from the focus of this paper. Yang at el. presents a probabilistic model to handle noise in mining strings [18]. However, Yang et al's work cannot be easily generalized to apply to the sequential data, sequences of sets, targeted in this paper. Furthermore, it does not address the issue of generating huge number of patterns that share significant redundancy.

There is a rich body of literature on string (simple ordered lists) analysis in computer science as well as computational biology that can be extended for this problem domain. In particular, multiple alignment has been studied extensively in computational biology [5, 6, 16] to find common patterns in a group of strings. In this paper, we have generalized string multiple alignment to find patterns in ordered lists of sets.

8 Conclusion and Future Works

Designing good operational definitions for patterns is the first step in mining useful and interesting patterns. Sequential patterns are commonly defined as frequent subsequences based on exact match. Noting some inherent problems in the frequent sequential pattern model, we present an entirely different model, *multiple alignment sequential pattern mining*. We performed a comparative study of the two models using a novel evaluation method utilizing the well-known IBM synthetic data generator [1]. The evaluation method comprehensively describes the quality of the patterns that emerge from certain definitions empirically.

When the evaluation method was applied to the support model for sequential pattern mining, it revealed that the model generates a huge number of redundant and spurious patterns in long sequences. These confounding patterns bury the true patterns in the mined result. Our theoretical analysis of the expected support of short patterns in long sequences confirm that many short patterns can occur frequently simply by chance. Hence, support alone cannot distinguish between statistically significant patterns and random occurrences. Furthermore, in the presence of noise in the data, the support model cannot detect the underlying patterns well because a sequence supports a pattern if, and only if, the pattern is fully contained in the sequence. Hence, noise in the data can cause the exact matching approach to miss general trends in the sequence database. This becomes especially problematic in real applications. Many customers may share similar buying habits, but few of them follow exactly the same buying patterns.

In contrast, the evaluation results demonstrate that, under all the circumstances we examined – including those where the support model fails, the alignment model is able to better recover the underlying patterns with little confounding information. ApproxMAP returns a succinct but accurate summary of the base patterns with few redundant or spurious patterns.

The evaluation method can be used in general to compare different sequential pattern definitions and mining models. This admittedly empirical method is crucial in understanding the quality of the mined results because often the models are too complicated for theoretical analysis. An empirical evaluation method is especially important for understanding the performance of approximate solutions.

Our initial work has clearly shown much promise in the alignment model. There are some interesting work remaining in optimizing the performance of ApproxMAP as well as fine tuning the evaluation method.

References

1. R. Agrawal and R. Srikant. Mining sequential patterns. In *Proc. of International Conference on Data Engineering (ICDE)*, pp. 3–14, Taipei, Taiwan, March 1995.
2. J. Ayres, J. Flannick, J. Gehrke, and T. Yiu. Sequential pattern mining using a bitmap representation. In *Proceedings of the ACM international conference on Knowledge discovery and data mining (SIGKDD)*, pp. 429–435, July 2002.
3. R. J. Bayardo. Efficiently mining long patterns from databases. In *Proc. of ACM International Conference On Management of Data (SIGMOD)*, pp. 85–93, June 1998.
4. S. Brin, R. Motwani, and C. Silverstein. Beyond market baskets: generalizing association rules to correlations. In *Proc. of ACM International Conference On Management of Data (SIGMOD)*, pp. 265–276, 1997.
5. O. Gotoh. Multiple sequence alignment: Algorithms and applications. In *Adv. Biophys.*, Vol. 36, pp. 159–206, 1999.
6. D. Gusfield. Algorithms on strings, trees, & sequences: Computer Science and Computational Biology. Cambridge Univ. Press, Cambridge, England. 1997.
7. H.C. Kum, J. Pei, W. Wang, and D. Duncan. ApproxMAP : Approximate Mining of Consensus Sequential Patterns. In *Third SIAM International Conference on Data Mining (SIAM-DM)*, pp. 311–315, San Fransico. CA, 2003.

8. F. Liang, S. Ma, and J. L. Hellerstein. Discovering Fully Dependent Patterns. In *Second SIAM International Conference on Data Mining (SIAM-DM)*, Arlington. VA, 2002.
9. D. Lin and Z. Keadem. Pincer-search: a new algorithm for discovering the maximum frequent set. In *Proc. 6th Intl. Conf Extending Database Technology (EDBT)*, pp. 105–119, 1998.
10. G. R. McPherson and S. DeStefano. Applied Ecology and Natural Resource Management. Cambridge University Press, Cambridge, England. 2002.
11. S. Ma and J. L. Hellerstein Mining Mutually Dependent Patterns. In *Proc. of International Conference on Data Mining (ICDM)*, pp. 409–461, San Jose, CA, Nov. 2001.
12. N. Pasquier, Y. Bastide, R. Taouil, and L. Lakhal. Discovering frequent closed itemsets for association rules. In *Proc. of the International Conference on Database Theory (ICDT)*, pp. 398–416, Jan. 1999.
13. J. Pei, J. Han, et al. PrefixSpan: Mining sequential patterns efficiently by prefix-projected pattern growth. In *Proc. of International Conference on Data Engineering (ICDE)*, pp. 215–224, April 2001.
14. M. Spiliopoulou. Managing interesting rules in sequence mining. In *Proc. European Conf. on Principles and Practice of Knowledge Discovery in Databases*, pp. 554–560, 1999.
15. R. Srikant and R. Agrawal. Mining sequential patterns: Generalizations and performance improvements. In *Proc. Intl. Conf Extending Database Technology (EDBT)*, pp. 3–17, Mar. 1996.
16. J. Thompson, F. Plewniak, and O. Poch. A comprehensive comparison of multiple sequence alignment programs. In *Nucleic Acids Research*. Vol. 27. No. 13: pp. 2682–2690. Oxford University Press. 1999.
17. X. Yan, J. Han, and R. Afshar. CloSpan: Mining Closed Sequential Patterns in Larege Datasets. In *Third SIAM International Conference on Data Mining (SIAM-DM)*, pp. 166–177, San Fransico. CA, 2003.
18. J. Yang, P. S. Yu, W. Wang, and J. Han. Mining long sequential patterns in a noisy environment. In *Proc. of ACM International Conference On Management of Data (SIGMOD)*, pp. 406–417, Madison, WI, June 2002.
19. M. J. Zaki. Efficient enumeration of frequent sequences. In *7th International Conference Information and Knowledge Management (CIKM)*, pp. 68–75, Nov. 1998.

A Appendix

A.1 Derivation of Expected Support on Random Data

The expression for the expected support of a subsequence of arbitrary length will be cumbersome. Instead, we present the derivation of the recursive procedure which can be used to calculate the expected support under the null hypothesis of complete independence. We provide explicit expressions for the expected support for subsequences of length one, two, and three only; however, the logic used to calculate the expected support for a subsequence of length three from that of length two can also be used to calculate successively longer subsequences. This should be implemented in a computer program, where a numerical value (rather than an analytical expression) is desired for the expected support of a particular subsequence.

Recall that the expected support (measured as a percentage) for a subsequence under our null hypothesis is equivalent to the probability that the subsequence appears at least once in an observed sequence. We consider first the expected support for a subsequence of length one. Let $p(\mathsf{A})$ be the probability of item A appearing in the sequence at any particular position. Then the expected support of A, ie. the probability that A will appear at least one time in a sequence of length, L is

$$\begin{aligned} E\{sup(\mathsf{A})\} &= Pr\{\mathsf{A} \text{ appears at least once}\} \\ &= 1 - Pr\{\mathsf{A} \text{ never appears}\} \\ &= 1 - [1 - p(\mathsf{A})]^L \end{aligned} \tag{10}$$

Now consider the calculation of $Pr\{seq_i\}$ where the subsequence length is two. Let A and B arbitrarily represent two items, which need not be distinct. Rather than directly calculate $Pr\{\mathsf{A..B}\}$, the probability that item A will be followed by item B at least one time in a sequence of length L, it is easier to instead calculate $Pr\{\sim(\mathsf{A..B})\}$, i.e. the probability of never seeing an A followed by a B. Then, $Pr\{\mathsf{A..B}\}$ – the quantity of interest – is found by subtracting $Pr\{\sim(\mathsf{A..B})\}$ from one.

The probability of $\sim(\mathsf{A..B})$ can be systematically calculated by dividing all the possible outcomes into the $L+1$ mutually exclusive sets, which will be denoted for convenience by "$first(A,i)$":

$$\begin{aligned} \mathsf{A} \text{ first appears in position } 1 &= first(\mathsf{A},1) \\ &\dots \\ \mathsf{A} \text{ first appears in position } j &= first(\mathsf{A},j) \\ &\dots \\ \mathsf{A} \text{ first appears in position } L &= first(\mathsf{A},L) \\ \mathsf{A} \text{ never appears} & \end{aligned}$$

The probability of each possible outcome is

$$Pr\{first(\mathsf{A},j)\} = [1 - p(\mathsf{A})]^{j-1} p(\mathsf{A}) \tag{11}$$

for $j = 1..L$. The probability that A never appears is:

$$Pr\{\mathsf{A} \text{ never appears}\} = [1 - p(\mathsf{A})]^L \tag{12}$$

Next, we condition $Pr\{\sim(\mathsf{A..B})\}$ upon the mutually exclusive event sets listed above:

$$\begin{aligned} Pr\{\sim(\mathsf{A..B})\} = \textstyle\sum_{j=1}^{L} & Pr\{\sim(\mathsf{A..B})|first(\mathsf{A},j)\} \cdot Pr\{first(\mathsf{A},j)\} \\ &+ Pr\{\sim(\mathsf{A..B})|\mathsf{A} \text{ never appears}\} \cdot Pr\{\mathsf{A} \text{ never appears}\} \end{aligned} \tag{13}$$

Where in general,

$$Pr\{\sim(\mathsf{A..B})|first(\mathsf{A},j)\} = Pr\{\mathsf{B}\text{ never appears in a sequence of length } L-j\} \\ = [1-p(\mathsf{B})]^{L-j} \tag{14}$$

By substituting equations 12, 11, and 14 into 13 and recognizing that we are certain not to observe A..B if A never appears, i.e.:

$$Pr\{\sim(\mathsf{A..B})|\mathsf{A}\text{ never appears}\} = 1$$

we have

$$Pr\{\sim(\mathsf{A..B})\} = p(\mathsf{A})\sum_{j=1}^{L}\{[1-p(\mathsf{B})]^{L-j}[1-p(\mathsf{A})]^{j-1}\} + [1-p(\mathsf{A})]^{L} \tag{15}$$

Then the expected support from a sequence of length L for A..B is

$$E\{sup(\mathsf{AB})\} = Pr\{(\mathsf{A..B})\} \\ = 1 - Pr\{\sim(\mathsf{A..B})\} \\ = 1 - \left(p(\mathsf{A})\sum_{j=1}^{L}\{[1-p(\mathsf{B})]^{L-j}[1-p(\mathsf{A})]^{j-1}\} + [1-p(\mathsf{A})]^{L}\right) \tag{16}$$

The probability of observing longer subsequences can now be calculated recursively. To determine $Pr\{\mathsf{A..B..C}\}$ we again first calculate $Pr\{\sim(\mathsf{A..B..C})\}$ and condition on A appearing first in either the jth position or not at all:

$$Pr\{\sim(\mathsf{A..B..C})\} = \sum_{j=1}^{L} Pr\{\sim(\mathsf{A..B..C})|first(\mathsf{A},j)\}\cdot Pr\{first(\mathsf{A},j)\} \\ +Pr\{\sim(\mathsf{A..B..C})|\mathsf{A}\text{ never appears}\}\cdot Pr\{\mathsf{A}\text{ never appears}\} \tag{17}$$

Note that the conditional probability

$$Pr\{\sim(\mathsf{A..B..C})|first(\mathsf{A},j)\} = Pr\{\sim(\mathsf{B..C})\text{ in a sequence of length } L-j\} \tag{18}$$

Applying (15) to a sequence of length $L-j$ we find that:

$$Pr\{\sim(\mathsf{B..C})\text{ in a sequence of length } L-j\} \\ = p(\mathsf{B})\sum_{k=1}^{L-j}\{[1-p(\mathsf{C})]^{L-j-k}[1-p(\mathsf{B})]^{k-1}\} + [1-p(\mathsf{B})]^{L-j} \tag{19}$$

Again we recognize that the probability of not seeing A..B..C given that A does not appear in the sequence must equal 1. Then substituting equations 18, 11, and 12 into 17 produces an expression for $Pr\{\sim(\mathsf{A..B..C})\}$:

$$Pr\{\sim(\mathsf{A..B..C})\} = [1-p(\mathsf{A})]^{L} + p(\mathsf{A})\cdot[1-p(\mathsf{A})]^{j-1} \\ \cdot\sum_{j=1}^{L}\left(p(\mathsf{B})\sum_{k=1}^{L-j}\{[1-p(\mathsf{C})]^{L-j-k}[1-p(\mathsf{B})]^{k-1}\} + [1-p(\mathsf{B})]^{L-j}\right) \tag{20}$$

from which $Pr\{\mathsf{A..B..C}\}$ can be found by subtraction:

$$E\{sup(\mathsf{ABC})\} = Pr\{(\mathsf{A..B..C})\} = 1 - Pr\{\sim (\mathsf{A..B..C})\} \tag{21}$$

The analytical expression for the probability of observing a subsequence at least once in a longer sequence quickly becomes cumbersome as the length of the subsequence increases, but may be readily computed in a recursive manner as illustrated above.

Designing Robust Regression Models

Murlikrishna Viswanathan and Kotagiri Ramamohanarao

Department of Computer Science and Software Engineering, University of Melbourne, Australia 3010
murli,rao@cs.mu.oz.au

1 Introduction

In this study we focus on the preference among competing models from a family of polynomial regressors. Classical statistics offers a number of well-known techniques for the selection of models in polynomial regression, namely, Finite Prediction Error (FPE) [1], Akaike's Information Criterion (AIC) [2], Schwartz's criterion (SCH) [10] and Generalized Cross Validation (GCV) [4]. Wallace's Minimum Message Length (MML) principle [16, 17, 18] and also Vapnik's Structural Risk Minimization (SRM) [11, 12] – based on the classical theory of VC-dimensionality – are plausible additions to this family of model-selection principles. SRM and MML are generic in the sense that they can be applied to any family of models, and similar in their attempt to define a trade-off between the complexity of a given model and its *goodness of fit* to the data under observation – although they do use different trade-offs, with MML's being Bayesian and SRM's being non-Bayesian in principle. Recent empirical evaluations [14, 15] comparing the performance of several methods for polynomial degree selection provide strong evidence in support of the MML and SRM methods over the other techniques.

We consider a simple domain where the x data are randomly selected from the interval $[-1, 1]$ and the y values are derived from a univariate function, $y = t(x)$, corrupted with Gaussian noise having a zero mean. Least-squares approximations of polynomials of degree up to 20 are derived from the data generated. These univariate polynomials of varying orders are then offered to the five model selection methods and the performance of the preferred polynomials are evaluated by their predictive accuracy on test data, similarly generated. This work supplements the recent studies by Wallace and Viswanathan [14, 15] by including an extensive empirical evaluation of five polynomial selection methods – FPE [1], SCH [10], GCV [4], MML and SRM. (In unreported results, we have found that AIC [2] performs almost identically to FPE.) We aim to analyze the behaviour of these five methods with respect

M. Viswanathan and K. Ramamohanarao: *Designing Robust Regression Models*, Studies in Computational Intelligence (SCI) **6**, 71–86 (2005)
www.springerlink.com

to variations in the number of training examples and the level of noise in the data.

2 Background

As mentioned in Sect. 1, the target problem is the selection of a univariate polynomial regression function. In this section we aim to summarize the two major approaches based on the minimum message length (MML) and structural risk minimization (SRM) principles. Let us assume that we have a finite number N of observations of a function $t(x)$ corrupted with additive noise ε,

$$y_i = t(x_i) + \varepsilon_i \quad \text{for} \quad i = 1, \ldots, N \ .$$

Our approximation of the target function is based on the training set of N observations, where the values, x_i, of the independent variable x are independently and uniformly distributed in the interval $[-1, 1]$ and the noise values $\varepsilon_i = y_i - t(x_i)$, are independently and identically distributed by a Normal density with zero mean and unknown variance. The framework of the problem follows Wallace [15] and Viswanathan et al. [14]. The values, x_i, of the independent variable x are randomly selected from the uniform distribution on the interval $[- - 1, 1]$.

The task then is to find some polynomial function, $f(x) = \hat{t}(x)$, of degree d that may be used to predict the value of $t(x)$ in the interval, $-1 \leq x \leq 1$. In our evaluation we only consider polynomials $f(\cdot)$ of degrees up to 20 and for any given degree d, we select the polynomial $f(d, x)$ that minimizes the squared error SE on the training data,

$$SE(f(d,x)) = \sum_{i=1}^{N} (y_i - f(d, x_i))^2 \tag{1}$$

The performance or *prediction risk* is the expected performance of the chosen polynomial for new (future) samples. This is measured by its Squared Prediction Error SPE, which is estimated using a simple Monte Carlo method:

$$SPE\,(f(d,x)) = \frac{1}{m} \sum_{i=1}^{m} (f(d, x_i) - t(x_i))^2 \tag{2}$$

where t(x) is the target function and m is max(N, 50) and the test data (i = 1 to m) are randomly selected from a uniform distribution in [-1,1].

2.1 Classical Methods for Model Selection

Among the classical methods compared in this paper two general approaches can be observed. While Generalized Cross-Validation (GCV) [4] is based on data re-sampling, Finite Prediction Error (FPE) [1] and Schwartz's Criterion (SCH) [10] attempt to penalize model complexity. The use of FPE as opposed to the Akaike Information Criterion (AIC) [2] is justified since FPE is specially derived under the assumption that the distributions of the predictors used in learning and prediction, is identical. Furthermore, FPE and AIC give almost the same inference for this class of problem [9, Sect. 8.4], as borne out by some unreported results of our own. As described in Wallace [15], the selection process for these classical methods is to choose the polynomial which minimizes

$$g(p, N) * SE(f(d, x))$$

where $p = (d+1)/N$, and $g(\cdot, \cdot)$ is a function characteristic of the given method of inference. The function $g(\cdot, \cdot)$ is known as a penalty function since it inflates the training error (average residual sum of squares). The following characteristic penalty functions are derived from the classical approaches :

1. Finite Prediction Error (FPE), $g(p, N) = (1+p)/(1-p)$
2. Schwartz's Criterion (SCH), $g(p, N) = 1 + 0.5\log(n) * p/(1-p)$
3. Generalized Cross Validation (GCV), $g(p, N) = 1/(1-p)^2$

2.2 VC Dimension and Structural Risk Minimization

As defined earlier in Sect. 2, the *prediction risk* is the expected performance of an estimator on new samples. The Vapnik-Chervonenkis theory [11, 12] provides non-asymptotic "guaranteed" bounds for the prediction risk of a model based on the concept of the VC-dimension. Generally speaking, the VC-dimension [11, 12] is a measure of model complexity. For a given set of functions the VC-dimension is the number of instances that can be "shattered" – i.e., all possible subsets of the instances being partitioned from their complement subset by functions from this set. For example, in the binary classification case, the VC-dimension is the maximum number of instances m which can be separated into two classes in all possible 2^m ways by using functions from the hypothesis space. The VC-dimension for the set of polynomial functions of degree d can be shown [12] to be equal to $(d+1)$.

The Structural Risk Minimization (SRM) principle [12, 13] is based on the well-known assumption that, in order to infer models with high generalization ability, we need to define a trade-off between the model complexity and goodness of fit to the data. Employing the VC-dimension as the measure of model complexity the SRM principle attempts to achieve this trade-off and avoid *over-fitting*.

According to Vapnik [12], in order to choose the polynomial $f(d, x)$ of the best degree d, one can minimize the following function based on the Structural Risk Minimization principle:

$$R(\underline{a}, d) \leq \frac{\frac{1}{N}\sum (y_i - f(d, x_i))^2}{(1 - c\sqrt{\xi_N})} \tag{3}$$

where

$$\xi_N = \frac{V\,(\log \frac{N}{V} + 1) - \log \eta}{N}$$

In the expression above, $R(\underline{a}, d)$ is the estimate of the prediction risk of a polynomial of degree d and coefficients $\underline{a}$. The numerator of the Right Hand Side of (3) is the average squared error, namely $(1/N)\sum_{i=1}^{N}(y_i - f(d, x_i))^2$, achieved by a polynomial of degree d and set of co-efficients $\underline{a} =< a_0, \ldots, a_d >$ on N training examples, and the denominator is $1 - c\sqrt{\xi_N}$. ξ_N is the error term where N is the number of examples, c is a constant that reflects the tails of the training error distribution and V is the VC dimension for $f(d, x)$. This approach also takes into account the confidence interval for the prediction. Specifically, it provides an upper bound for the estimate, $R(\underline{a}, d)$, of the prediction risk. The inequality in (3) then holds with probability (1-η), where η represents a confidence interval for the prediction [12]. In our empirical evaluation we have used $V = (d + 1)$, since the VC dimension for a polynomial of degree d is (d + 1) and, as suggested in [13], we have elected to use $c = 1$. In this evaluation we employ the error term ξ_N as described in (3) with $\eta = 0.125$ (this value of η was derived from Vapnik's book [12]). An older error term was investigated in empirical comparisons done by Wallace [15] where the confidence interval η was implemented as a function of the sample size ($\eta = \frac{1}{\sqrt{N}}$). The recent version [13] from (3) employs a fixed user-defined value. This application of a fixed confidence interval improves the predictive performance of the older approach [12, 14, 15] at least in specific cases.

2.3 Minimum Message Length Principle

Minimal Length Encoding techniques like the Minimum Message Length (MML) [16, 17, 18] and Minimum Description Length (MDL) [6, 7, 8] principles have been popular due to their successful application in model selection. MML is an invariant Bayesian principle based on the information-theoretic assertion that the best model for a given set of data is one that allows the shortest joint encoding of the model and the data, given the model. MML seeks to minimize the length of a "message" which encodes the data (in this problem the training y-values) by first stating a probabilistic model for the data, and then encoding the data using a code which would be optimal were the stated model true. MML has been shown to provide robust and highly competitive model selection in comparison to various classical model selection criteria (including AIC) within the polynomial degree selection framework [3], although we improve this even further here in Sect. 2.3 by our use of an orthonormal basis.

In our case, since the model is assumed to be a polynomial function with Gaussian noise, the model description need only specify the degree of the polynomial, the co-efficients of the polynomial, and the estimated variance (or, equivalently, SD) of the noise. In the general case, the best MML polynomial for any degree is the one that has the shortest two-part message length, of which the first part describes the polynomial in terms of its degree d, coefficients $\underline{a}$ and estimated variance v, while the second part describes the data (via the ε_i) using the given polynomial. An important point to note is that our encoding system uses the degree of the polynomial rather than the number of non-zero coefficients. One reason for this is that the number of non-zero coefficients will depend upon whether we choose the basis $1, x, x^2, x^3, \ldots$, the orthonormal basis for integration over the region $[-1, 1]$ or another basis. However, the degree of the polynomial will remain the same regardless of this choice of basis.

In the current experiment, the coefficients are estimated using the maximum likelihood technique, since the difference from MML estimation is small for this problem and the other methods all advocate using the maximum likelihood estimate. (For examples of problems where MML and maximum likelihood estimation are substantially different, see, e.g., Wallace and Freeman, 1992 and Wallace, 1995.) The following sections provide details of the actual MML encoding scheme.

Encoding the Model with Prior Beliefs

MML is a Bayesian principle and thus requires the formal specification of prior beliefs. We start by considering the degree of our polynomial models. All degrees from 0 to 20 are considered equally likely a priori, so each degree is coded with a code word of length log (21) *nits*, or natural bits. The coding of the model degree has no influence on the choice of model as all degrees have the same code length.

In coding estimates of the noise variance v, and the polynomial coefficients $\underline{a}$, some scale of magnitude may be assumed. Here, we use the second order sample moment of the given y-values to determine such a scale by defining

$$\vartheta = \frac{1}{N} \sum_{i=1}^{N} (y_i)^2 \tag{4}$$

In encoding a polynomial model of degree d, we suppose that the noise and each of the $(d+1)$ degrees of freedom of the polynomial may be expected a priori to contribute equally (in very rough terms) to the observed variance ϑ of the y-values. Defining

$$u = \sqrt{\left(\frac{\vartheta}{d+2}\right)} = \sqrt{\frac{\sum_{i=1}^{N} y_i^2}{N(d+2)}} \tag{5}$$

we assume the Standard Deviation s, of the "noise" v, where $s = \sqrt{v}$, has a Negative Exponential prior density with mean u, and that each of the coefficients $(a_0, \ldots, a_j, \ldots, a_d)$ of the polynomial has independently a $N(0, u^2)$

prior density. If the coefficients were the usual coefficients of the successive powers of x, the latter assumption would be unreasonable, and highly informative. Instead, we represent a d-degree polynomial as

$$f(d,x) = \sum_{j=0}^{d} (a_j Q_j(x)) \tag{6}$$

where the set $Q_j(\cdot) : j = 0 \ldots d$ is a set of polynomials, $Q_j(\cdot)$ being of degree j, which are orthonormal under integration on the interval $[-1,1]$. The orthonormal polynomials represent effectively independent modes of contributing to the variance of $f(d,\cdot)$, and therefore it seems reasonable to assume independent prior densities for the co-efficients $\{a_j : j = 0,\ldots,d\}$. With these assumptions, the overall prior density for the unknown parameters $\{s,\{a_j\}\}$ is

$$h(s,\underline{a}) = (1/u)^{(-s/u)} * \prod_{j=0}^{d} \left(\frac{1}{\sqrt{2\pi u}}\right) e^{(-a_j^2/(2u^2))} \tag{7}$$

The amount of information needed to encode the estimates s and $\underline{a}$ depends on the precision to which these are stated in the "message". Specifying the estimates with high precision leads to an increase in the model part of the message length while lower precision leads to lengthening of the data part. The optimum precision, as described in [18], is inversely proportional to the square root of the *Fisher information* $F(s,\underline{a})$, which in this case is given by

$$F(s,\underline{a}) = 2\left(\frac{N}{s^2}\right)^{(d+2)} |M| \tag{8}$$

where M is the co-variance matrix of the orthonormal polynomials evaluated at the given x-values:

$$M[j,k] = (1/N)\sum_{i=1}^{N} Q_j(x_i)\, Q_k(x_i) \tag{9}$$

Encoding the Data

Once the model polynomial $f(d,\cdot)$ and the noise standard deviation s have been stated, the given y-values can be coded simply by their differences from $f(d,\cdot)$, these differences being coded as random values from the Normal density $N(0,v)$. The message length required for this is then given by

$$\begin{aligned} DataMessLen = {} & \left(\frac{N}{2}\right)\log(2\pi v) \\ & + (1/2v) * SE(f(d,\cdot)) \end{aligned} \tag{10}$$

The Total Message Length

The total message length is approximately given by

$$\begin{aligned} MessLen = & -\log h(s, \underline{a}) \\ & + 0.5 \log F(s, \underline{a}) + DataMessLen \\ & - ((d+2)/2) \log(2\pi) + 0.5 \log((d+2)\pi) \end{aligned} \quad (11)$$

where the last two terms arise from the geometry of an optimum quantizing lattice in $(d+2)$–space. The noise variance $v = s^2$ is estimated as

$$\hat{v}_{MML} = \left(\frac{1}{(N-d-1)}\right) \sum_{i=1}^{N} (f(d, x_i) - y_i)^2 \quad (12)$$

and the co-efficients of $\underline{a}$ are estimated by conventional least-squares fit. These estimates do not *exactly* minimize the message length *MessLen*, but, as in Sect. 2.3, for this problem the difference is small. The MML method selects that degree d which minimizes *MessLen* as calculated above in (11). Recent theory [5] suggests that the MML estimator will closely approximate the estimator minimizing the expected Kullback-Leibler distance.

3 Experimental Evaluation

The following discussion outlines the experimental procedure used in this empirical evaluation also employed in Wallace [15] and Viswanathan et al. [14]. For each experiment a target function is selected in the required interval [-1,1]. The noise is defined in terms of the "signal-to-noise ratio" (SNR), where SNR is defined as the second moment of the target function, $t(x)$, about zero, divided by the noise variance v. The number of training data points N and the number of evaluations (training and test runs) are specified.

An experiment consists of (averaging over) 10000 evaluations. In each evaluation or "case", N training examples and $m = \max(N, 50)$ test examples are generated. For each "case", all least-squares polynomial approximations of degrees up to 20 are found by standard techniques, and the training error, $SE(d)$, computed for each of the 21 polynomials. These training error values are then given to each of the model selection methods being compared and each method selects its preferred choice among the polynomials.

The prediction risk for a method in the current case is then the average squared prediction error (SPE) achieved by the polynomial chosen by the method on the test data. Note again that all selection methods must choose from the same set of 21 polynomials, and their choices are evaluated on the same set of test data. Thus if two selection methods choose the same degree in some case, they are choosing the same polynomial and will incur the same SPE for this case.

After the 10000 cases of an experiment have been completed, we obtain for each selection method:

- its Squared Prediction Error (SPE), averaged over all cases (10000); and,
- the Standard Deviation of its SPE.

The five selection methods – namely, MML, SRM, FPE, SCH and GCV – were evaluated on six target functions $t(x)$ in the interval [-1,1]. The methods were tested under different scenarios. First, the noise level was kept constant with varying numbers of training points and, then the noise levels were varied with a fixed number of training points. For each method evaluated under a particular scenario, comparisons were made on the basis of squared prediction error (SPE) and standard deviation of the SPE. As an artificial yardstick, we also include a method called "BEST". The "BEST" polynomial is obtained by selecting from the 21 candidate least-square fit polynomials (from Sect.1) the polynomial which would give the smallest SPE as calculated on the test data. Thus, BEST shows the best performance which could possibly be obtained in principle, but these results are of course unrealizable, as they are based on knowledge of the noise-free values of the target function at the test x_i's.

3.1 Target Functions

In evaluating the five polynomial selection methods we consider the following two polynomial and four non-polynomial target functions.

1. **A Higher-order Polynomial**: $y = 0.623x^{18} - 0.72x^{15} - 0.801x^{14} + 9.4x^{11} - 5.72x^9 + 1.873x^6 - 0.923x^4 + 1.826x - 21.45$;
2. **A Lower-order Polynomial**: $y = 9.72x^5 + 0.801x^3 + 9.4x^2 - 5.72x - 136.45$;

Non-polynomial Target Functions

It is interesting to observe the performance of the different methods when the target function are not polynomials. The task here is challenging as we seek the best polynomial approximation. The following target functions were utilized in the comparison, with $-1 \leq x \leq 1$.

1. **SIN**: $y = (\sin(\pi(x+1)))^2$;
2. **LOG**: $y = \log(x + 1.01)$;
3. **FABS**: $y = |x + 0.3| - 0.3$; and,
4. **DISC**: if $(x < 0.0)$ $y = 0.1$; else $y = 2x - 1.0$.

4 Analysis of Results

In the experimental evaluation we consider two sets of target functions. In the former the target functions belong to the family of polynomials of maximum degrees of 20, while the latter set consists of non-polynomials. Figs. 1–6 include plots of results from experiments with six target functions. Each figure presents two scenarios. In the first case the sample size is increased and the standard deviation of the noise is kept constant, and vice versa in the second case. Some of the prediction methods can be seen to have a squared prediction error typically of the order of 100 times that of MML. Due to this

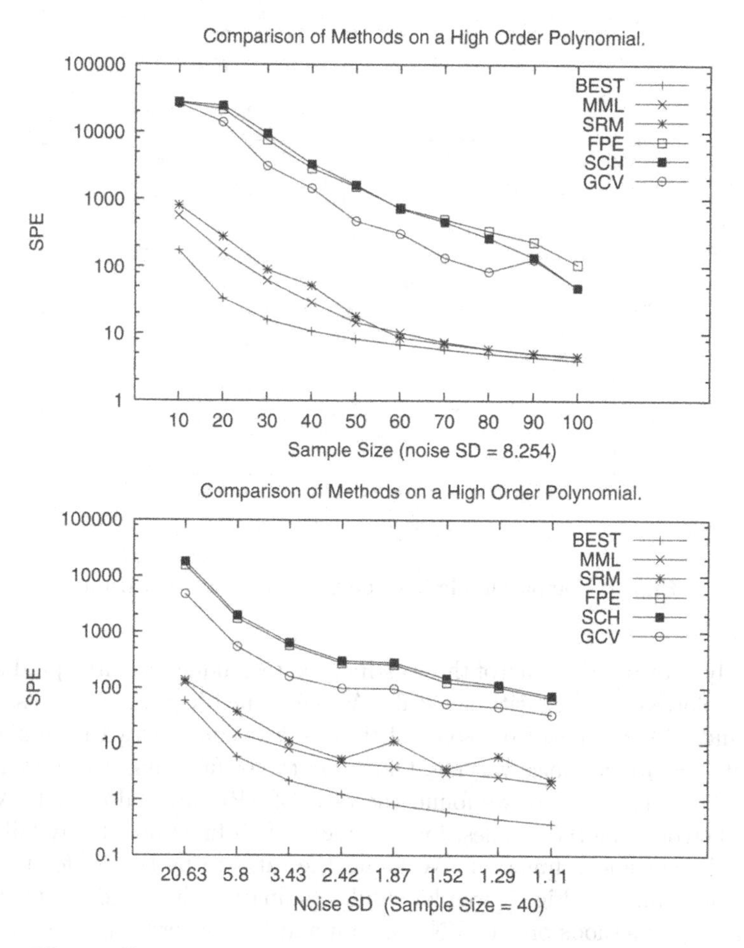

Fig. 1. Comparing methods on Squared Prediction Error (SPE)

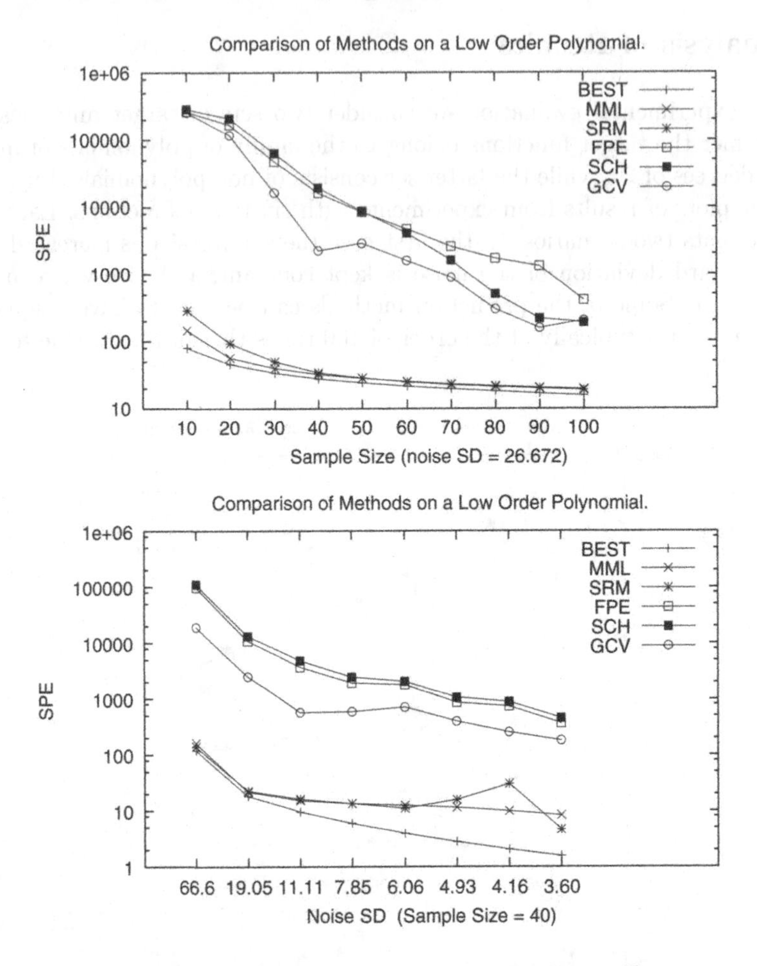

Fig. 2. Comparing Methods on Squared Prediction Error

relatively poor performance of those estimation techniques, squared prediction error is plotted on a logarithmic scale. As mentioned in Sect. 3, we also plot the standard deviations from some of these simulation runs with sample size fixed at $N = 40$ and noise varying. Of the six target functions considered (see Sect. 3.1), when $N = 40$, we found the ratio of FPE squared error to MML squared error to be the smallest for the smooth SIN function, closely followed by the discontinuous function. We found the ratio to be largest for the low order polynomial. In Fig. 7, we plot results comparing the standard deviation of SPE of all methods on the SIN function and the low-order polynomial.

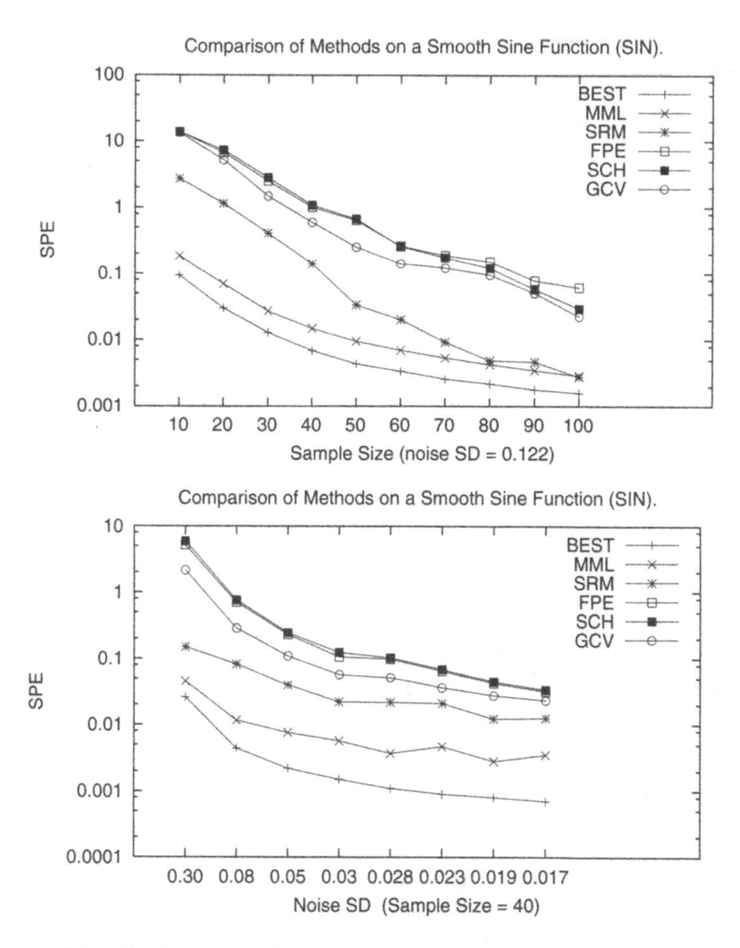

Fig. 3. Comparing Methods on Squared Prediction Error

In general, from the examples included in this paper and other tests it can be observed that MML and SRM give lower errors than other methods in most cases. The results clearly show that none of the methods FPE, SCH, or GCV is competitive with SRM or MML, except under conditions of low noise and large N, when all methods give similar results. An interesting observation is that the SRM method is based on theory which does not assume that the target function belongs to the model family from which the approximation is drawn (in this case, the polynomials). MML, however, is in part motivated by theory which does make this assumption. It is curious that the only target/test conditions under which SRM performs comparably with MML are

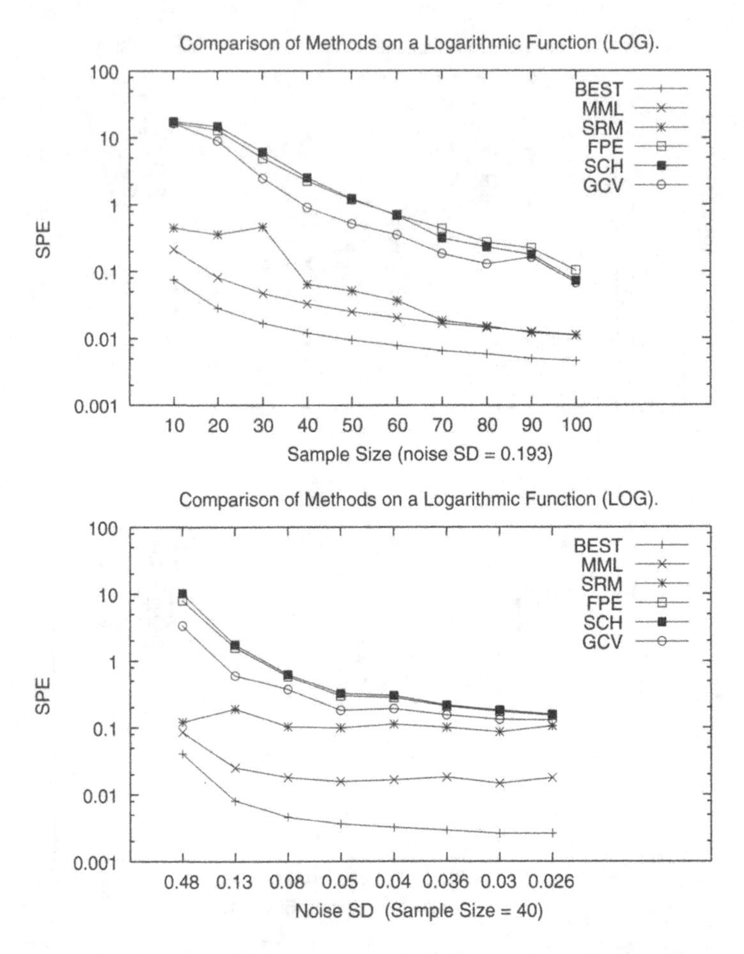

Fig. 4. Comparing Methods on Squared Prediction Error

ones where this assumption is largely valid. Furthermore, an examination of the percentiles of the error distribution suggests that the MML and SRM-based methods usually have similar median errors. The high SPE of the SRM method seems to be the result of occasional very large errors using the selected model. Evidence of this is given in Fig. 8.

Finally, we find that, in its application to the current model selection framework, the SRM principle is not paying any attention to the variance of the selected model. On an examination of the cases where the performance of the SRM-based models is worse than usual we find that the estimated variance of the approximating polynomial is orders of magnitude larger than the second moment of the training samples. Thus the SRM principle is accepting some

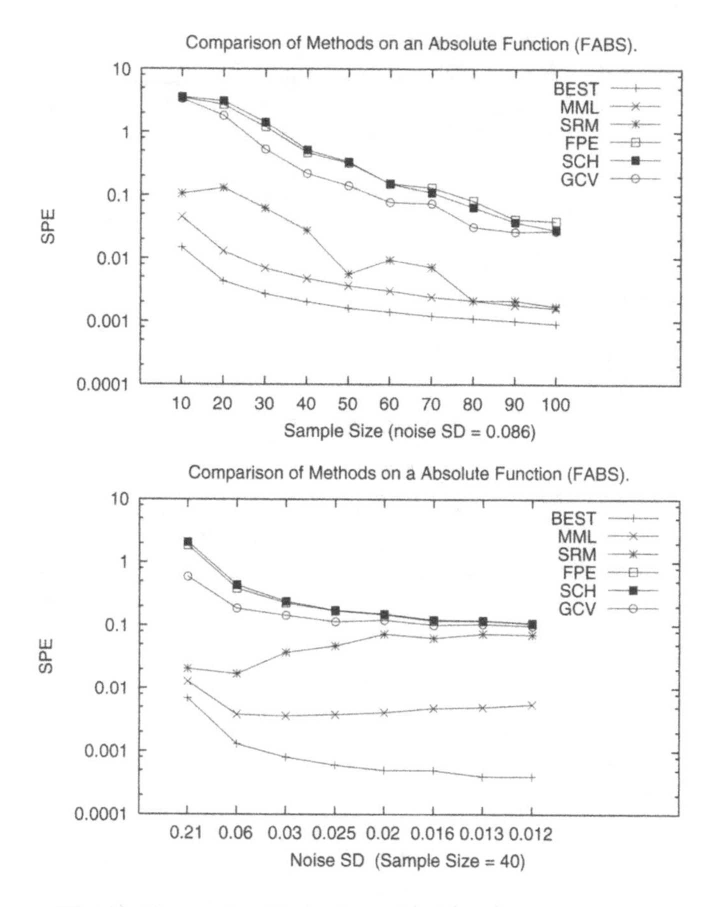

Fig. 5. Comparing Methods on Squared Prediction Error

clearly unreasonable models. As an example, the plots in Fig. 8 present a comparison of the variance of the worst model (in terms of SPE) selected by the SRM and MML methods over the 10,000 runs with their average training errors. The model selection methods were evaluated on the sine function from Sect. 3.1 with a fixed SNR ratio and variable sample size.

5 Conclusions

In this paper we present a detailed technique for modeling noisy data with univariate polynomials. The Minimum Message Length methodology is used to develop a coding scheme to encode these univariate polynomials which are estimated using a maximum likelihood technique. The model that minimizes this MML-based encoding scheme is preferred. We compared our approach to several well-known techniques by a detailed experimental evaluation involving

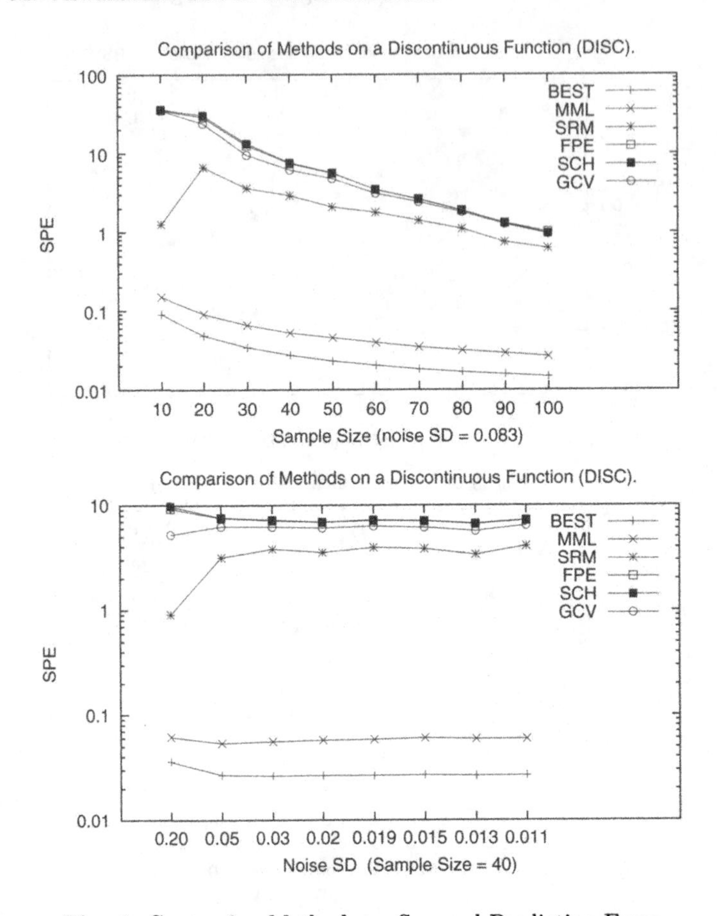

Fig. 6. Comparing Methods on Squared Prediction Error

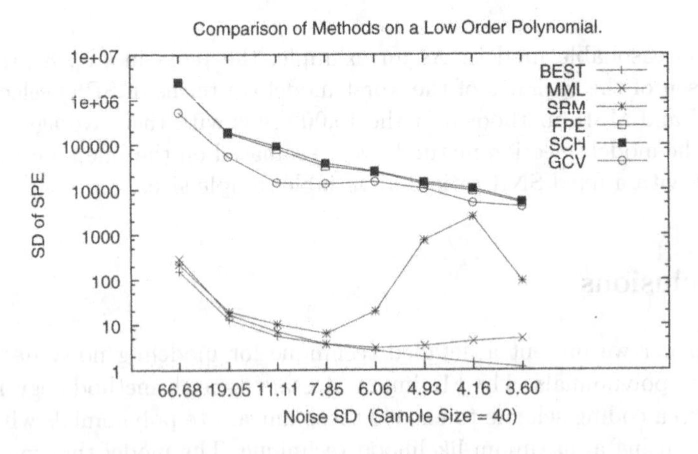

Fig. 7. Standard Deviation of Squared Prediction Error

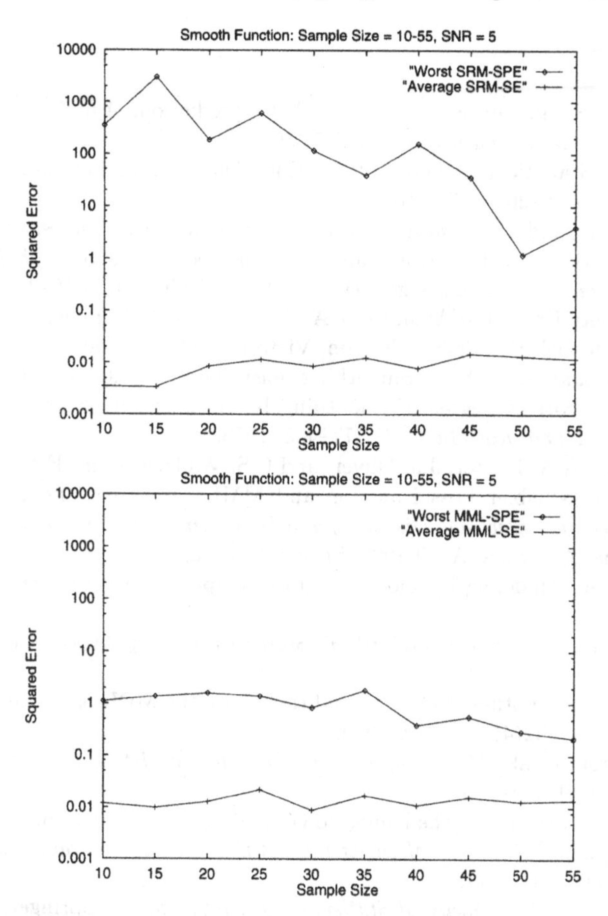

Fig. 8. A Comparison of the SPE (test data) and SE (training data) of the Worst Models Selected by SRM and MML

several model selection methods (SRM, FPE, SCH and GCV) on both continuous and discontinuous target functions. The empirical evaluation consisted of averaging errors from 10000 cases of an experiment with a specific sample size and noise level. In general models based on the MML methodology provided excellent predictive accuracy in the presence of noise with improved quality of prediction. One of our research efforts in progress is to extend the current univariate regression problem to a multivariate design, thus enabling direct comparisons with classical multivariate techniques. In future we also aim to analyse the performance of our approach on other target functions.

References

1. H. Akaike. Fitting autoregressive models for prediction. *Annals of the Institute of Statistical Mathematics*, 21:243–247, 1969.
2. H. Akaike. Statistical predictor information. *Annals of the Institute of Statistical Mathematics*, 22:203–217, 1970.
3. R.A. Baxter and D.L. Dowe. Model selection in linear regression using the MML criterion. In J.A. Storer and M. Cohn, editors, *Proc. 4'th IEEE Data Compression Conference*, page 498, Snowbird, Utah, March 1994. IEEE Computer Society Press, Los Alamitos, CA. Also TR 276 (1996), Dept. of Computer Science, Monash University, Clayton, Victoria 3168, Australia.
4. P. Craven and G. Wahba. Smoothing noisy data with spline functions: Estimating the correct degree of smoothing by the method of generalized cross-validation. *Numerical Math.*, 31:377–403, 1979.
5. D.L. Dowe, R.A. Baxter, J.J. Oliver, and C.S. Wallace. Point Estimation using the Kullback-Leibler Loss Function and MML. In *Proc. 2nd Pacific Asian Conference on Knowledge Discovery and Data Mining (PAKDD'98)*, pp. 87–95, Melbourne, Australia, April 1998. Springer Verlag.
6. J. Rissanen. Modeling by shortest data description. *Automatica*, 14:465–471, 1978.
7. J. Rissanen. *Stochastic Complexity in Statistical Inquiry.* World Scientific, N.J., 1989.
8. J. Rissanen. Hypothesis selection and testing by the MDL principle. *The Computer Journal*, 42(4):260–269, 1999.
9. Y. Sakamoto et al. *Akaike information criterion statistics*, pp. 191–194. KTK scientific publishers, 1986.
10. G. Schwarz. Estimating the Dimension of a Model. *Ann. Stat.*, 6:461–464, 1978.
11. V. N. Vapnik. *Estimation of Dependencies Based on Empirical Data.* Springer-Verlag, New York, 1982.
12. V. N. Vapnik. *The Nature of Statistical Learning Theory.* Springer, New York, 1995.
13. V. N. Vapnik. *Computational Learning and Probabilistic Reasoning*, chapter Structure of Statistical Learning Theory. Wiley and Sons, 1996.
14. M. Viswanathan and C. Wallace. A note on the comparison of polynomial selection methods. In *Proc. 7th Int. Workshop on Artif. Intell. and Stats.*, pp. 169–177. Morgan Kauffman, January 1999.
15. C.S. Wallace. On the selection of the order of a polynomial model. Technical report, Royal Holloway College, London, 1997.
16. C.S. Wallace and D.M. Boulton. An information measure for classification. *Computer Journal*, 11(2):195–209, 1968.
17. C.S. Wallace and D.L. Dowe. Minimum message length and Kolmogorov complexity. *Computer Journal*, 42(4):270–283, 1999.
18. C.S. Wallace and P.R. Freeman. Estimation and inference by compact coding. *J. R. Statist. Soc B*, 49(3):240–265, 1987.

A Probabilistic Logic-based Framework for Characterizing Knowledge Discovery in Databases

Ying Xie and Vijay V. Raghavan

The Center for Advanced Computer Studies, University of Louisiana at Lafayette
{yxx2098,raghavan}@cacs.louisiana.edu

In order to further improve the KDD process in terms of both the degree of automation achieved and types of knowledge discovered, we argue that a formal logical foundation is needed and suggest that Bacchus' probability logic is a good choice. By completely staying within the expressiveness of Bacchus' probability logic language, we give formal definitions of "pattern" as well as its determiners, which are "previously unknown" and "potentially useful". These definitions provide a sound foundation to overcome several deficiencies of current KDD systems with respect to novelty and usefulness judgment. Furthermore, based on this logic, we propose a logic induction operator that defines a standard process through which all the potentially useful patterns embedded in the given data can be discovered. Hence, general knowledge discovery (independent of any application) is defined to be any process functionally equivalent to the process specified by this logic induction operator with respect to the given data. By customizing the parameters and providing more constraints, users can guide the knowledge discovery process to obtain a specific subset of all previously unknown and potentially useful patterns, in order to satisfy their current needs.

1 Introduction

The knowledge Discovery in Databases (KDD) process is defined to be the non-trivial extraction of implicit, previously unknown and potentially useful patterns from data [3]. This concise definition implies the basic capabilities that a KDD system should have. In reality, however, the user of current KDD systems is only provided limited automation: on the one hand, he is required to specify in advance exactly which type of knowledge is considered potentially useful and thus need to be discovered; on the other hand, often, he has to make the novelty judgment from a glut of patterns generated by KDD systems.

Y. Xie and V.V. Raghavan: *A Probabilistic Logic-based Framework for Characterizing Knowledge Discovery in Databases*, Studies in Computational Intelligence (SCI) **6**, 87–100 (2005)
www.springerlink.com

In other words, the KDD system can neither support in the specification of knowledge types that are potentially useful nor that a specific piece of discovered knowledge is previously unknown. Thus, it cannot guarantee what it discovers is previously unknown and potentially useful. This deficiency offers big room for the improvement of KDD process in terms of both the degree of automation and ability to discover a variety of knowledge types. In the following subsections, we will show in detail that in order to achieve such improvement, a formal theoretical framework of KDD is needed, and suggest that Bacchus' probabilistic logic provides a good foundation for us to start from.

1.1 On "Previously Unknown Pattern"

One can imagine that without a specification of what is "already known", there is no way for KDD systems to judge what is "previously unknown". Nevertheless, the view of a "real world" of current KDD systems is just a set of facts or data, so that any discovered pattern with parameters higher than thresholds will be deemed novel by the KDD systems, though they may not be so to human users. This is one of the major reasons why identifying and eliminating uninteresting discovered patterns is always a hot topic in the practice of data mining. In other words, the lack of the ability to model "already known patterns" burdens the user with the task of novelty judgment. In order to improve the degree of automation of the knowledge discovery process, we need a comprehensive model of the "real world", which requires a unified formalism to represent both facts and patterns.

However, the formalization of the representation alone is not enough to solve the novelty judgment problem. The inference ability among patterns is also required. For example, assume that we only have the following two "already known patterns": 1) More than 80% of the students didn't get A in the test; 2) Every student living on campus got A. Now suppose we obtain another pattern through some KDD process: more than 80% of the students didn't live on campus. Does it belong to "previously unknown pattern" or not? Most of us may agree that it does not, because this pattern can be easily deduced from the other two. Therefore, in order to effectively identify "previously unknown pattern", the deductive ability of the formalism that helps in recognizing relationships among patterns (statistical assertions) is also a necessity.

1.2 On "Potentially Useful Pattern"

In almost all previous literature, "potential usefulness" was viewed as a user-dependent measure. This popular opinion implies that the user has already known or currently has the ability to know which type of pattern will be potentially useful. If we cannot say that there exists, to some degree, a paradox in the way the terms of "previously unknown" and "potentially useful" are handled, we can at least say that current practices ignore a group of patterns that are indeed potentially useful. This group of patterns have the potential to lead to some useful actions that are not currently realized by the user.

In practice, a pattern with conditional probability (or, confidence) of 30% will be thought to be uninteresting. However, consider the following situation: the conditional probability of A given B is 1% or 80% (that is, either it is too low or too high) and the conditional probability of A given $B \wedge C$ is 30%, where A is some kind certain type of disease; B denotes some population and $B \wedge C$ is a subpopulation of B. Now one will feel the latter pattern is potentially useful. Conversely, a pattern with conditional probability 90% will be always thought useful. However, consider the scenario: the conditional probability of A given $B \wedge C$ is 90%, and the conditional probability of A given B equals to 91.5%. For this case, obviously, it is not $B \wedge C$ that implies A.

Therefore, on the one hand, what we can get, with current practices, is just a subset of all potentially useful patterns; on the other hand, what we really get may include those that are not potentially useful. In order to solve this problem, we need a more sophisticated formulation of the criteria for the notion of "potential usefulness".

One of the major purposes of KDD is to utilize the statistical knowledge extracted from given data to help one's decision making. In decision theory, propositional probability, whose statements represent assertions about the subjective state of one's beliefs, is always employed as the guide to action in the face of uncertainty [1]. However, the subjective state of one's belief is always coming from or affected by the objective statistical state of the world, which is described by statistical probability. For verification, one can think about some examples: recommender system, classification system, actuarial reasoning, medical expert system and so on. Therefore, direct inference, which deals with the inference of propositional probability from statistical assertion in AI, provides us the clue of how to define "potential usefulness". Roughly speaking, the statistical pattern that can affect one's propositional assertion, when one faces some particular situation, will be potentially useful.

1.3 Bacchus' Probabilistic Logic

Given that the scope of our discussion is limited to statistical knowledge discovery, which is one of the most active themes in KDD, we found that Bacchus's probabilistic logic [1], which is an augmentation of the first order logic and provides a unified formalism to represent and reason with both the statistical probability and propositional probability, provides a good foundation for formalizing the KDD process. For example, its first order logic part can be applied to model facts and relational structures, while the statistical probability part of its language can be used to represent and reason with patterns. Moreover, it integrates an elegant direct inference mechanism, based on which we can give a formal definition of "potential usefulness". Furthermore, the logic itself is sound.

1.4 Contributions and Outline of this Paper

Based upon Bacchus's probabilistic logic, this paper proposes a theoretical framework for (statistical) knowledge discovery in databases: first, a logical definition of "previously unknown, potentially useful pattern" are given by completely remaining within the framework of Bacchus's logic language, which provides a sound foundation to solve several deficiencies of current KDD systems with respect to novelty and usefulness judgment. Secondly, we propose a logic induction operator that defines a standard process through which all the potentially useful patterns embedded in the given data can be discovered. This logic induction operator formally characterizes the discovery process. Therefore, general knowledge discovery (independent of any application) is defined to be any process functionally equivalent to the process determined by this logic induction operator for a given data. Finally, by showing that the following knowledge types: association rule, negative association rule and exception rule are all special cases of our potentially useful patterns, user-guided knowledge discovery is defined to be discovery of a subset of all previously unknown and potentially useful patterns, discoverable via general knowledge discovery, in order to satisfy a particular user's current needs.

The rest of this paper will be organized as follows: A brief review of Bacchus' probability logic will be presented in Sect. 2. In Sect. 3, we will provide the logical model of a "real world" including both the data (facts) and "already known knowledge". Based on this model, in Sect. 4, formal definitions of pattern as well as its determiners "previously unknown" and "potentially useful" will be given. Then, in Sect. 5, the logical induction operator will be introduced based on these definitions, and several existing knowledge types will be analyzed within the theoretical framework that we propose for KDD. Finally, we will provide our conclusions in Sect. 6.

2 Review of Bacchus' Probability Logic

In this section, we briefly review the syntax, semantics and direct inference mechanism provided by Bacchus' probability logic. For a detailed discussion of its axiom system, the analysis of soundness and completeness, please refer to [1].

2.1 Syntax

The syntax defines the allowed symbols and specifies the strings of symbols that are well-formed formulas.

Allowed Symbols

- A set of n-ary function symbols (*f, g, h, ...*), including two types: *object function symbols* and *numeric function symbols*;
- A set of n-ary predicate symbols (*P, Q, R, ...*), including two types: *object predicate symbols* and *numeric predicate symbols*;
- The binary object predicate symbol =;
- The binary numeric predicate symbol < and = ($\leq, \geq$ and can be derived from them); the binary numeric function symbols + and $\times$; The numeric constants -1,1,0;
- A set of numeric variables and a set of object variables;
- The connectives $\wedge$, $\neg$; the qualifier $\forall$;
- The propositional probability operator **prob**;
- The statistical probability operator [and].

Formulas and Terms

- T0: A single object variable or constant is an *o-term*; a single numeric variable or constant is an *f-term*;
- T1: If *f* is an n-ary object function symbol and *t1, ..., tn* are o-term, then *f*(*t1, ..., tn*) is an o-term; if **f** is an n-ary numeric function symbol and *t1, ..., tn* are f-terms, then **f**(*t1, ...,tn*) is an f-term;
- F1: If *P* is an n-ary object predicate symbol and *t1, ..., tn* are o-terms, *P*(*t1, ..., tn*) is a formula;
- F2: If **P** is an n-ary numeric predicate symbol and *t1, ..., tn* are f-terms, then **P**(*t1, ..., tn*) is a formula;
- F3: If α is a formula, so is $\neg\alpha$;
- F4: If α, β are formulas, so is $\alpha \wedge \beta$;
- F5: If α is a formula, and x is variable, then $\forall x.\alpha$ is a formula;
- T2a: If α is a formula, then $\mathbf{prob}(\alpha)$ is an f-term;
- T2b: If α is a formula and $\mathbf{x}$ is a vector of n object variables $\langle x1, \ldots, xn\rangle$, then $[\alpha]_{\mathbf{x}}$ is an f-term.

Definition Extentions

- Propositional Conditional Probability:
  $prob(\beta) \neq 0 \rightarrow prob(\alpha|\beta) \times prob(\beta) = prob(\alpha \wedge \beta) \wedge$
  $prob(\beta) = 0 \rightarrow prob(\alpha|\beta) = 0$
- Statistical Conditional Probability:
  $[\beta]_{\mathbf{x}} \neq 0 \rightarrow [\alpha|\beta]_{\mathbf{x}} \times [\beta]_{\mathbf{x}} = [\alpha \wedge \beta]_{\mathbf{x}} \wedge$
  $[\beta]_{\mathbf{x}} = 0 \rightarrow [\alpha|\beta]_{\mathbf{x}} = 0$

2.2 Semantics

The logic semantic structure $\mathbf{M} = \langle \mathcal{O}, \mathcal{S}, \mathcal{V}, \mu_{\mathcal{O}}, \mu_{\mathcal{S}} \rangle$ is provided to interpret the symbols and formulas of Bacchus' probability logic, where

- $\mathcal{O}$ is a set of objects of the domain of interest;
- $\mathcal{S}$ is the set of possible worlds;
- $\mathcal{V}$ is a function that associates an interpretation of the language with each world; For every $s \in \mathcal{S}$, $\mathcal{V}(s)$ assigns to every n-ary object predicate symbol a subset of $\mathcal{O}^n$, to every n-ary object function symbol a function from $\mathcal{O}^n$ to $\mathcal{O}$, to every n-ary numeric predicate symbol a subset of $\mathbf{R}^n$, and to every n-ary numeric function symbol a function from $\mathbf{R}^n$ to $\mathbf{R}$.
- $\mu_{\mathcal{O}}$ is a discrete probability function over $\mathcal{O}$; for $A \subseteq \mathcal{O}$, $\mu_{\mathcal{O}}(A) = \Sigma_{o \in A} \mu_{\mathcal{O}}(o)$ and $\mu_{\mathcal{O}}(\mathcal{O}) = 1$.
- $\mu_{\mathcal{S}}$ is a discrete probability function over $\mathcal{S}$; for $B \subseteq \mathcal{S}$, $\mu_{\mathcal{S}}(B) = \Sigma_{s \in B} \mu_{\mathcal{S}}(s)$ and $\mu_{\mathcal{S}}(\mathcal{S}) = 1$.

In addition, a variable assignment function v is needed to map each variable to a particular individual from the proper domain. Hence, the truth assignment of a formula and the interpretation of a term will depend on three factors: the logic semantic structure $\mathbf{M}$, the current world s, and variable assignment function v. If a formula α is assigned true by the triple $(\mathbf{M}, s, v)$, it is denoted as $(\mathbf{M}, s, v) \models \alpha$; while the interpretation of term t is denoted as $t^{(\mathbf{M},s,v)}$. In details:

- T0: If x is a variable, then $x^{(\mathbf{M},s,v)} = v(x)$, which is independent of current world s.
- T1: If f is an n-ary function symbol (of either type) and $t1, \ldots, t2$ are terms of the same type, then

$$(f(t1, \ldots, tn))^{(\mathbf{M},s,v)} = f^{\mathcal{V}(s)}(t1^{(\mathbf{M},s,v)}, \ldots, t1^{(\mathbf{M},s,v)})$$

- F1: If P is an n-ary predicate symbol (of either type) and $t1, \ldots, t2$ are terms of the same type, then

$$(\mathbf{M}, s, v) \models P(t1, \ldots, tn) \textit{ iff } < t1^{(\mathbf{M},s,v)}, \ldots, t1^{(\mathbf{M},s,v)} > \in P^{\mathcal{V}(s)}$$

- F1=: If s and t are terms of the same type, then

$$(\mathbf{M}, s, v) \models (s = t) \textit{ iff } s^{(\mathbf{M},s,v)} = t^{(\mathbf{M},s,v)}$$

- F2: For every formula α, $(\mathbf{M}, s, v) \models \neg\alpha$ *iff* $(\mathbf{M}, s, v) \not\models \alpha$
- F3: For every formulas α, β, $(\mathbf{M}, s, v) \models (\alpha \wedge \beta)$ *iff* $(\mathbf{M}, s, v) \models \alpha$ and $(\mathbf{M}, s, v) \models \beta$.
- F4a: For every formula α and object variable x,

$$(\mathbf{M}, s, v) \models \forall x.\alpha \textit{ iff } (\mathbf{M}, s, v[x/o]) \models \alpha \text{ for all } o \in \mathcal{O},$$

  where $v[x/o]$ is identical to v except that it maps x to the individual o.
- F4b: For every formula α and numeric variable x,

$$(\mathbf{M}, s, v) \models \forall x.\alpha \textit{ iff } (\mathbf{M}, s, v[x/r]) \models \alpha \text{ for all } r \in \mathbf{R} \ ,$$

where $v[x/r]$ is identical to v except that it maps x to the individual r.

- T2a: For every formula α, $(\mathbf{prob}(\alpha))^{(\mathbf{M},s,v)} = \mu_{\mathcal{S}}\{s' \in \mathcal{S} : (\mathbf{M}, s', v) \models \alpha\}$.
- T2b: For every unary formula α, $([\alpha]_x)^{(\mathbf{M},s,v)} = \mu_{\mathcal{O}}\{o \in \mathcal{O} : (\mathbf{M}, s, v[x/o]) \models \alpha\}$.

2.3 Direct Inference

Direct inference deals with the inference of a propositional probability from a statistical probability. In terms of knowledge discovery, as discussed before, it provides a mechanism to formalize the decision making process based on discovered statistical knowledge. In the next section, we will define the notion of "potential usefulness" of a discovered pattern based on this mechanism. Generally speaking, direct inference always adopts the *reference class* principle that says the probability of an individual having a property should be equal to the relative frequency of that property among the *proper reference class* to which that individual belonged. Intuitively, the mechanism of choosing proper reference class used by Bacchus' logic can be described as follows: Unless we have knowledge that the statistics over a narrower reference class differ from the wider reference class, the wider reference class will be chosen as the proper one. Please refer to [1] for the formal description.

3 Logical Model of a "Real World"

Due to the advantages of Bacchus' probability logic in knowledge representation and reasoning, in this section, we will utilize it to model a "real world" consisting of an information table (database) and the background knowledge (previously known knowledge). Based on this model, in Sect. 4, formal definitions of "pattern" and its determiners "previously unknown" and "potentially useful" are provided. In Sect. 5, a logic induction operator is introduced in order to formalize the knowledge discovery process.

In this paper, a "real world" $\mathcal{W}$ is a tuple (I, KB_0), where I denotes an information table transformed from a database, while KB_0 is the background knowledge base. Therefore, logically modeling a "real world" includes building the logic semantic structure from the information table and representing the background knowledge base with logic formulas. In this subsection, we mainly talk about the first task.

Because all the facts in the "real world" should remain true across all the possible worlds, the logic semantic structure $\mathbf{M} = \langle \mathcal{O}, \mathcal{S}, \mathcal{V}, \mu_{\mathcal{O}}, \mu_{\mathcal{S}} \rangle$ of the "real world" will reduce to a triple $\langle \mathcal{O}, \mathcal{V}, \mu = \mu_{\mathcal{O}} \rangle$, and the truth assignment of a formula and the interpretation of a term will only depend on $\mathbf{M}$ and the variable assignment function v.

Now, given an information table $I = (U, A, V_a | a \in A, I_a | a \in A)$, where U is a set of objects of the interesting domain; A is the set of attributes; V_a is called the value set of attribute $a \in A$ and $I_a : U \rightarrow V_a$ is an information function, we will build the logic semantic structure $\mathbf{M} = \langle \mathcal{O}, \mathcal{V}, \mu \rangle$ through the following steps:

- $\mathcal{O} = U$;
- For each $V_a \in \{V_a | a \in A\}$
  - For each $va \in V_a$, va is a unary object predicate symbol, and interpretation function $\mathcal{V}$ assigns to va the set of objects $\{o \in \mathcal{O} | I_a(o) = va\}$, i.e., $va^{\mathcal{V}} = \{o \in \mathcal{O} | I_a(o) = va\}$;
- Two specific object predicate symbols are defined, $\emptyset$ and F, such that $\emptyset^{\mathcal{V}} = \emptyset$, $F^{\mathcal{V}} = \mathcal{O}$;
- μ is a discrete probability function, such that for any object $o \in \mathcal{O}, \mu(o) = 1/|\mathcal{O}|$, and for any subset $A \subseteq \mathcal{O}, \mu(A) = \Sigma_{o \in A} \mu(o)$ and $\mu(\mathcal{O}) = 1$;
- Interpretation function $\mathcal{V}$ assigns to every n-ary numeric predicate symbol a subset of $\mathbf{R}^n$, for example, 2-ary predicate symbol "=" will be assigned the set $\{(1/8, 1/8), (2/8, 2/8), \ldots\}$; and $\mathcal{V}$ assigns to every n-ary numeric function symbol a function from $\mathbf{R}^n$ to $\mathbf{R}$, which gives the proper meanings to symbols $+, -, \times$, min, max and so on.

Example 1. The student information table (shown in Table 1) is given by $I = (U, A, \{V_a | a \in A\}, \{I_a | a \in A\})$, where

- $U = \{S1, S2, \ldots, S8\}$;
- $A = \{Grad./Under., Major, GPA\}$;
- $V_{Grad./Under.} = \{Gr, Un\}, V_{Major} = \{Sc, En, Bu\}, V_{GPA} = \{Ex, Gd, Fa\}$;
- $I_{Grad./Under.} = U \rightarrow V_{Grad./Under.}, I_{major} = U \rightarrow V_{Major}, I_{GPA} = U \rightarrow V_{GPA}$.

Hence, the corresponding semantic structure for this information table is $M = \langle \mathcal{O}, \mathcal{V}, \mu \rangle$, where

- $\mathcal{O} = U = \{S1, S2, \ldots, S8\}$;
- The set of unary object predicate symbols is {Gr, Un, Sc, En, Bu, Ex, Gd, Fa}; and by interpretation function $\mathcal{V}, Gr^{\mathcal{V}} = \{S1, S2, S3\}; Un^{\mathcal{V}} = \{S4, S5, S6, S7, S8\}; Sc^{\mathcal{V}} = \{S1, S3, S5\}$ and so on;
- By discrete probability function μ, for every student $S_i \in \mathcal{O}, \mu(S_i) = 1/8$; and for any subset $A \subseteq \mathcal{O}, \mu(A) = \Sigma_{S_i \in A} \mu(S_i)$ and $\mu(\mathcal{O}) = 1$;
- By interpretation function $\mathcal{V}$, numeric predicate symbols such as $>, =, <$, and numeric function symbols such as +, -, $\times$, min, max get the proper meaning.

Now, based on the semantic structure $\mathbf{M}$ we build, some truth assignments of formulas and interpretations of terms will be shown as follow:

- $(\mathbf{M}, v) \models Gr(S1)$, because $S1 \in Gr^{\mathcal{V}}$;
- $(\mathbf{M}, v) \not\models En(S1)$, because $S1 \notin En^{\mathcal{V}}$;

Table 1. Student Information Table

Student	Grad./Under.	Major	GPA
S1	Gr	Sc	Ex
S2	Gr	En	Ex
S3	Gr	Sc	Gd
S4	Un	Bu	Ex
S5	Un	Sc	Gd
S6	Un	Bu	Fa
S7	Un	En	Fa
S8	Un	En	Gd

- $(\mathbf{M}, v) \not\models \emptyset(Si)$, for any $Si \in \mathcal{O}$;
- $(\mathbf{M}, v) \models F(Si)$, for any $Si \in \mathcal{O}$;
- $[Gr]_x = \mu\{Si : (\mathbf{M}, v[x/Si]) \models Gr(Si)\} = \mu\{S1, S2, S3\} = 3/8$;
- $(\mathbf{M}, v) \models [Gr]_x = 3/8$.

In this paper, we will not discuss the detail of how to represent background knowledge using probabilistic logic formula. However, it should be mentioned that if two object predicate symbols P and Q (e.g. Gr and Un in Table 1) are coming from the same domain, we need put the following formulas into $KB_0 : [P|Q]_x = 0; [Q|P]_x = 0$.

4 Previously Unknown and Potentially Useful Pattern

Based on the logical representation of a "real world", in this section, a logical definition of "previously unknown, potentially useful pattern" is provided by completely remaining within the framework of Bacchus's logic language.

In terms of statistical knowledge, a pattern can be viewed as the quantitative description of concepts and the relationships between concepts. Therefore, a logic definition of pattern will be based on the logic definition of concept.

Definition 1. *Concept: If P is an object predicate symbol, P is a concept; it is also called atomic concept, if $P \neq F$ and $P \neq \emptyset$. both $\emptyset$ and F are called special concepts; If P, Q are concepts, $P \wedge Q$ is also a concept, where binary operator $\wedge$ is defined as follows:*

- $(P \wedge Q)^{\mathcal{V}} = P^{\mathcal{V}} \cap Q^{\mathcal{V}}$;
- $P \wedge P = P$;
- $P \wedge Q = Q \wedge P$;
- $P \wedge (Q \wedge R) = (P \wedge Q) \wedge R$;
- $P \wedge F = P$;
- $P \wedge \emptyset = \emptyset$;

If a concept P is neither atomic concept nor special concept, it is called composite concept.

Now, let $C = P_1, P_2, \ldots, P_n, F, \emptyset$ denote all the atomic concepts and special concepts defined on $\mathcal{O}$. The $\wedge$-closure C^* is defined to be the minimum set containing all the concepts in C and is closed under $\wedge$. Now we define the binary relation $\leq$ on C^*, such that for any pair of concepts $Qi, Qj \in C^*$, we have:

$$Q_i \leq Q_j \Leftrightarrow Q_i = Q_j \wedge P\,, \tag{1}$$

where P can be any concept. The tuple $\langle C^*, \leq \rangle$ is a complete lattice that we call concept lattice.

Example 2. Let's continue with the logic representation built in example 1. The set of all the atomic concepts is: $C = \{Gr, Un, Sc, En, Bu, Ex, Gd, Fa\}$. And we have the following concept lattice $\langle C^*, \leq \rangle$ (Fig. 1), each node of which represents a concept.

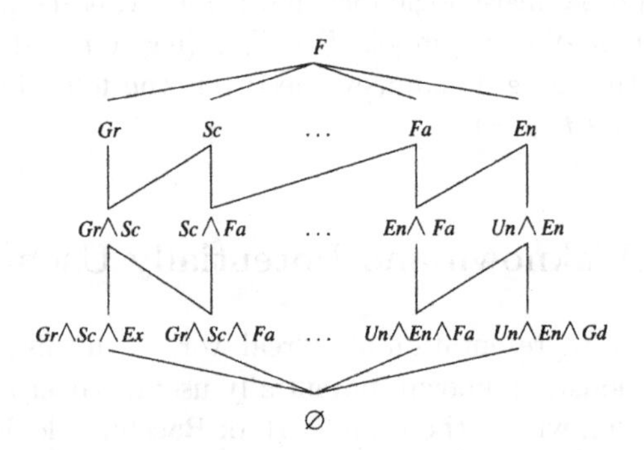

Fig. 1. Concept Lattice

Now, we define *inside-attribute* for each concept as follows: For concept F, every atomic concept is called *inside-attribute* of F; For concept $\emptyset$, *inside-attribute* has no definition; For any other concept $\wedge_{i=1}^{k} Q_i$, where $k \geq 1$, each atomic concept $Q_i (1 \leq i \leq k)$ is called *defining-attribute* of this concept; any other atomic concetp is called *inside-attribute* of this concept.

Example 3. For the concept lattice shown in Fig. 1,

- Concept F's inside-attribute includes: $Gr, Un, Sc, En, Bu, Ex, Gd, Fa$;
- Concept Gr's inside-attribute includes: $Un, Sc, En, Bu, Ex, Gd, Fa$;
- Concept $Gr \wedge Sc$'s inside-attribute includes: Un, En, Bu, Ex, Gd, Fa;
- Concept $Gr \wedge Sc \wedge Ex$'s inside-attribute includes:Un, En, Bu, Gd, Fa.

Definition 2. *Pattern: Let P and Q be two concepts, and $r \in \mathbf{R}$. We call formula $[P|Q]_x = r$ a pattern, iff $P = \wedge_k a_k$ and $[Q]_x > n$, where $k \geq 1$, each a_k is an inside attribute of Q, and parameter n is called noise controller.*

Definition 3. *Validity, Implicitness of pattern: A pattern $\mathcal{P}$ is valid by $\mathbf{M}$ iff $(\mathbf{M}, v) \models \mathcal{P}$. If formula $[P|Q]_x = r$ is a pattern, it is called implicit Pattern by $\mathbf{M}$ iff it is valid by $\mathbf{M}$ and $Q \neq F$.*

Intuitively, an implicit pattern cannot be obtained by just querying the database using only one standard SQL statement.

As mentioned in Sect. 1, if a pattern can be easily deduced from the previously known ones, it should not be called previously unknown pattern. The fact that the previously known knowledge base can be represented formally using Bacchus' logic formula, not only makes it possible to give a formal definition of what is previously unknown pattern, but also provides a mechanism by which the KDD system can judge the novelty of a pattern to a greater degree of precision.

Definition 4. *Previously Unknown pattern: Assume Pr is an efficient enough deductive program based on Bacchus's probability logic. For any implicit pattern $[P|Q]_x = r$, if no formula like $[P|Q]_x = r1 \wedge (r - e < r1) \wedge (r1 < r + e)$, where $r1$ can be any number that belongs to $\mathbf{R}$, can be deduced from KB_0 by Pr within c steps, this implicit pattern is called previously unknown pattern, which is denoted as $KB_0 \not\vdash_{Pr(c)} [P|Q]_x = r$. Parameter c is called complexity controller, which is utilized to balance the use of deduction and induction processes. Parameter e is called error controller, which is utilized to identify similar or nearly identical knowledge.*

Now, it is time for us to formalize the notion of usefulness of patterns. Intuitively, our idea of "potential usefulness", which is based on the direct inference mechanism of Bacchus's probability logic, can be described as follows: if a narrower reference class provides much more knowledge than any of the wider reference about a property, the statistical information of this narrower reference class will be potentially useful in deciding the probability of any individual, which belongs to these reference classes, having that property.

Definition 5. *Potentially Useful Pattern: Let Q be any concept except F and $\emptyset$ on the concept lattice, and let concepts $P1, P2, \ldots, Pj$ be all the parent concepts of Q. Let $A = \wedge_k a_k$, where a_k is an inside-attribute of Q, (of course a_k is also inside-attribute of all Q's parent concepts). Now a valid pattern $[A|Q]_x = r$ is called positive pattern by $\mathbf{M}$, if $(\mathbf{M}, v) \models r - max_{1 \leq i \leq j}[A|P_i]_x > s$; and it is called negative pattern by $\mathbf{M}$, if $(\mathbf{M}, v) \models min_{1 \leq i \leq j}[A|P_i]_x - r > s$, where $(\mathbf{M}, v) \models (0 < s) \wedge (s \leq 1)$. Both positive and negative patterns are potentially useful patterns. Parameter s is called significance controller.*

5 Logical Induction Operator $\Vdash$

In order to formalize the knowledge discovering process using Bacchus' logic language, we define a logic induction operator $\Vdash$ that allows to formally characterize a discovery process, based on the concept lattice, through which all the potentially useful patterns can be found.

Definition 6. *Logic Induction Operator $\Vdash$: Given the logic semantic structure $\mathbf{M}$, assume that $\langle C^*, \leq \rangle$ is the concept lattice decided by $\mathbf{M}$. Let n be the noise controller. The Logic Induction Operator, $\Vdash$, determines the following inductive process to get a set of potentially useful patterns, which is denoted as PUP: traverse the concept lattice from F to $\emptyset$, for each concept node P (except F and $\emptyset$), if $[P]_x > n$, check all its offspring nodes with the form $P \wedge \gamma$, where γ can be either atomic concept or composite concept. If $[\gamma|P]_x = r$ is a potentially useful pattern, put it into PUP. This process is denoted as $(\mathbf{M}, v) \Vdash PUP$.*

Theorem 1. *Completeness of Logic Induction Operator $\Vdash$: Given $(\mathbf{M}, v) \Vdash PUP$, a pattern $\mathcal{P}$ is a potentially useful pattern by $\mathbf{M}$ iff $\mathcal{P} \in PUP$.*

Proof: Given $(\mathbf{M}, v) \Vdash PUP$ and n is the noise controller, 1) according to the definition of logic induction operator $\Vdash$, all patterns put into PUP are potentially useful patterns; 2) if $\mathcal{P}$ is a potentially useful pattern by $\mathbf{M}$, $\mathcal{P}$ must have the form $[Q|P]_x = r$, where $[P]_x > n$. Therefore, concept P must be checked by operator $\Vdash$. When checking P, operator $\Vdash$ will examine all its offspring concepts including $P \wedge Q$. Thus, if $[Q|P]_x = r$ is potentially useful pattern, it will be put into PUP.

Definition 7. *General Knowledge Discovery: Given a information Table I and KB_0, a general knowledge discovery is any efficient inductive process that extracts a set of patterns $\mathbf{P}$, such that $\mathcal{P} \in \mathbf{P}$ iff $(\mathbf{M}, v) \Vdash PUP$, $\mathcal{P} \in PUP$ and $KB_0 \not\models_{Pr(c)} \mathcal{P}$, where $\mathbf{M}$ is the logic semantic structure built from I.*

Developing such an efficient inductive process will be the topic of our future work. Users can guide the knowledge discovery process by providing more constrains to get a subset of PUP that satisfies their current needs. Several existing statistical knowledge discovery types can be viewed as user-guided knowledge discovery.

Definition 8. *User-guided Knowledge Discovery: Given a information Table I and KB_0, a user-guided knowledge discovery is any efficient inductive process that extracts a set of patterns $\mathbf{P}$, such that $\mathbf{P} \subset PUP$, and for any $\mathcal{P} \in \mathbf{P}$, $KB_0 \not\models_{Pr(c)} \mathcal{P}$, where $\mathbf{M}$ is the logic semantic structure built from I and $(\mathbf{M}, v) \Vdash PUP$.*

The following existing statistical knowledge types can be viewed as special subsets of PUP:

- Association rule $P \Rightarrow Q$ [8]: $[Q|P]_x = r$ is a positive pattern; extra constrains: $r > threshold$;
- Negative association rule $P \Rightarrow \neg Q$ [9]: $[Q|P]_x = r$ is a negative pattern; extra constrains: $r < threshold$;
- Exception rule $P \Rightarrow Q, P \wedge P^* \Rightarrow \neg Q$ [10]: $[Q|P]_x = r_1$ is a positive pattern and $[Q|P \wedge P^*]_x = r_2$ is a negative pattern; extra constrains: $r_1 > threshold1$ and $r_2 < threshold2$;

6 Conclusions

In order to further improve the KDD process in terms of both the degree of automation achieved and types of knowledge discovered, we propose a theoretical framework for (statistical) knowledge discovery in database based on Bacchus's probability logic. Within this framework, formal definitions of "previously unknown, potentially useful pattern" are proposed, which provide a sound solution to overcome several deficiencies of current KDD systems with respect to novelty and usefulness judgment. Furthermore, a logic induction operator that describes a standard knowledge discovering process is proposed, through which all potentially useful pattern can be discovered. This logic induction operator provides a formal characterization of the "discovery" process itself. Based on this logical foundation, general knowledge discovery (independent of any application) is defined to be any efficient process functionally equivalent to the process determined by this logic induction operator for given data. Due to the fact that several existing knowledge types are special cases of our potentially useful patterns, users can, by customizing the parameters and providing more constraints, guide the knowledge discovery process to obtain a specific subset of the general previously unknown and potentially useful patterns in order to satisfy their current needs. Our future work will focus on designing scalable algorithm that is functionally equivalent to the process determined by this logic induction operator for the given data.

References

1. Bacchus, F. (1990) Representing and Reasoning With Probabilistic Knowledge. MIT-Press, Cambridge, MA.
2. Pawlak, Z. (1991) Rough Sets, Theoretical Aspects of Reasoning about Data. Kluwer Academic Publishers, Dordrecht.
3. Frawley, W., Piatetsky-Shapiro G., Matheus C. (1991) Knowledge Discovery In Databases: An Overview. In: Piatetsky-Shapiro, G. Frawley, W. (eds) Knowledge Discovery In Databases. AAAI Press/MIT Press, Cambridge, MA.
4. Yao, Y. (2001) On Modeling Data Mining with Granular Computing. In: Proceeding of the 25th Annual International Computer Software and Applications Conference, Chicago, IL.

5. Lin, T., Louie, E. (2001) Modeling the Real World for Data Mining: Granular Computer Approach. In: Proceeding of IFSA/NAFIPS, Vancouver, Canada.
6. Louie, E., Lin, T. (2002) Semantics Oriented Association Rules. In: Proceeding of FUZZ-IEEE Conference IEEE World Congress on Computational Intelligence, Honolulu, HI.
7. Murai, T., Murai, M., and Sato Y. (2001) A Note on Conditional Logic and Association Rules. In: Proceeding of JSAI International Workshop on Rough Set Theory and Granular Computing, Matsue, Japan.
8. Agrawal, R., Imielinski, T., and Swami, A. (1993) Mining association rules between sets of items in large databases. In Proceeding of ACM-SIGMOD International Conference on Management of Data, Washington, DC.
9. Savasere, A., Savasere, and E., Navathe, S. (1998) Mining for Strong Negative Associations in a Large Database of Customer Transactions. In: Proceeding of the 14th International Conference on Data Engineering, Orlando, Florida.
10. Suzuki, E. (1997) Autonomous Discovery of Reliable Exception Rules. In: Proceeding of the 3th International Conference on Knowledge Discovery and Data Mining (KDD-97), Newport Beach, California.
11. Zhong, N., Yao, Y., and Ohsuga, S. (1999) Peculiarity Oriented Multi-database Mining. In: Proceeding of the 5th European Conference on Principles and Practice of Knowledge Discovery in Databases, Freiburg, Germany.
12. Bacchus, F. (1990) Lp, A Logic for Representing and Reasoning with Statistical Knowledge. Computational Intelligence 6:209–231.

A Careful Look at the Use of Statistical Methodology in Data Mining

Norman Matloff

Department of Computer Science, University of California, Davis, CA 95616 USA
matloff@cs.ucdavis.edu

Summary. Knowledge discovery in databases (KDD) is an inherently statistical activity, with a considerable literature drawing upon statistical science. However, the usage has typically been vague and informal at best, and at worst of a seriously misleading nature. In addition, much of the classical statistical methodology was designed for goals which can be very different from those of KDD. The present paper seeks to take a first step in remedying this problem by pairing precise mathematical descriptions of some of the concepts in KDD with practical interpretations and implications for specific KDD issues.

1 Introduction

The field of KDD has made extensive use of statistical methodology. Such methodology is clearly of great potential, but is also fraught with a myriad of pitfalls. A lack of insight into how the methods actually work may result in unnecessarily weak KDD machinery. Moreover, naive "transplantation" of many statistical methods to KDD arenas for which the methods were not designed may result in poor or even misleading analyses.

The remedy is for KDD practitioners to on the one hand to gain a better, more precise mathematical understanding of the statistical methodology, and on the other hand to develop a better intuitive understanding of what the methodology does.

In this paper, we will encourage KDD practitioners to

- devise simple mathematical models which will facilitate precise statements of the problems at hand and deepen intuitive insight into possible solutions
- take a close look at the goals of the statistical methodology they use, assessing how well those methods fit the given KDD application

We will present a simple framework, consisting of some simple mathematical constructs motivated by intuitive notions tied to the actual practice of

N. Matloff: *A Careful Look at the Use of Statistical Methodology in Data Mining*, Studies in Computational Intelligence (SCI) **6**, 101–117 (2005)
www.springerlink.com

KDD. It is important to note that the latter, i.e. the intuitive "philosophical" issues, will play an integral role here.

It is assumed here that the reader has at least a first-level knowledge of standard statistical methods, e.g. hypothesis testing, confidence intervals and regression, and a basic background in probabilistic constructs such as random variables, expected value and so on. We first develop some simple infrastructure, some of which will be at least somewhat familiar to many readers, and then move to detailed worked-out examples which illustrate the issues.

2 Statistical Sampling

As is common in theoretical treatments, we will phrase the issues in terms of a statistical prediction problem. This is not to say we consider KDD to be limited to prediction settings, but rather that such settings are among the most common KDD applications. We depart from tradition, though, by engaging in an explicit discussion of the practical interpretation of what we mean by "statistical."

2.1 Notation

Denote our attribute set by $X^{(1)}, \ldots, X^{(d)}$. It is assumed that our database constitutes a *statistical sample* of n observations on these attributes; the ith observation on the jth attribute from this sample is denoted by $X_i^{(j)}$, $i = 1, \ldots, n, j = 1, \ldots, d$. We need to spend some time here on the question of what this really means.

To make things concrete—again, this is one of our principle aims—let's consider the well-known KDD "market basket" example. Each row in the database corresponds to some individual consumer. The jth attribute might record whether the consumer bought a given item (1 for yes, 0 for no).[1] We wish to predict the event $X^{(i)} = 1$ from several other simultaneous events $X^{(j)} = 1$. In other words, we wish to know whether a consumer's purchase of one item is related to his/her possible interest in buying a related item. For example, in an online book sales site, if a consumer purchases a certain book, the site may then suggest to the customer that he/she consider buying other related books. Note that this is unlike typical statistical contexts, in which we would be interested in predicting from events of the form $X^{(j)} = 0$ as well.

The vector $(X_i^{(1)}, \ldots, X_i^{(d)})$, representing the values of all our attributes in the ith observation will be denoted by X_i. In relational database terms, this vector is the ith row in our relation.

[1] In other KDD contexts, some of the attributes may be characteristics of a consumer, say age, income or gender (say 1 for male, 0 for female), and so on.

2.2 Sampling from Populations, Real or Conceptual

In considering our database to be a "statistical sample," we mean that it is a sample from some "population." This interpretation is, in our view, key.

The population may be tangible, as in the "basket" example, where we are sampling from the population of all customers of this business. Or, the population may be more conceptual in nature. A database consisting of students in a new curriculum in a university could be considered as a sample from the conceptual population of all students at this university who *might* be in this major. If for example we imagine the university overall enrollment had been 20 percent larger this year, with no change in demographic or other makeup of the enrollment, then some of the increased overall enrollment would have been students choosing this major. This population is then conceptual. Similar remarks hold when considering potential future students in the major.

Here is an example of a "population" which is even more conceptual in nature. Consider the subject of quadratic equations, studied in school algebra classes:

$$ax^2 + bx + c = 0 \tag{1}$$

The students learn that this equation has a real root if and only the *discriminant* $b^2 - 4ac$ is nonnegative. Suppose one did not know this rule, and tried to find it using KDD.

This sounds like an inherently non-statistical problem. Yet one could convert it to a statistical problem in the following way. One could sample randomly from $a/b/c$ space, according to a distribution of one's choice, and for each sample triplet from this space, determine somehow (say by graphing the quadratic polynomial) whether a real root exists. One could then apply various *statistical regression models* (see below), trying to predict the 0–1 variable w from a, b and c, where w is 1 if there are real roots and 0 otherwise. In this manner, we might possibly stumble onto the discriminant rule.

2.3 Relation to Probability Distributions

It is important to relate the abstract mathematical variables to the population being studied. When we speak of the distribution of the random variable $X^{(j)}$, what we really mean is the distribution of that attribute in the population. Say $X^{(1)}$ is age of the customer. When we say, for instance, that $P(X^{(1)} > 32) = 0.22$, we mean that 22 percent of all customers in this population are older than 32.

A similar point holds for expected value. Some KDD practitioners with an engineering or physical science background might be accustomed to interpreting $E(X^{(1)})$ in terms of the physics metaphor of center of gravity. Yet for statistical applications such as KDD, the relevant interpretation of this quantity is as the mean age of all customers in this population. This interpretation is especially important when considering sample-distribution issues such as bias and variance, as we will see.

3 Prediction

As we have noted, our focus in the statistical nature of KDD is on prediction. For notational convenience, in the remainder of this paper, let us suppose that we are using $X^{(1)}, \ldots, X^{(d-1)}$ to predict $X^{(d)}$, and rename the latter attribute Y.

Our focus here will largely be on predicting *dichotomous*, i.e. 0/1-valued, attributes Y. (We do not make this restriction on the attributes $X^{(j)}$.) However, as will be seen, most of the issues which arise also occur in the case of continuous-valued Y.

3.1 Statement of the Problem

Suppose for the moment that we know the population distributions of the attributes, and we wish to minimize the overall probability of misclassification.[2] Suppose that we observe $X^{(j)}$ to have the value v_j, $j = 1, \ldots, d-1$. Then we would guess Y to be either 0 or 1, according to whether the quantity

$$
\begin{aligned}
r(v) &= r(v_1, \ldots, v_{d-1}) \\
&= P(Y = 1 | X^{(1)} = v_1, \ldots, X^{(d-1)} = v_{d-1}) \quad (2)
\end{aligned}
$$

is less than 0.5 or greater than 0.5, respectively, where $v = (v_1, \ldots, v_{d-1})$).[3] Let us then denote this guess as $g[r(v)]$, where $g(u) = \text{floor}(2u)$ for u in [0,1].

Note that in theory $r(v)$ should converge to 1 or 0 (depending on v) as $d \to \infty$. In other words, if you know enough about the situation, you can always predict correctly! This of course is a very big "if", but it puts in perspective notions such as that of *unexpected rules* in [10]. Nothing is "unexpected," strictly speaking; we simply lack data. This issue will become relevant in our discussion of the bias/variance tradeoff and Simpson's Paradox later.

3.2 Classification Vs. Regression

Some authors, e.g. Han [5] consider the case of dichotomous Y to be a conceptually separate case from that of continuous Y, and refer to it as *classification*

[2] The latter would not be the case if we assigned different costs to different types of errors. It may be more costly to falsely guess Y to be 1 than to falsely guess it to be 0, for example.

[3] Here and below, will write this function simply as $r()$, suppressing the dependence on d, i.e. not writing it as r_{d-1}. The dependence will be clear from the number of arguments.

instead of prediction. However, mathematically it is the same problem, in the following sense.

Classically, the problem of predicting a general attribute Y from a vector of attributes $X = (X^{(1)}, \ldots, X^{(d-1)})$ is posed as finding a function h() that minimizes

$$E[(Y - h(X))^2] \tag{3}$$

One can easily show that the minimizing solution is the *regression function*,

$$h(t) = E(Y|X = t) \tag{4}$$

Now, if Y is dichotomous, i.e. Y takes on the values 0 and 1, then

$$E(Y|X = t) = w \cdot 1 + (1 - w) \cdot 0 = r(t) \tag{5}$$

where $w = P(Y = 1|X = t)$.

In other words, the general formulation of the prediction problem yields the regression function $r()$ anyway.

Thus the classification and regression problems are the same. This is not just a semantic issue. There are indeed some aspects of the classification problem which differ from the regression setting, but there is a great amount of commonality. A vast literature exists on the general regression problem, with much material relevant to the dichotomous case,[4] and it would be a loss not to draw upon it.[5]

3.3 The Function r() Must Be Estimated from Sample Data

The situation is complicated by the fact that we do not know the population distributions of the attributes, as assumed in the previous few paragraphs. We thus do not know the function $r()$ above, and need to estimate it from the observations in our database.

The estimated function, $\hat{r}(v)$, is obtained either by parametric or nonparametric means. A common parametric approach, for instance, uses the logistic regression model, which postulates that $r(v)$ has the form

$$r(v_1, \ldots, v_{d-1}) = \frac{1}{1 + \exp[-(\beta_0 + \beta_1 v_1 + \ldots + \beta_{d-1} v_{d-1})]}$$

The parameters β_j are estimated from our sample data $X_i^{(j)}$, yielding the estimated parameters $\hat{\beta}_j$ and the estimated $r(v)$:[6]

[4] A large separate literature on the classification problem has also been developed, but much of it draws upon the material on regression.

[5] By the way, some of these points are also noted (albeit rather abstractly) in [3].

[6] See for example the **lrm** procedure in the R statistical package [11].

$$\hat{r}(v) = \frac{1}{1 + \exp[-(\hat{\beta}_0 + \hat{\beta}_1 v_1 + \ldots + \hat{\beta}_{d-1} v_{d-1})]}$$

Many nonparametric method have been used in KDD for estimating r(v), such as CART [1].

4 Over/underfitting

There is a lot of talk about "noise mining," "overfitting" and the like in the KDD literature, but again this is rarely precisely defined.

4.1 Bias and Variance

In some theoretical papers, the literature does at least point out that the "average" discrepancy between $\hat{r}(v)$ and $r(v)$ can be shown to consist of two components—(the square of) a *bias* component,

$$E\hat{r}(v) - r(v) \tag{6}$$

and a *variance* component,

$$E[(\hat{r}(v) - E\hat{r}(v))^2] \tag{7}$$

Note that v is fixed here, not random. Instead, the randomness involves the fact that these expected values are averages over all possible n-observation samples from the given population.[7] This interpretation is very important when one is assessing various competing types of prediction methodology, and especially important in understanding the bias/variance problem.

A large bias is due to using too simple a model in the parametric case, or to using too coarse a granularity in the nonparametric case (e.g. leaf nodes too large in CART). In both cases, one common source of the problem is that we are using too few predictor attributes.

However, any efforts to reduce the bias will increase the variance, i.e. increase the amount of "noise." This is due to having an insufficient sample size n for the given model. In CART, for example, if we use smaller hyper-rectangles in order to reduce bias, a given hyper-rectangle might contain very few observations, thus causing $\hat{r}()$ to have a large variance within the rectangle. The same rectangle, applied to a larger sample from the same population, might work fine. In the case of a logistic model, if we add more predictor variables in order to reduce bias, then for each fixed j, $Var(\hat{\beta}_j)$ will tend to increase.[8]

[7] We could also make v random. i.e. replace it by X in the expressions above, so that the averaging is being done both over all possible n-observation samples and over all values of v.

[8] See [14] for an analytical proof of this in the linear regression setting.

These phenonema arise in the "market basket" setting as well. If confidence and support levels (see Sect. 6.1) are set too low, a bias problem occurs in much the same way as it does if we have too large a rectangle in CART. If the confidence and support levels are set too high, it becomes a variance problem.

4.2 Illustrative Model

Recall that our theme here has been that empirical research work in KDD should include a mathematically precise statement of the problem, and should present mathematical treatment of at least a small but illustrative model of the effects being studied. In that light, we now present such a model of the "noise fitting" problem. We ask the reader to keep in mind our specific goal here in devising this model—we desire to devise a simple but precise model in which the roles of both d and n in the bias/variance tradeoff are explicitly visible in the model's results.[9] It is our hope that KDD practitioners will often engage in setting up such simple models in order to gain insight into specific applications.

Continue to assume the setting described at the beginning of Sect. 3, but with the additional specialization that all the predictor attributes $X^{(j)}$, $j = 1, \ldots, d-1$ are dichotomous.

Suppose that $X^{(j)}$, $j = 1, \ldots, d-1$ are all "coin tosses," i.e. have probability 0.5 of taking on the value 1 and are statistically independent. Suppose in addition that $P(Y = 1|X^{(1)} = v_1)$ is equal to 0.6 for $v_1 = 1$ and equal to 0.4 for $v_1 = 0$, and that $X^{(j)}$, $j = 2, \ldots, d-1$ have no predictive power for Y at all, i.e.

$$r(v_1, v_2, \ldots, v_{d-1}) = P(Y = 1|X^{(1)} = v_1) \tag{8}$$

independent of $v_2, \ldots, v_{d-1}$.

But we would not have this information, since we would not have the population data. We would have only *sample* estimates of $r(v)$ to work with, $\hat{r}(v)$. The point then is that that estimate will be subject to bias/variance issues. We discuss the variance issue first, and focus our attention on the estimation of $r(1, 1, \ldots, 1)$.

One decision we would need to make is which of the attributes $X^{(j)}$ to use as predictors. Let us compare the effects of using just $X^{(1)}$ alone to predict Y, versus using $X^{(1)}, X^{(2)}, \ldots, X^{(d-1)}$ for that prediction. In the former situation, note again that we would be modeling r(v) to be a function which does not depend on $X^{(2)}, \ldots, X^{(d-1)}$ (see (8)). Again, this modeling assumption would be correct, but we would not know this.

Suppose we are not using a parametric model, and instead are simply using straight sample proportions to estimate $r()$. Then if we use only $X^{(1)}$ as our predictor, our estimate of $r(1, 1, \ldots, 1)$ would be the proportion of records in our database for which $Y = 1$, among those for which $X^{(1)} = 1$, i.e.

[9] A number of much more complex examples of this tradeoff for various kinds of estimators are presented (without derivation) in [6].

$$\hat{r}(1,1,\ldots,1) = \frac{\sum_i X_i^{(1)} X_i^{(d)}}{\sum_i X_i^{(1)}} = \frac{T_1}{U_1} \tag{9}$$

Recalling that $r(1,1,\ldots,1) = 0.6$, the question at hand is, "What is the probability that $\hat{r}(1,1,\ldots,1)$ will make the right decision for us in this situation, which is to guess that $Y = 1$?"[10] Well, this is

$$P(\hat{r}(1,1,\ldots,1) > 0.5) = P(T_1 > 0.5U_1) \tag{10}$$

To evaluate this probability, note first that T_1 and U_1, are binomially distributed.[11] Thus they have approximate normal distributions. But in addition, their bivariate distribution approximates that of a bivariate normal.[12] The means and variances of T_1 and U_1 are then np, nq, $np(1-p)$ and $nq(1-q)$, where $p = P(X^{(1)} = X^{(d)} = 1) = 0.3$ and $q = P(X^{(1)} = 1) = 0.5$. Their covariance is

$$\begin{aligned} &Cov(T_1, U_1) \\ &= n[E(X^{(1)}X^{(d)}X^{(1)}) - E(X^{(1)}X^{(d)}) \cdot EX^{(1)}] \\ &= np(1-q) \end{aligned} \tag{11}$$

Any linear combination of T_1 and U_1, say $aT_1 + bU_1$, then has an approximate normal distribution with mean n(ap+bq), and variance

$$n[a^2Var(T_1) + b^2Var(U_1) + 2abCov(T_1, U_1)] \tag{12}$$

In our case here, $a = 1$ and $b = -0.5$. After doing the calculations we find that $E(T_1 - 0.5U_1) = 0.05n$ and $Var(T_1 - 0.5U_1) = 0.1225n$, and thus

$$P(T_1 > 0.5U_1) \approx 1 - \Phi(-0.14\sqrt{n}) \tag{13}$$

where Φ is cumulative distribution function of a standard $N(0,1)$ variate.

So, (13) is the probability that we make the right decision if we predict Y from only $X^{(1)}$. Let's see how that probability changes if we predict Y from $X^{(1)}, \ldots, X^{(d-1)}$.

In this setting, (2) again reverts to (8), and (9) becomes

$$\hat{r}(1,1,\ldots,1) = \frac{\sum_i X_i^{(1)} X_i^{(2)} \ldots X_i^{(d-1)} X_i^{(d)}}{\sum_i X_i^{(1)} X_i^{(2)} \ldots X_i^{(d-1)}} = \frac{T_{d-1}}{U_{d-1}} \tag{14}$$

The analog of (13) is then (after a bit of algebraic approximation)

[10] Note that the term "the right decision" means the decision we would make if we had full knowledge of the population distributions, rather than just sample estimates. It does not mean that our guess for Y is guaranteed to be correct.

[11] The variable $B = X^{(1)}X^{(d)}$ is 0-1 valued, and the terms are independent, so the sum T_1 is binomial.

[12] This stems from the fact the vector form of the Central Limit Theorem.

$$P(T_{d-1} > 0.5U_{d-1}) \approx 1 - \Phi\left(-0.28 \cdot \sqrt{\frac{n}{2^d}}\right) \tag{15}$$

Compare (13) and (15), focusing on the roles of d and n. They are both of the form $P(Z > c)$ for a negative c, and the algebraically smaller (i.e. more negative) c is, the better. So, for fixed n, the larger d is, the worse is our predictive ability for Y, if we use all $d-1$ predictors.

Now, remember the context: We devised a model here in which $X_i^{(2)}$ $\ldots X_i^{(d-1)}$ had no predictive ability at all for Y in the population distribution, though the analyst would not know this. *In other words, not only will the analyst not gain predictive ability by using these attributes, he/she would actually lose predictive power by using them, i.e. "overfit."*

So, this is the variance side of the bias/variance tradeoff. The number of records in our sample which have $X^{(1)} = 1, X^{(2)} = 1, \ldots, X^{(d-1)} = 1$ will be very small for large d (similar to having a small leaf node in CART), leading to a high variance for $\hat{r}(1, 1, \ldots, 1)$.[13]

Equation (15) also shows the role of n in the overfitting issue: For fixed d, as n increases the harmful effect of overfitting will diminish.

Now, what about the bias side of the bias/variance tradeoff? Suppose we are considering using $X^{(1)}, X^{(2)} \ldots, X^{(k)}$ as our predictor attributes. Due to the nature of the model here, i.e. the fact that $X^{(2)} \ldots, X^{(k)}$ have no predictive power, the bias in using any k in the range $1 \leq k < d-1$ is 0.[14] So, if we use k greater than 1, we are incurring the problem of increasing variance without reducing bias, a pure loss.

On the other hand, using $k = 0$ would produce a bias, since $X^{(1)}$ does have some predictive value for Y: If k were taken to be 0, then the population value of $r(1, 1, \ldots, 1)$ would reduce to the unconditional probability $P(Y = 1) = 0.5$, rather than the achievable value 0.6.

Again, our point in devising this model here is to illustrate our theme that even empirical KDD research should anchor its presentation with (a) a precise mathematical statement of the problem being studied, and (b) a simple mathematical model which explicitly illustrates the issues.

The word *explicitly* in (b) should be emphasized. Equation (15) explicitly shows the roles of d and n. One sees that for a fixed value of n, use of a larger d increases the variance. As d increases, at some point our predictive ability based on sample data will begin to diminish, i.e. we will overfit. One also sees, though, that for a larger value of n, that crossover point will occur for a larger d, i.e. we can use more attributes as our predictors.

---

[13] We did not directly calculate that variance here, showing the variance effects only indirectly. However, $Var(\hat{r})$ could be calculated by using the *delta method*, applied to the function $f(t, u) = t/u$ [13].

[14] Here we are using the term *bias* only informally, not quite as in (6). To make it formal, we would have to make the expected value conditional on U_k, the latter defined analogously to U_1 and U_{d-1} for the case of k predictors. The technical reason for this is that U_k may be 0, and thus $\hat{r}(1, 1, \ldots, 1)$ would be undefined.

5 Attribute Selection

As we have seen, there is a tradeoff between bias and variance for fixed n. As finer models are fitted, with more attributes, the bias is reduced (or stays the same) but the variance increases. If too much attention is paid to minimizing bias rather than variance, the decision rules found from the analysis may be spurious, hence the term *noise mining.*

The problem of finding the attribute set which maximizes predictive ability, i.e. finding the optimal point in the bias/variance tradeoff spectrum, is as old as the field of statistics itself. It must be emphasized in the strongest possible terms that this is still an unsolved problem, in spite of numerous papers in the KDD literature which report on "promising" solutions.

5.1 The Use of Hypothesis Testing

We wish to emphasize also the importance of phrasing the problem directly in terms of the goals of the KDD settings being analyzed. For example, the classical statistical method for selecting predictor attributes, hypothesis testing, is of questionable propriety. In the case of a logistic model, say, this approach would involve testing the hypothesis

$$H_0 : \beta_j = 0 \tag{16}$$

and then either including or excluding the attribute $X^{(j)}$ in our predictor set, depending on whether the hypothesis is rejected or accepted. This procedure is often applied sequentially, one potential predictor attribute at a time, to determine which attributes to use; this algorithm is called *stepwise variable selection.*

Yet the classic use of hypothesis testing in regression analysis is largely aimed at descriptive, rather than predictive, types of applications. An example of descriptive use is the identification of risk factors for a certain disease. The attributes having large β_j are considered to be the more important factors. There the goal is more to understand how a disease arises in the population at large, rather than to actually predict whether a particular individual develops the disease. By contrast, in many KDD settings one really does want to predict, and thus one should be hesitant to apply classical variable-selection algorithms in the KDD context.

For example, in the classical use of regression, the hypotheses are typically tested at traditional significance levels such as $\alpha = 0.05$. Yet some studies (e.g. [7]) have found that for the prediction of continuous variables, the best values of α are in the range 0.25 to 0.40.[15]

Some readers will notice this as a statistical *power* issue; the term refers to the probabilities of rejecting the null hypothesis under various scenarios of

[15] The picture is further muddled by the fact that the stated α value is nominal anyway, due to issues such as *multiple inference*, discussed in Sect. 6.

the alternative hypothesis. As such, it is not a new concept, since the original usage of hypothesis testing in the early 20th century did assume that analysts would carefully balance the values of α and power in a manner suitable to the application. However, modern usage has institutionalized the value of α to be set to 0.05 or 0.01, so much so that a "star" notation has become standard in research in the social and life sciences (a statistic is adorned with one or two asterisks, depending on whether α is 0.05 or 0.01).[16] Power is rarely discussed, let alone calculated.

Again, in KDD the goals may be very different from classical statistical usage, and thus that our analyses must not blindly mimic that usage. In this case, the point is that if one does use hypothesis testing for model selection, power considerations are crucial. Of course, it is not always clear how best to use power analyses in a given situation, or even how to calculate it in many cases, but it is certainly clear that classical values of α are not the best.

6 The Multiple Inference Problem

If hypothesis testing is used for attribute selection, there is not only the problem of considering power levels, but also the issue of accuracy of the α and power levels. First there is the problem that even though each test has the specified significance level, the collective significance level of all the tests may be much greater. In addition, many attribute selection algorithms, e.g. stepwise selection, are *adaptive*, and thus even the individual significance levels may have values quite different from their nominal values.

The problem of the collective significance level being much greater than the level applied to each individual test can be addressed by the use of *multiple inference methods*, which allow one to set an overall significance level for multiple tests. In this light, a worthy future research project would be to revisit past dismissals of the use of multiple inference for rule finding [8]. Earlier authors had found such approaches to be too conservative, finding too few rules. However, the point we made above in Sect. 5.1 suggests that with a larger value of overall α, such methodology may work well.

6.1 Illustrative Example

Consider the market basket problem with two attributes, $X^{(1)}$ and $X^{(2)}$. Typically one is interested in finding all attributes for which the *confidence*, say

$$P(X^{(2)} = 1 | X^{(1)} = 1) \tag{17}$$

is greater than a given level c, and the *support*, say

[16] Even if one's goal is descriptive rather than predictive, the usage of these institutionalized values is questionable [9].

$$P(X^{(2)} = X^{(1)} = 1) \tag{18}$$

is a above s. If both conditions are satisfied, we will use the rule $X^{(1)} \Rightarrow X^{(2)}$.

(For the sake of simplicity, we are using only two attributes in this example. Typically there are many more than two attributes, in which case combinations of attributes are considered. With three attributes, for instance, we would assess not only potential rules such as $X^{(1)} \Rightarrow X^{(2)}$ but also some like $X^{(1)}, X^{(3)} \Rightarrow X^{(2)}$. In the latter case, quantities such as $P(X^{(2)} = 1 | X^{(1)} = X^{(3)} = 1)$ would be checked.)

Let $p_{ij} = P(X^{(1)} = i, X^{(2)} = j)$, i,j = 0,1. Then to determine whether to use the rule $X^{(1)} \Rightarrow X^{(2)}$, we might test the hypothesis

$$H_0 : p_{11} \leq s \; or \; \frac{p_{11}}{p_{11} + p_{10}} \leq c \tag{19}$$

and then use the rule if the hypothesis is rejected. But for mathematical tractability here, let us treat (19) as two separate hypotheses. Accounting also for the possible rule $X^{(2)} \Rightarrow X^{(1)}$, we have a total of three hypotheses to test in all.[17] We will now investigate how well we can assess the two rules with a given value of α. We will calculate $E(K)$, where K is the number of hypotheses in which we make the correct decision; K varies from 0 to 3.

As a test case, let us take the matrix $p = (p_{ij})$ to be

$$\begin{pmatrix} 0.15 & 0.45 \\ 0.10 & 0.30 \end{pmatrix} \tag{20}$$

and take $s = 0.35$, $c = 0.60$. In this setting, the potential rules have confidence and support as shown in Table 1. Then in this simple example,[18] of the three hypothesis tests to be performed, ideally two of them should be rejected and one accepted. $E(K)$ will be the sum of the probabilities of the two rejections and one acceptance.

Table 1. Rule Results

poss. rule	conf.	supp.
$X^{(1)} \Rightarrow X^{(2)}$	0.75	0.30
$X^{(2)} \Rightarrow X^{(1)}$	0.40	0.30

A standard method for multiple inference on a small number of tests involves use of the *Bonferroni Inequality*.[19] If one is performing k tests and wishes an overall significance level of at most α, then one sets the individual significance level of each test at α/k. Here, to achieve an overall significance

[17] The two potential rules have one support test in comon.

[18] A much more general study is being conducted for a separate paper.

[19] A reference on this and many other multiple inference methods is [12].

level of 0.05, we use a level of $0.05/3 = 0.017$ for each of the three tests. For a one-sided test, this corresponds to a "Z value" of 2.12 in the normal distribution, i.e. $1 - \Phi(2.12) = 0.017$.

To test the potential rule $X^{(1)} \Rightarrow X^{(2)}$ for our confidence level 0.35, we reject if

$$\frac{\hat{p}_{11} - 0.35}{\sqrt{\hat{Var}(\hat{p}_{11})}} > 2.12 \tag{21}$$

where

$$\hat{Var}(\hat{p}_{11}) = \frac{\hat{p}_{11}(1 - \hat{p}_{11})}{n} \tag{22}$$

Thus we need to compute

$$P\left(\frac{\hat{p}_{11} - 0.35}{\sqrt{\hat{Var}(\hat{p}_{11})}} < 2.12\right) \tag{23}$$

in the setting $p_{11} = 0.30$. This probability is computed (approximately) via standard normal distribution calculations as seen earlier in Sect. 4.2.[20]

For testing whether the potential rule $X^{(1)} \Rightarrow X^{(2)}$ meets the confidence criterion, note that

$$\frac{p_{11}}{p_{11} + p_{10}} > c \tag{24}$$

holds if and only if

$$(1 - c)p_{11} - cp_{10} > 0 \tag{25}$$

Thus the probability of making the correct decision regarding the confidence level of $X^{(1)} \Rightarrow X^{(2)}$ is

$$P\left(\frac{(1 - c)\hat{p}_{11} - c\hat{p}_{10}}{\sqrt{\hat{Var}[(1 - c)\hat{p}_{11} - c\hat{p}_{10}]}} > 2.12\right) \tag{26}$$

where, using (12) and the fact that $Cov(\hat{p}_{11}, \hat{p}_{10}) = -p_{11}p_{10}/n$,

$$\sigma^2 = Var[(1 - c)\hat{p}_{11} - c\hat{p}_{10}] =$$

$$\frac{1}{n}[(1 - c)^2 p_{11}(1 - p_{11}) + c^2 p_{10}(1 - p_{10}) + 2c(1 - c)p_{11}p_{10}]$$

The probability of the correct decision is then

$$1 - \Phi\left(2.12 - \frac{(1 - c)p_{11} - cp_{10}}{\sigma}\right) \tag{27}$$

[20] Here, though, one uses the exact variance in the denominator in (23), i.e. $p_{11}(1 - p_{11})/n = 0.18/n$.

The probability of the correct decision for $X^{(2)} \Rightarrow X^{(1)}$ (that the confidence level is *not* met) is computed similarly.

After doing all this computation, we found that the value of $E(K)$ for an α of 0.05 turns out to be 2.70. After some experimentation, we found that a very large level of 0.59 improves the value of $E(K)$ to 2.94. Though this amount of improvement is modest, it does show again that the best choice of significance level in KDD settings may be quite different from those used classically in statistical applications. Moreover, it shows that multiple inference methods may have high potential in KDD after all, if only one considers nontraditional significance levels. And as mentioned, the multiple-inference approach dismissed in [8] appear to be worth revisiting.

The fact that the relevant hypotheses involve linear combinations of the p_{ij}, as in (25), suggests that the Scheffe' method of multiple inference could be used. That method can simultaneously test all linear combinations of the p_{ij}. Those rules for which the confidence and support tests are both rejected would be selected. Again, it may be the case that with suitable significance levels, this approach would work well. As noted earlier, this is under investigation.

7 Simpson's Paradox Revisited

A number of KDD authors have cautioned practitioners to be vigilant for *Simpson's Paradox* [4]. Let us first couch the paradox in precise mathematical terms, and then raise the question as to whether, for predictive KDD settings, the "paradox" is such a bad thing after all.

Suppose each individual under study, e.g. each customer in the market basket setting, either possesses or does not possess traits A, B and C, and that we wish to predict trait A. Let $\bar{A}$, $\bar{B}$ and $\bar{C}$ denote the situations in which the individual does not possess the given trait. Simpson's Paradox then describes a situation in which

$$P(A|B) > P(A|\bar{B}) \tag{28}$$

and yet

$$P(A|B, C) < P(A|\bar{B}, C) \tag{29}$$

In other words, the possession of trait B seems to have a positive predictive power for A by itself, but when in addition trait C is held constant, the relation between B and A turns negative.

An example is given in [2], concerning a classic study of tuberculosis mortality in 1910. Here the attribute A is mortality, B is city (Richmond, with $\bar{B}$ being New York), and C is race (African-American, with $\bar{C}$ being Caucasian). In probability terms, the data show that:[21]

- P(mortality | Richmond) = 0.0022

[21] These of course are sample estimates.

- P(mortality | New York) = 0.0019
- P(mortality | Richmond, black) = 0.0033
- P(mortality | New York, black) = 0.0056
- P(mortality | Richmond, white) = 0.0016
- P(mortality | New York, white) = 0.0018

The data also show that

- P(black | Richmond) = 0.37
- P(black | New York) = 0.002

a point which will become relevant below.

At first, New York looks like it did a better job than Richmond. However, once one accounts for race, we find that New York is actually worse than Richmond. Why the reversal? The answer stems from the fact that racial inequities being what they were at the time, blacks with the disease fared much worse than whites. Richmond's population was 37% black, proportionally far more than New York's 0.2%. So, Richmond's heavy concentration of blacks made its overall mortality rate look worse than New York's, even though things were actually much worse in New York.

But is this "paradox" a problem? Some statistical authors say it merely means that one should not combine very different data sets, in this case white and black. But is the "paradox" really a problem in KDD contexts?

The authors in [2] even think the paradox is something to be exploited, rather than a problem. Noting that many KDD practitioners are interested in finding "surprising" rules (recall [10]), the authors in [2] regard instances of Simpson's Paradox as generally being surprising. In other words, they contend that one good way to find surprising rules is to determine all instances of Simpson's Paradox in a given data set. They then develop an algorithm to do this.

That is interesting, but a different point we would make (which, to our knowledge has not been made before) is that the only reason this example (and others like it) is surprising is that the predictors were used in the wrong order. As noted in Sect. 5, one normally looks for predictors (or explanatory variables, if the goal is understanding rather than prediction) one at a time, first finding the best single predictor, then the best pair of predictors, and so on. If this were done on the above data set, the first predictor variable chosen would be race, not city. In other words, the sequence of analysis would look something like this:

- P(mortality | Richmond) = 0.0022
- P(mortality | New York) = 0.0019
- P(mortality | black) = 0.0048
- P(mortality | white) = 0.0018
- P(mortality | black, Richmond) = 0.0033
- P(mortality | black, New York) = 0.0056

- P(mortality | white, Richmond) = 0.0016
- P(mortality | white, New York) = 0.0018

The analyst would have seen that race is a better predictor than city, and thus would have chosen race as the best single predictor. The analyst would then investigate the race/city predictor pair, and would never reach a point in which city alone were in the selected predictor set. Thus no anomalies would arise.

8 Discussion

We have, in the confines of this short note, endeavored to argue for the need for more mathematical content in empirically-oriented KDD research. The mathematics should be kept simple and should be carefully formulated according to the goals of the research. We presented worked-out examples of how a simple mathematical model could be used to illustrate, and hopefully gain insight into, the issues at hand.

We wish to reiterate, on the other hand, that very theoretical treatments, written by and for mathematical statisticians, are generally inaccessible to empirical KDD researchers and KDD practitioners. We hope that theoretical work be made more intuitive and tied to practical interpretations.

References

1. L. Breiman, J. Friedman, R. Olshen and C. Stone. *Classification and Regression Trees*. Wadsworth Publishers, 1984.
2. C.C. Fabris and A.A. Freitas. Discovering Surprising Patterns by Detecting Occurrences of Simpson's Paradox. In *Research and Development in Intelligent Systems XVI (Proc. ES99, The 19th SGES Int. Conf. on Knowledge-Based Systems and Applied Artificial Intelligence)*, 148–160. Springer-Verlag, 1999.
3. J. Friedman. On Bias, Variance, 0/1-Loss, and the Curse of Dimensionality, *Data Mining and Knowledge Discovery*, 1, 55–77, 1997.
4. C. Glymour, D. Madigan, D. Pregibon and P. Smyth. Statistical Themes and Lessons for Data Mining, *Data Mining and Knowledge Discovery*, 1, 25–42, 1996.
5. J. Han and M. Kamber. *Data Mining: Concepts and Techniques*, Morgan Kaufman Publishers, 2000.
6. T. Hastie, R. Tibshirani and J. Friedman. *The Elements of Statistical Learning: Data Mining, Inference, and Prediction*. Springer, 2001.
7. R. Bendel and A. Afifi. Comparison of Stopping Rules in Forward Stepwise Regression, Joint ASA/IMS Meeting, St. Louis, 1974.
8. P. Domingos. E4—Machine Learning. http://citeseer.nj.nec.com/205450.html.
9. N. Matloff. Statistical Hypothesis Testing: Problems and Alternatives. *Journal of Economic Entomology*, 20, 1246–1250, 1991.
10. B. Padmanabhan and A. Tuzhilin. Finding Unexpected Patterns in Data, in *Data Mining, Rough Sets and Granular Computing*, T.Y. Lin, Y.Y. Yao, L. Zadeh (eds), Physica-Verlag, 2001.

11. R. statistical package home page, http://www.r-project.org.
12. Y. Hochberg and A. Tamhane. *Multiple Comparison Procedures*, Wiley, 1987.
13. R.J. Serfling. *Approximation Theorems of Mathematical Statistics*, Wiley, 1980.
14. R. Walls and D. Weeks. A Note on the Variance of a Predicted Response in Regression, *The American Statistician*, 23, 24–26, 1969.

Justification and Hypothesis Selection in Data Mining

Tuan-Fang Fan[1,3], Duen-Ren Liu[1] and Churn-Jung Liau[2]

[1] Institute of Information Management, National Chiao-Tung University, Hsinchu, Taiwan
tffan.iim92g@nctu.edu.tw, dliu@iim.nctu.edu.tw
[2] Institute of Information Science, Academia Sinica, Taipei, Taiwan
liaucj@iis.sinica.edu.tw.
[3] Department of Information Engineering, National Penghu Institute of Technology, Penghu, Taiwan
dffan@npit.edu.tw

Summary. Data mining is an instance of the inductive methodology. Many philosophical considerations for induction can also be carried out for data mining. In particular, the justification of induction has been a long-standing problem in epistemology. This article is a recast of the problem in the context of data mining. We formulate the problem precisely in the rough set-based decision logic and discuss its implications for the research of data mining.

1 The Circular Justification of Induction

Induction is widely accepted as a cornerstone of the modern scientific methodology. It regards the systematic use of past experiences in the prediction of the future. According to Hume's epistemology, our beliefs are established on the basis of observation and can be divided into two cases:

1. Observed matters of fact.
2. Unobserved matters of fact.

While the beliefs of observed matters of fact are based directly on observation, unobserved matters of fact can only be known indirectly on the basis of observation by means of inductive argument.

An argument is called inductive if it passes from singular statements regarding the results of observations to universal statements, such as hypotheses or theories. The inductive argument can be formulated deductively as follows: Let $\varphi(x)$ denote a hypothesis regarding the individuals x we are interested in. In general, $\varphi(x)$ is in the form of a universal conditional sentence "$P(x) \supset Q(x)$", however it is not necessarily so. Let Φ and Ψ denote the following sentences

T.-F. Fan et al.: *Justification and Hypothesis Selection in Data Mining*, Studies in Computational Intelligence (SCI) **6**, 119–130 (2005)
www.springerlink.com

Φ : $\varphi(x)$ holds for all observed x.
Ψ: $\varphi(x)$ holds for all x (or at least the next observed x).

Then the inductive argument is just an application of the modus ponens rule and can be expressed as

$$\frac{\begin{array}{c}\Phi\\ \Phi \supset \Psi\end{array}}{\Psi} \tag{1}$$

where $\Phi \supset \Psi$ is called "The Principle of the Uniformity of Nature" (PUN).

If we can assume that the observation is noiseless, then the validity of the conclusion of the inductive argument will rely completely on that of PUN. However, since PUN represents unobserved portions of nature, it cannot be known directly by observation. Thus PUN, if known at all, must be known on the basis of induction. This means that to justify PUN, we must first justify the validity of the inductive argument which in turn relies on the validity of PUN. This kind of justification is thus circular and cannot serve as a proper justification. This leads to Hume's skepticism on induction [16].

Does the same kind of inductive skepticism arise in the data mining context? Let us answer this question by considering the mining of rules from data tables in the framework of Pawlak's decision logic [22].

2 Decision Logic

Definition 1 *A data table*[4] *is a triplet* $T = (U, A, \{a_T \mid a \in A\})$ *such that*

- U *is a nonempty finite set, called the universe,*
- A *is a nonempty finite set of primitive attributes, and*
- *for each* $a \in A$, $a_T : U \to V_a$ *is a total function, where* V_a *is the domain of values for* a_T. *Usually, we will simply write* a *instead of* a_T *for the functions.*

Given a data table T, we will denote its universe U and set of attributes A by $Uni(T)$ and $Att(T)$ respectively.

In [22], a decision logic(DL) is proposed for the representation of the knowledge discovered from data tables. The logic is called decision logic because it is particularly useful in a special kind of data table, called *decision table.* A decision table is a data table $T = (U, C \cup D, \{a_T \mid a \in C \cup D\})$, where $Att(T)$ can be partitioned into two sets C and D, called condition attributes and decision attributes respectively. By data analysis, decision rules relating the condition and the decision attributes can be derived from the table. A rule is then represented as an implication between formulas of the logic. However, for a general data table, the acronym DL can also denote *data logic.*

[4] Also commonly known as information table, knowledge representation system, information system, attribute-value system, or bag relation in the literature.

The basic alphabet of a DL consists of a finite set of attribute symbols $\mathcal{A}$ and for $a \in \mathcal{A}$, a finite set of value symbols $\mathcal{V}_a$. The syntax of DL is then defined as follows:

Definition 2

1. *An atomic formula of DL is a descriptor* (a, v), *where* $a \in \mathcal{A}$ *and* $v \in \mathcal{V}_a$.
2. *The well-formed formulas (wff) of DL is the smallest set containing the atomic formulas and closed under the Boolean connectives* $\neg, \wedge,$ *and* $\vee$.

A data table $T = (U, A, \{a_T \mid a \in A\})$ is an interpretation of the given DL if there is a bijection $f : \mathcal{A} \to A$ such that for every $a \in \mathcal{A}$, $V_{f(a)} = \mathcal{V}_a$. Thus, by somewhat abusing the notation, we will usually denote an atomic formula as (a, v), where $a \in A$ and $v \in V_a$, if the data tables are clear from the context. Intuitively, each element in the universe of a data table corresponds to a data record and an atomic formula, which is in fact an attribute-value pair, describes the value of some attribute in a data record. Thus the atomic formulas (and likely the wffs) can be verified or falsified in each data record. This gives rise to a satisfaction relation between the universe and the set of wffs.

Definition 3 *Given a DL and a DT* $T = (U, A, \{a_T \mid a \in A\})$ *for it, the satisfaction relation* $\models$ *between* $x \in U$ *and wffs of DL is defined inductively as follows:*

1. $(T, x) \models (a, v)$ *iff* $a(x) = v$
2. $(T, x) \models \neg\varphi$ *iff* $(T, x) \not\models \varphi$
3. $(T, x) \models \varphi \wedge \psi$ *iff* $(T, x) \models \varphi$ *and* $(T, x) \models \psi$
4. $(T, x) \models \varphi \vee \psi$ *iff* $(T, x) \models \varphi$ *or* $(T, x) \models \psi$

If φ *is a DL wff, the set* $m_T(\varphi)$ *defined by:*

$$m_T(\varphi) = \{x \in U \mid (T, x) \models \varphi\} ,$$

is called the meaning of the formula φ *in* T. *If* T *is understood, we simply write* $m(\varphi)$.

A formula φ is said to be valid in a data table T, written $T \models \varphi$, if and only if $m(\varphi) = U$. That is, ϕ is satisfied by all individuals in $Uni(T)$.

3 Circularity of the Justification

Obviously, a data table consists of the observed matters of fact in some domain. To cover the unobserved part of the world, we slightly extend the definition of data tables to that of situation tables. A *situation table*, just like a data table, is a triplet $S = (U, A, \{a_S \mid a \in A\})$ except that the finiteness assumption of the universe U is relaxed, so that U may be finite or infinite. Thus, a data table is just a special kind of situation table. Given two situation tables $S_1 = (U_1, A_1, \{a_{S_1} \mid a \in A_1\})$ and $S_2 = (U_2, A_2, \{a_{S_2} \mid a \in A_2\})$, S_1 is said to be a sub-table of S_2, denoted by $S_1 \sqsubseteq S_2$, if $U_1 \subseteq U_2$, $A_1 = A_2$, and $a_{S_1}(u) = a_{S_2}(u)$ for all $a \in A_1$ and $u \in U_1$.

Let S be a situation table, $T \sqsubseteq S$ a data table, and φ a DL wff, then the inductive argument (1) can be reformulated in the data mining context as

$$\frac{\begin{array}{l} T \models \varphi \\ T \models \varphi \supset S \models \varphi \end{array}}{S \models \varphi} \tag{2}$$

or

$$\frac{\begin{array}{l} T \models \varphi \\ T \models \varphi \supset (S, u) \models \varphi \end{array}}{(S, u) \models \varphi} \tag{3}$$

where $u \in Uni(S) - Uni(T)$, i.e., u is the next observed individual.

The data mining schema (4) and (5) can be respectively called the generalization and the prediction version of data mining. As in the case of inductive schema, we have to verify the validity of the PUN "$T \models \varphi \supset S \models \varphi$" or "$T \models \varphi \supset (S, u) \models \varphi$". Let us consider the case for the predication version of data mining. The generalization version can be done analogously. A *data mining task* is characterized by (T, φ, S, u) which are respectively a data table, a DL wff, a situation table, and a test example. To ensure that the PUN holds for any data mining task, we can use a (meta-) data table specified by $T^{dm} = (U, A, \{a_{T^{dm}} \mid a \in A\})$ such that

- $U = \{t_1, t_2, \cdots, t_n\}$ is the finite set of data mining tasks which have been performed so far,
- $A = \{Tr, Fm, Te, Si, PUN\}$
- For each task $t \in U$, $Tr(t)$ is a data table consisting of the training samples, $Fm(t)$ is a DL wff denoting the rule to be mined in the task, $Te(t)$ is a test example not observed yet in $Tr(t)$, $Si(t)$ is a situation table containing, among others, all training samples in $Tr(t)$ and the test example $Te(t)$ and PUN is a binary attribute such that $PUN(t) = 1$ iff $Tr(t) \models Fm(t)$ implies $(Si(t), Te(t)) \models Fm(t)$.

Let S^{dm} be the situation table consisting of all possible data mining tasks, either performed or not, such that the attribute PUN is defined in all tasks t exactly as above. In other words, $PUN(t) = 1$ iff $Tr(t) \models Fm(t)$ implies $(Si(t), Te(t)) \models Fm(t)$ for any $t \in Uni(S^{dm})$. Thus T^{dm} is a sub-table of S^{dm}. We can then define a new task $t_0 = (T^{dm}, (PUN, 1), S^{dm}, t_0)$ in $Uni(S^{dm})$ but not in $Uni(T^{dm})$. Intuitively, t_0 is the data mining task to infer the PUN on itself by the observations in T^{dm}. By instantiating it into the inductive schema (5), we have

$$\frac{\begin{array}{l} T^{dm} \models (PUN, 1) \\ T^{dm} \models (PUN, 1) \supset (S^{dm}, t_0) \models (PUN, 1) \end{array}}{(S^{dm}, t_0) \models (PUN, 1)} \tag{4}$$

We can assume that $T^{dm} \models (PUN, 1)$ holds, namely, all data mining tasks in T^{dm} have made successful predications on their test examples. Thus the truth of $(S^{dm}, t_0) \models (PUN, 1)$ will depend on that of $T^{dm} \models (PUN, 1) \supset (S^{dm}, t_0) \models (PUN, 1)$, which by the definition of S^{dm}, holds exactly when $PUN(t_0) = 1$ in S^{dm} and is equivalent to $(S^{dm}, t_0) \models (PUN, 1)$ by the semantics of DL. In summary, the justification of $(S^{dm}, t_0) \models (PUN, 1)$ depends on itself, so it is circular as in Hume's argument.

4 Infinite Regress of the Justification

The self-referential character of the circular justification of data mining is apparent. First, t_0 occurs in the definition of t_0 itself. Second, the situation table S^{dm} contains t_0 with its value of attribute Si being S^{dm} itself. Someone may feel uncomfortable with this kind of self-reference and stick to stratifying the data tables into levels. We can do this in the following way. Let us call integers, strings, real numbers, DL wffs, other nominal values, etc. *0th order elements* and a situation table $S = (U, A, \{a_S \mid a \in A\})$ is called first order if for each $a \in A$ and $x \in U$, $a_S(x)$ is a 0th order element. And generally, a situation table $S = (U, A, \{a_S \mid a \in A\})$ is called kth order if for each $a \in A$ and $x \in U$, $a_S(x)$ is a 0th order element or jth order situation table for some $j < k$.

To validate the data mining schema (5) for first-order situation tables, we have to justify the following first-order PUN:

$$\text{PUN}_1 =_{def} T_1 \models \varphi \supset (S_1, u) \models \varphi$$

where T_1 and S_1 are ranged over first-order situation tables. As in the preceding subsection, we can construct a second-order data table T_2^{dm} for the justification of PUN_1. The T_2^{dm} is essentially the same as T^{dm} except in the following two aspects. First, the data mining tasks in the universe of T_2^{dm} are restricted to those based on first-order situation tables. Second, the attribute PUN is replaced by PUN_1, though its dependency on the other attributes remains unchanged. Let S_2^{dm} be a situation table for all first-order data mining tasks so that T_2^{dm} is its sub-table. Then, even though all data mining tasks have made successful predictions, so that $T_2^{dm} \models (PUN_1, 1)$, the validity of PUN_1 still depends on the following

$$T_2^{dm} \models (PUN_1, 1) \supset (S_2^{dm}, t) \models (PUN_1, 1)$$

which is an instance of the second-order PUN

$$\text{PUN}_2 =_{def} T_2 \models \varphi \supset (S_2, u) \models \varphi$$

where T_2 and S_2 are ranged over second-order situation tables. In this way, the justification of data mining schema (3) for first-order situation tables will in turn depend on PUN_3, PUN_4,..., and so on, which are defined in analogy of PUN_1 and PUN_2. Therefore, this leads to the infinite regress of the justification.

5 Implications

In the preceding discussions, we have formulated the circularity and infinite regress problems of the justification for data mining in decision logic. In fact, it is even argued that such circularity or infinite regress is inevitable in any attempt to the justification of induction. For example, one of the most important philosophers in the 20th century, Karl Popper, wrote in ([24], P.29):

> *...For the principle of induction must be a universal statement in its turn. Thus if we try to regard its truth as known from experience, then the very same problems which occasioned its introduction will arise all over again. To justify it, we should have to employ inductive inferences; and to justify these we should have to assume an inductive principle of a higher order; and so on. Thus the attempt to base the principle of induction on experience breaks down, since it must lead to an infinite regress.*

It seems that this argument also applies to data mining analogously according to our decision logic formulation.

This argument clearly shows the anti-inductivist position of Popper. It also leads him to refuse any form of verificationism of scientific theory, which claims that a scientific theory is supported by a collection of verifying instances. Instead, he proposes falsification as the methodology of science. In the falsification philosophy of science, a scientific theory is not confirmed by observational instances but corroborated by passing the test of possibly falsifying instances. As explained in [26]:

> *According to Popper, a scientific theory is posed without any prior justification. It gains its confirmation, or "corroboration", which is Popper's preferred term in later writings, not by experimental evidence demonstrating why it should hold, but by the fact that all attempts at showing why it does not hold have failed.*

In the 1960's, the verificationists have developed theories of confirmation [2, 12, 14]. These theories suggest a qualitative relation of confirmation between observations and scientific hypotheses or a quantitative measure for the degree of confirmation of a hypothesis by an observation. It seems that the most frequently used measures in data mining, i.e., the support and the confidence, are based on such verificationism philosophy. For example, it is not unusual in the data mining community to consider the confidence of a mined rule as the probability that the data confers upon the rule.

However, by the justification problem of induction, Popper claims that the attempt to develop a theory of confirmation for scientific theories is simply mistaken, so he proposed an alternative measure, called degree of corroboration, to compare competitive scientific hypotheses. Though we do not advocate such a radical position against verificationism philosophy, we indeed feel that data mining researchers can learn from the falsification methodology in the design of measures for mined rules which are conformed to the notion of corroboration. Therefore, our decision logic formulation for the justification problem of data mining will motivate the technical development of new kinds of measures for the strength of mined rules. A technical development of Popper's concept of corroboration for inductive inference has been carried out and applied to machine learning in [30]. It would be worthwhile to try the same thing in the data mining context. This kind of attempt represents a flow of ideas from the philosophy of science to data mining research.

In [31], the dynamic interactions between machine learning and the philosophy of science has been exemplified by the research of Bayesian networks, where a dynamic interaction between two disciplines is characterized by the flow of ideas in both directions. We have seen above that the justification problem of data mining may motivate the flow of ideas from the philosophy of science, in particular, the Popper's methodology, to data mining research, so we could also ask what impacts the data mining research has on the philosophy of science.

The answer to this question may not be too difficult if data mining is considered as a sub-discipline of machine learning. The viewpoint that machine learning is computational philosophy of science is not unusual. The ideas in the philosophy of science can be implemented and tested by machine learning algorithms, so machine learning techniques can help the philosophy of science to form new methodological rules. Let us quote the observation in [31]:

> *Machine learning has influenced the controversy in the philosophy of science between inductivism and falsificationism: Gillies argues that the success of the GOLEM machine learning program at learning a scientific hypothesis provides evidence for inductivism[7]. Proponents of falsificationism remain sceptical about machine learning[1], but further successes in the automation of scientific hypothesis generation may dampen their ardour. Machine learning has also proven invaluable to the philosophy of science as a means of testing formal models of scientific reasoning, including inductive reasoning[5, 20], abductive reasoning[4, 6, 23], coherence-based reasoning[27, 28, 29], analogical reasoning[15], causal reasoning, and theory revision[3].*

Of course, data mining can also help in the same way, though it is perhaps in a narrower scope. Therefore, a challenging research programme is to investigate to what extent data mining can serve as a computational tool for testing formal models of scientific reasoning in the philosophy of science.

Though it seems that the justification problem of data mining has a negative result, this does not render data mining useless. From a practical aspect, mined rules may not be justified scientific laws, but they may be very useful for a decision maker. Indeed, in many application domains, simply the summarization of the relationship between data is very helpful for knowledge management and decision making. In such cases, we are only concerned with the regularity within the scope of the existing data.

However, once we want to apply the mined rules beyond this scope, the justification problem could become critical. The result above implies that it is not justified to predict the future occurrence of an event by the regularity discovered from the past data. In fact, for a finite data table, a data mining algorithm can discover a large number of rules valid in that table. These rules may be mutually incompatible in the future instances to be observed. This kind of problem has been known as Goodman paradox or inductive consistencies in the philosophy of science[8, 13]. The implications of inductive consistencies for data mining will be discussed further in the extended version of this paper.

Last, but not the least, the skepticism on induction has lead Popper and the falsificationists to claim that all observations must be based on some background theory. In data mining context, this means that to discover interesting rules, we have to resort to a large amount of background knowledge. Background knowledge makes it possible to find not only the relations among the data but also the phenomena that give rise to data. This is exactly the direction of phenomenal data mining explored in [19].

6 Hypothesis Selection

From the arguments above, we can see that it is extremely implausible (if not impossible) to justify the validity of rules mined from a data table in a logical ground. Hence, a mined rule is simply a hypothesis from a logical viewpoint. In general, a large number of competitive hypotheses can be mined from a data set. Thus, an important question in data mining is to select and/or evaluate competitive hypotheses. In this section, we discuss some possible criteria for hypothesis selection based on the interpretation of data as models or as theories. In both cases, we assume that a logical language $\mathcal{L}$ is used for the representation of hypotheses (i.e. the mined rules). The decision logic introduced in Sect. 2 is an example of such representation languages, however, our discussion applies to any general logic languages.

6.1 Data as Models

In this interpretation, each data item is considered as a model of the language $\mathcal{L}$. Hence, a hypothesis φ in $\mathcal{L}$ is verified or falsified in a data item d, denoted respectively by $d \models \varphi$ and $d \not\models \varphi$. In such interpretation, a data mining context is a pair $(\mathcal{L}, \mathcal{D})$, where $\mathcal{L}$ is the representation language for hypotheses and $\mathcal{D}$ is a finite set of models of $\mathcal{L}$.

A naive criterion for evaluating a hypothesis is the number (or proportion) of models verifying the hypothesis. In other words, we can select a hypothesis φ that maximizes $|m_{\mathcal{D}}(\varphi)|$ or $\frac{|m_{\mathcal{D}}(\varphi)|}{|\mathcal{D}|}$, where $m_{\mathcal{D}}(\varphi) = \{d \in \mathcal{D} \mid d \models \varphi\}$ is the meaning set of φ with respect to $\mathcal{D}$.

Someone may argue that not all data items in $\mathcal{D}$ is relevant to the evaluation of a hypothesis φ, so we should only concentrate on the relevant data items. This leads to the most popular criteria of hypothesis selection in data mining, i.e. support and confidence. Let us denote $d \not\perp \varphi$ if d is relevant to the evaluation of φ and define $r_{\mathcal{D}}(\varphi) = \{d \in \mathcal{D} \mid d \not\perp \varphi\}$, then the support and confidence of φ is respectively defined as

$$supp_{\mathcal{D}}(\varphi) = |r_{\mathcal{D}}(\varphi) \cap m_{\mathcal{D}}(\varphi)|$$

$$conf_{\mathcal{D}}(\varphi) = \frac{|r_{\mathcal{D}}(\varphi) \cap m_{\mathcal{D}}(\varphi)|}{|r_{\mathcal{D}}(\varphi)|} .$$

The usual practice in data mining is to set a threshold, and then evaluate the hypotheses with supports above the threshold according to their confidence values.

We must remark that the acceptance of a hypothesis based on a threshold suffers from the lottery paradox [18]. The resolution of the paradox and its implications on epistemology have been extensively discussed in philosophy [11, 10, 21], however, it seems that this problem has never received enough attention yet in data mining research.

6.2 Data as Theories

In this interpretation, each data item is considered as a sentence of the language $\mathcal{L}$. Hence, a data set $\mathcal{D}$ is a finite subset of $\mathcal{L}$. We also assume that there is a consistent background theory $\mathcal{T} \subset \mathcal{L}$. Thus, a data mining context is a triplet $(\mathcal{L}, \mathcal{D}, \mathcal{T})$. We further assume a logical consequence relation $\models$ exists between sets of sentences and sentences of $\mathcal{L}$, so that $S \models \varphi$ means that $\varphi \in \mathcal{L}$ is a logical consequence of $S \subseteq \mathcal{L}$.

In such interpretation, a prerequisite for a hypothesis to be accepted is that it does not contradict with the background theory. That is, we consider a hypothesis φ only if $\mathcal{T} \not\models \neg\varphi$, where $\neg\varphi$ is the Boolean negation of φ. The sets of data items explained by φ, contradicted with φ, and neutral with φ are respectively defined as

$$e_{\mathcal{D}}(\varphi) = \{\alpha \in \mathcal{D} \mid \mathcal{T} \cup \{\varphi\} \models \alpha\} ,$$

$$c_{\mathcal{D}}(\varphi) = \{\alpha \in \mathcal{D} \mid \mathcal{T} \cup \{\varphi\} \models \neg\alpha\} ,$$

and

$$n_{\mathcal{D}}(\varphi) = \{\alpha \in \mathcal{D} \mid \mathcal{T} \cup \{\varphi\} \not\models \alpha \wedge \mathcal{T} \cup \{\varphi\} \not\models \neg\alpha\} .$$

Given these three sets, we can select hypotheses that explain the most data items and contradict the least data items. Sometimes, the importance of different data items may be different. If such information of importance is available, we can associate a weight to each data item in $\mathcal{D}$, and respectively define the explanatory and contradictory scores of a hypothesis φ as

$$es_{\mathcal{D}}(\varphi) = \frac{\sum_{\alpha \in e_{\mathcal{D}}(\varphi)} w_\alpha}{\sum_{\alpha \in \mathcal{D}} w_\alpha}$$

and

$$cs_{\mathcal{D}}(\varphi) = \frac{\sum_{\alpha \in c_{\mathcal{D}}(\varphi)} w_\alpha}{\sum_{\alpha \in \mathcal{D}} w_\alpha},$$

where w_α is the weight associated to a data item α.

To evaluate hypotheses, we can rank the hypotheses according to a lexicographical ordering on the pair $(es_{\mathcal{D}}(\varphi), 1 - cs_{\mathcal{D}}(\varphi))$, if the explanatory power of a hypothesis is emphasized, or a lexicographical ordering on the pair $(1 - cs_{\mathcal{D}}(\varphi), es_{\mathcal{D}}(\varphi))$, if the consistency property has the higher priority. Other intermediate combinations of these two criteria are possible by employing multi-criteria decision making techniques [17].

7 Concluding Remarks

In addition to the criteria mentioned above, there are two general criteria for hypothesis selection, i.e. generality and simplicity.

1. Generality: a general hypothesis is usually preferred than the specific ones. In the logical sense, a hypothesis φ is more general than another hypothesis ψ if $\varphi \models \psi$ (or $\mathcal{T} \cup \{\varphi\} \models \psi$), but not vice versa.
2. Simplicity: for the economic purpose, a simple hypothesis is preferred than the complex ones. There is not a common agreement regarding the formal definition of simplicity yet. The criterion is related to the principle of the Occam's razor. A form of simplicity is formalized as *minimal description length principle* [25]. The space limit prevents us from presenting the technical detail of the principle. We refer the reader to a recent publication in this topic [9].

Furthermore, domain-specific criteria may be also helpful in the selection of hypotheses. For a data mining problem, these criteria of hypothesis selection may be in conflict with each other. Again, we can employ the techniques of multi-criteria decision making [17] to find overall optimal hypotheses.

Acknowledgement

The authors would like to thank T.Y. Lin for useful comments. This work has been partly supported by NSC of Taiwan (91-2213-E-001-024).

References

1. J.A. Allen. "Bioinformatics and discovery: induction beckons again". *BioEssays*, 23:104–107, 2001.

2. R. Carnap. *Logical Foundations of Probability.* The University of Chicago Press, Routledge, 1962.
3. L. Darden. "Anomaly-driven theory redesign: computational philosophy of science experiments". In T.W. Bynum and J.H. Moor, editors, *The Digital Phoenix: How Computers are Changing Philosophy*, pp. 62–78. Blackwell Publishers, 1998.
4. Y. Dimopoulos and A. Kakas. "Abduction and inductive learning". In L. De Raedt, editor, *Advances in Inductive Logic Programming*, pp. 144–171. IOS Press, 1996.
5. P. Flach. *An Inquiry Concerning the Logic of Induction.* ITK Dissertation Series, 1995.
6. P. Flach and A. Kakas. "On the relation between abduction and inductive learning". In D.M. Gabbay and R. Kruse, editors, *Handbook of Defeasible Reasoning and Uncertainty Management Systems Vol. 4: Abductive Reasoning and Learning*, pp. 1–33. Kluwer Academic Publishers, 2000.
7. D. Gillies. *Artificial Intelligence and Scientific Method.* Oxford University Press, 1996.
8. N. Goodman. *Fact, Fiction, and Forecast.* Bobbs-Merrill, Indianapolis, third edition, 1971.
9. P. Grünwald, I.J. Myung and M. Pitt. *Advances in Minimum Description Length: Theory and Applications.* The MIT Press, 2004.
10. J. Hawthorne. *Knowledge and Lotteries.* Oxford University Press, 2003.
11. J. Hawthorne and L. Bovens. "The preface, the lottery, and rational belief". *Mind*, 108:241–264, 1999.
12. C.G. Hempel. "Studies in the logic of confirmation". In C. Hempel, editor, *Aspects of Scienfific Explanation and Other Essays in the Philosophy of Science*, pp. 3–46. Free Press, New York, 1965.
13. C.G. Hempel. "Studies in the logic of confirmation". In C. Hempel, editor, *Aspects of Scienfific Explanation and Other Essays in the Philosophy of Science*, pp. 53–79. Free Press, New York, 1965.
14. J. Hintikka and P. Suppes editors. *Aspects of Inductive Logic.* North Holland, Amsterdam, 1967.
15. D. Hofstadter and Fluid Analogies Research Group. *Fluid Concepts and Creative Analogies: Computer Models of the Fundamental Mechanisms of Thought.* Penguin, 1998.
16. D. Hume. *A Treatise of Human Nature.* Clarendon Press, Oxford, 1896.
17. C.L. Hwang and K. Yoon. *Multiple Attribute Decision Making : Methods and Applications.* Springer-Verlag, 1981.
18. H. Kyburg. *Probability and the Logic of Rational Belief.* Wesleyan University Press, Middletown, 1961.
19. J. McCarthy. Phenomenal data mining. *Communications of the ACM*, 43(8):75–79, 2000.
20. S. Muggleton and L. de Raedt. "Inductive logic programming: theory and methods". *Journal of Logic Programming*, 19/20:629–679, 1994.
21. D.K. Nelkin. "The lottery paradox, knowledge, and rationality". *The Philosophical Review*, 109(3):373–409, 2000.
22. Z. Pawlak. *Rough Sets–Theoretical Aspects of Reasoning about Data.* Kluwer Academic Publishers, 1991.

23. D. Poole. "Learning, Bayesian probability, graphical models, and abduction". In P. Flach and A. Kakas, editors, *Abduction and Induction: Essays on Their Relation and Integration.* Kluwer Academic Publishers, 1998.
24. K. R. Popper. *The Logic of Scientific Discovery.* Harper & Row, New York, 1968.
25. J. Rissanen. "Modeling by shortest data description". *Automatica*, 14:465–471, 1978.
26. P. Schroeder-Heister. "Popper, Karl Raimund". In *International Encyclopedia of the Social and Behavioral Sciences.* Elsevier, 2001.
27. P. Thagard. *Computational Philosophy of Science.* The MIT Press, 1988.
28. P. Thagard. *Conceptual Revolutions.* PrincetonUniversity Press, 1992.
29. P. Thagard. "Probabilistic networks and explanatory coherence". *Cognitive Science Quarterly*, 1:91–144, 2000.
30. P. Watson. "Inductive learning with corroboration". In O. Watanabe and T. Yokomori, editors, *Proceedings of the 10th International Conference on Algorithmic Learning Theory*, LNAI 1720, pp. 145–156. Springer-Verlag, 1999.
31. J. Williamson. Machine learning and the philosophy of science: a dynamic interaction. In *Proceedings of the ECML-PKDD-01 Workshop on Machine Leaning as Experimental Philosophy of Science*, 2001.

On Statistical Independence in a Contingency Table

Shusaku Tsumoto

Department of Medical Informatics, Shimane University, School of Medicine,
89-1 Enya-cho Izumo 693-8501 Japan
tsumoto@computer.org

Summary. This paper gives a proof showing that statistical independence in a contingency table is a special type of linear independence, where the rank of a given table as a matrix is equal to 1.0. Especially, the equation obtained is corresponding to that of projective geometry, which suggests that a contingency matrix can be interpreted in a geometrical way.

Key words: Statistical Independence, Linear Independence, Contingency Table, Matrix Theory

1 Introduction

Statistical independence between two attributes is a very important concept in data mining and statistics. The definition $P(A, B) = P(A)P(B)$ show that the joint probability of A and B is the product of both probabilities. This gives several useful formula, such as $P(A|B) = P(A)$, $P(B|A) = P(B)$. In a data mining context, these formulae show that these two attributes may not be correlated with each other. Thus, when A or B is a classification target, the other attribute may not play an important role in its classification.

Although independence is a very important concept, it has not been fully and formally investigated as a relation between two attributes.

In this paper, a statistical independence in a contingency table is focused on from the viewpoint of granular computing.

The first important observation is that a contingency table compares two attributes with respect to information granularity. It is shown from the definition that statistifcal independence in a contingency table is a special form of linear depedence of two attributes. Especially, when the table is viewed as a matrix, the above discussion shows that the rank of the matrix is equal to 1.0. Also, the results also show that partial statistical independence can be observed.

S. Tsumoto: *On Statistical Independence in a Contingency Table*, Studies in Computational Intelligence (SCI) **6**, 131–141 (2005)
www.springerlink.com

The second important observation is that matrix algebra is a key point of analysis of this table. A contingency table can be viewed as a matrix and several operations and ideas of matrix theory are introduced into the analysis of the contingency table.

The paper is organized as follows: Section 2 discusses the characteristics of contingency tables. Section 3 shows the conditions on statistical independence for a 2×2 table. Section 4 gives those for a $2 \times n$ table. Section 5 extends these results into a multi-way contingency table. Section 6 discusses statistical independence from matrix theory. Finally, Sect. 7 concludes this paper.

2 Contingency Table from Rough Sets

2.1 Rough Sets Notations

In the subsequent sections, the following notations is adopted, which is introduced in [7]. Let U denote a nonempty, finite set called the universe and A denote a nonempty, finite set of attributes, i.e., $a : U \to V_a$ for $a \in A$, where V_a is called the domain of a, respectively. Then, a decision table is defined as an information system, $A = (U, A \cup \{\mathcal{D}\})$, where $\{\mathcal{D}\}$ is a set of given decision attributes. The atomic formulas over $B \subseteq A \cup \{\mathcal{D}\}$ and V are expressions of the form $[a = v]$, called descriptors over B, where $a \in B$ and $v \in V_a$. The set $F(B, V)$ of formulas over B is the least set containing all atomic formulas over B and closed with respect to disjunction, conjunction and negation. For each $f \in F(B, V)$, f_A denote the meaning of f in A, i.e., the set of all objects in U with property f, defined inductively as follows.

1. If f is of the form $[a = v]$ then, $f_A = \{s \in U | a(s) = v\}$
2. $(f \wedge g)_A = f_A \cap g_A$; $(f \vee g)_A = f_A \vee g_A$; $(\neg f)_A = U - f_a$

By using this framework, classification accuracy and coverage, or true positive rate is defined as follows.

Definition 1. *Let R and D denote a formula in $F(B, V)$ and a set of objects whose decision attribute is given as $\lceil$, respectively. Classification accuracy and coverage(true positive rate) for $R \to \mathcal{D}$ is defined as:*

$$\alpha_R(D) = \frac{|R_A \cap D|}{|R_A|} (= P(D|R)), \text{ and } \kappa_R(D) = \frac{|R_A \cap D|}{|D|} (= P(R|D)) ,$$

where $|A|$ denotes the cardinality of a set A, $\alpha_R(D)$ denotes a classification accuracy of R as to classification of $\mathcal{D}$, and $\kappa_R(D)$ denotes a coverage, or a true positive rate of R to $\mathcal{D}$, respectively.

2.2 Two-way Contingency Table

From the viewpoint of information systems, a contingency table summarizes the relation between two attributes with respect to frequencies. This viewpoint has already been discussed in [10, 11]. However, this study focuses on more statistical interpretation of this table.

Definition 2. *Let R_1 and R_2 denote binary attributes in an attribute space A. A contingency table is a table of a set of the meaning of the following formulas:* $|[R_1 = 0]_A|$,$|[R_1 = 1]_A|$, $|[R_2 = 0]_A|$,$|[R_1 = 1]_A|$, $|[R_1 = 0 \wedge R_2 = 0]_A|$,$|[R_1 = 0 \wedge R_2 = 1]_A|$, $|[R_1 = 1 \wedge R_2 = 0]_A|$,$|[R_1 = 1 \wedge R_2 = 1]_A|$, $|[R_1 = 0 \vee R_1 = 1]_A|(= |U|)$. *This table is arranged into the form shown in Table 1, where:* $|[R_1 = 0]_A| = x_{11} + x_{21} = x_{\cdot 1}$, $|[R_1 = 1]_A| = x_{12} + x_{22} = x_{\cdot 2}$, $|[R_2 = 0]_A| = x_{11} + x_{12} = x_{1\cdot}$, $|[R_2 = 1]_A| = x_{21} + x_{22} = x_{2\cdot}$, $|[R_1 = 0 \wedge R_2 = 0]_A| = x_{11}$, $|[R_1 = 0 \wedge R_2 = 1]_A| = x_{21}$, $|[R_1 = 1 \wedge R_2 = 0]_A| = x_{12}$, $|[R_1 = 1 \wedge R_2 = 1]_A| = x_{22}$, $|[R_1 = 0 \vee R_1 = 1]_A| = x_{\cdot 1} + x_{\cdot 2} = x_{\cdot\cdot}(= |U|)$.

Table 1. Two way Contingency Table

	$R_1 = 0$	$R_1 = 1$	
$R_2 = 0$	x_{11}	x_{12}	$x_{1\cdot}$
$R_2 = 1$	x_{21}	x_{22}	$x_{2\cdot}$
	$x_{\cdot 1}$	$x_{\cdot 2}$	$x_{\cdot\cdot}$ $(= \vert U \vert = N)$

From this table, accuracy and coverage for $[R_1 = 0] \to [R_2 = 0]$ are defined as:

$$\alpha_{[R_1=0]}([R_2 = 0]) = \frac{|[R_1 = 0 \wedge R_2 = 0]_A|}{|[R_1 = 0]_A|} = \frac{x_{11}}{x_{\cdot 1}},$$

and

$$\kappa_{[R_1=0]}([R_2 = 0]) = \frac{|[R_1 = 0 \wedge R_2 = 0]_A|}{|[R_2 = 0]_A|} = \frac{x_{11}}{x_{1\cdot}}.$$

2.3 Multi-way Contingency Table

Two-way contingency table can be extended into a contingency table for multinominal attributes.

Definition 3. *Let R_1 and R_2 denote multinominal attributes in an attribute space A which have m and n values. A contingency tables is a table of a set of the meaning of the following formulas:* $|[R_1 = A_j]_A|$, $|[R_2 = B_i]_A|$, $|[R_1 = A_j \wedge R_2 = B_i]_A|$, $|[R_1 = A_1 \wedge R_1 = A_2 \wedge \cdots \wedge R_1 = A_m]_A|$, $|[R_2 = B_1 \wedge R_2 = A_2 \wedge \cdots \wedge R_2 = A_n]_A|$ *and* $|U|$ $(i = 1, 2, 3, \cdots, n$ *and* $j = 1, 2, 3, \cdots, m)$.

This table is arranged into the form shown in Table 1, where: $|[R_1 = A_j]_A| = \sum_{i=1}^{m} x_{1i} = x_{\cdot j}$, $|[R_2 = B_i]_A| = \sum_{j=1}^{n} x_{ji} = x_{i\cdot}$, $|[R_1 = A_j \wedge R_2 = B_i]_A| = x_{ij}$, $|U| = N = x_{\cdot\cdot}$ $(i = 1, 2, 3, \cdots, n$ *and* $j = 1, 2, 3, \cdots, m)$.

Table 2. Contingency Table ($m \times n$)

	A_1	A_2	$\cdots$	A_n	Sum
B_1	x_{11}	x_{12}	$\cdots$	x_{1n}	$x_{1\cdot}$
B_2	x_{21}	x_{22}	$\cdots$	x_{2n}	$x_{2\cdot}$
$\vdots$	$\vdots$	$\vdots$	$\ddots$	$\vdots$	$\vdots$
B_m	x_{m1}	x_{m2}	$\cdots$	x_{mn}	$x_{m\cdot}$
Sum	$x_{\cdot 1}$	$x_{\cdot 2}$	$\cdots$	$x_{\cdot n}$	$x_{\cdot\cdot} = \lvert U\rvert = N$

3 Statistical Independence in 2 × 2 Contingency Table

Let us consider a contingency table shown in Table 1. Statistical independence between R_1 and R_2 gives:

$$\begin{aligned} P([R_1 = 0], [R_2 = 0]) &= P([R_1 = 0]) \times P([R_2 = 0]) \\ P([R_1 = 0], [R_2 = 1]) &= P([R_1 = 0]) \times P([R_2 = 1]) \\ P([R_1 = 1], [R_2 = 0]) &= P([R_1 = 1]) \times P([R_2 = 0]) \\ P([R_1 = 1], [R_2 = 1]) &= P([R_1 = 1]) \times P([R_2 = 1]) \end{aligned}$$

Since each probability is given as a ratio of each cell to N, the above equations are calculated as:

$$\begin{aligned} \frac{x_{11}}{N} &= \frac{x_{11} + x_{12}}{N} \times \frac{x_{11} + x_{21}}{N} \\ \frac{x_{12}}{N} &= \frac{x_{11} + x_{12}}{N} \times \frac{x_{12} + x_{22}}{N} \\ \frac{x_{21}}{N} &= \frac{x_{21} + x_{22}}{N} \times \frac{x_{11} + x_{21}}{N} \\ \frac{x_{22}}{N} &= \frac{x_{21} + x_{22}}{N} \times \frac{x_{12} + x_{22}}{N} \end{aligned}$$

Since $N = \sum_{i,j} x_{ij}$, the following formula will be obtained from these four formulae.

$$x_{11}x_{22} = x_{12}x_{21} \text{ or } x_{11}x_{22} - x_{12}x_{21} = 0$$

Thus,

Theorem 1. *If two attributes in a contingency table shown in Table 1 are statistical indepedent, the following equation holds:*

$$x_{11}x_{22} - x_{12}x_{21} = 0 \quad (1)$$

□

It is notable that the above equation corresponds to the fact that the determinant of a matrix corresponding to this table is equal to 0. Also, when these four values are not equal to 0, the (1) can be transformed into:

$$\frac{x_{11}}{x_{21}} = \frac{x_{12}}{x_{22}}.$$

Let us assume that the above ratio is equal to $C(constant)$. Then, since $x_{11} = Cx_{21}$ and $x_{12} = Cx_{22}$, the following equation is obtained.

$$\frac{x_{11} + x_{12}}{x_{21} + x_{22}} = \frac{C(x_{21} + x_{22})}{x_{21} + x_{22}} = C = \frac{x_{11}}{x_{21}} = \frac{x_{12}}{x_{22}}. \tag{2}$$

This equation also holds when we extend this discussion into a general case. Before getting into it, let us cosndier a 2×3 contingency table.

4 Statistical Independence in 2 × 3 Contingency Table

Let us consider a 2×3 contingency table shown in Table 3. Statistical independence between R_1 and R_2 gives:

$$\begin{aligned}
P([R_1 = 0], [R_2 = 0]) &= P([R_1 = 0]) \times P([R_2 = 0]) \\
P([R_1 = 0], [R_2 = 1]) &= P([R_1 = 0]) \times P([R_2 = 1]) \\
P([R_1 = 0], [R_2 = 2]) &= P([R_1 = 0]) \times P([R_2 = 2]) \\
P([R_1 = 1], [R_2 = 0]) &= P([R_1 = 1]) \times P([R_2 = 0]) \\
P([R_1 = 1], [R_2 = 1]) &= P([R_1 = 1]) \times P([R_2 = 1]) \\
P([R_1 = 1], [R_2 = 2]) &= P([R_1 = 1]) \times P([R_2 = 2])
\end{aligned}$$

Since each probability is given as a ratio of each cell to N, the above equations are calculated as:

Table 3. Contingency Table (2 × 3)

	$R_1 = 0$	$R_1 = 1$	$R_1 = 2$	
$R_2 = 0$	x_{11}	x_{12}	x_{13}	$x_{1\cdot}$
$R_2 = 1$	x_{21}	x_{22}	x_{23}	$x_{2\cdot}$
	$x_{\cdot 1}$	$x_{\cdot 2}$	$x_{\cdots 3}$	$x_{\cdot\cdot}$ $(= \lvert U \rvert = N)$

$$\frac{x_{11}}{N} = \frac{x_{11}+x_{12}+x_{13}}{N} \times \frac{x_{11}+x_{21}}{N} \quad (3)$$
$$\frac{x_{12}}{N} = \frac{x_{11}+x_{12}+x_{13}}{N} \times \frac{x_{12}+x_{22}}{N} \quad (4)$$
$$\frac{x_{13}}{N} = \frac{x_{11}+x_{12}+x_{13}}{N} \times \frac{x_{13}+x_{23}}{N} \quad (5)$$
$$\frac{x_{21}}{N} = \frac{x_{21}+x_{22}+x_{23}}{N} \times \frac{x_{11}+x_{21}}{N} \quad (6)$$
$$\frac{x_{22}}{N} = \frac{x_{21}+x_{22}+x_{23}}{N} \times \frac{x_{12}+x_{22}}{N} \quad (7)$$
$$\frac{x_{23}}{N} = \frac{x_{21}+x_{22}+x_{23}}{N} \times \frac{x_{13}+x_{23}}{N} \quad (8)$$
$$(9)$$

From equation (3) and (6),

$$\frac{x_{11}}{x_{21}} = \frac{x_{11}+x_{12}+x_{13}}{x_{21}+x_{22}+x_{23}}$$

In the same way, the following equation will be obtained:

$$\frac{x_{11}}{x_{21}} = \frac{x_{12}}{x_{22}} = \frac{x_{13}}{x_{23}} = \frac{x_{11}+x_{12}+x_{13}}{x_{21}+x_{22}+x_{23}} \quad (10)$$

Thus, we obtain the following theorem:

Theorem 2. *If two attributes in a contingency table shown in Table 3 are statistical indepedent, the following equations hold:*

$$x_{11}x_{22} - x_{12}x_{21} = x_{12}x_{23} - x_{13}x_{22} = x_{13}x_{21} - x_{11}x_{23} = 0 \quad (11)$$

□

It is notable that this discussion can be easily extended into a $2xn$ contingency table where $n > 3$. The important equation 10 will be extended into

$$\frac{x_{11}}{x_{21}} = \frac{x_{12}}{x_{22}} = \cdots = \frac{x_{1n}}{x_{2n}} = \frac{x_{11}+x_{12}+\cdots+x_{1n}}{x_{21}+x_{22}+\cdots+x_{2n}} = \frac{\sum_{k=1}^{n} x_{1k}}{\sum_{k=1}^{n} x_{2k}} \quad (12)$$

Thus,

Theorem 3. *If two attributes in a contingency table* $(2 \times k(k = 2, \cdots, n))$ *are statistical indepedent, the following equations hold:*

$$x_{11}x_{22} - x_{12}x_{21} = x_{12}x_{23} - x_{13}x_{22} = \cdots = x_{1n}x_{21} - x_{11}x_{n3} = 0 \quad (13)$$

□

It is also notable that this equation is the same as the equation on collinearity of projective geometry [2].

5 Statistical Independence in $m \times n$ Contingency Table

Let us consider a $m \times n$ contingency table shown in Table 2. Statistical independence of R_1 and R_2 gives the following formulae:

$$P([R_1 = A_i, R_2 = B_j]) = P([R_1 = A_i])P([R_2 = B_j]) \quad (i = 1, \cdots, m, j = 1, \cdots, n).$$

According to the definition of the table,

$$\frac{x_{ij}}{N} = \frac{\sum_{k=1}^{n} x_{ik}}{N} \times \frac{\sum_{l=1}^{m} x_{lj}}{N}. \tag{14}$$

Thus, we have obtained:

$$x_{ij} = \frac{\sum_{k=1}^{n} x_{ik} \times \sum_{l=1}^{m} x_{lj}}{N}. \tag{15}$$

Thus, for a fixed j,

$$\frac{x_{i_a j}}{x_{i_b j}} = \frac{\sum_{k=1}^{n} x_{i_a k}}{\sum_{k=1}^{n} x_{i_b k}}$$

In the same way, for a fixed i,

$$\frac{x_{i j_a}}{x_{i j_b}} = \frac{\sum_{l=1}^{m} x_{l j_a}}{\sum_{l=1}^{m} x_{l j_b}}$$

Since this relation will hold for any j, the following equation is obtained:

$$\frac{x_{i_a 1}}{x_{i_b 1}} = \frac{x_{i_a 2}}{x_{i_b 2}} \cdots = \frac{x_{i_a n}}{x_{i_b n}} = \frac{\sum_{k=1}^{n} x_{i_a k}}{\sum_{k=1}^{n} x_{i_b k}}. \tag{16}$$

Since the right hand side of the above equation will be constant, thus all the ratios are constant. Thus,

Theorem 4. *If two attributes in a contingency table shown in Table 2 are statistical indepedent, the following equations hold:*

$$\frac{x_{i_a 1}}{x_{i_b 1}} = \frac{x_{i_a 2}}{x_{i_b 2}} \cdots = \frac{x_{i_a n}}{x_{i_b n}} = const. \tag{17}$$

for all rows: i_a and i_b ($i_a, i_b = 1, 2, \cdots, m$).

□

6 Contingency Matrix

The meaning of the above discussions will become much clearer when we view a contingency table as a matrix.

Definition 4. *A corresponding matrix $C_{T_{a,b}}$ is defined as a matrix the element of which are equal to the value of the corresponding contingency table $T_{a,b}$ of two attributes a and b, except for marginal values.*

Definition 5. *The rank of a table is defined as the rank of its corresponding matrix. The maximum value of the rank is equal to the size of (square) matrix, denoted by r.*

The contingency matrix of Table 2 ($T(R_1, R_2)$) is defined as $C_{T_{R_1,R_2}}$ as below:

$$\begin{pmatrix} x_{11} & x_{12} & \cdots & x_{1n} \\ x_{21} & x_{22} & \cdots & x_{2n} \\ \vdots & \vdots & \ddots & \vdots \\ x_{m1} & x_{m2} & \cdots & x_{mn} \end{pmatrix}$$

6.1 Independence of 2 × 2 Contingency Table

The results in Section 3 corresponds to the degree of independence in matrix theory. Let us assume that a contingency table is given as Table 1. Then the corresponding matrix ($C_{T_{R_1,R_2}}$) is given as:

$$\begin{pmatrix} x_{11} & x_{12} \\ x_{21} & x_{22} \end{pmatrix},$$

Then,

Proposition 1. *The determinant of $det(C_{T_{R_1,R_2}})$ is equal to $x_{11}x_{22} - x_{12}x_{21}$,*

Proposition 2. *The rank will be:*

$$rank = \begin{cases} 2, & if\ det(C_{T_{R_1,R_2}}) \neq 0 \\ 1, & if\ det(C_{T_{R_1,R_2}}) = 0 \end{cases}$$

From Theorem 1,

Theorem 5. *If the rank of the corresponding matrix of a 2times2 contingency table is 1, then two attributes in a given contingency table are statistically independent. Thus,*

$$rank = \begin{cases} 2, & dependent \\ 1, & statistical\ independent \end{cases}$$

This discussion can be extended into $2 \times n$ tables. According to Theorem 3, the following theorem is obtained.

Theorem 6. *If the rank of the corresponding matrix of a $2 \times n$ contigency table is 1, then two attributes in a given contingency table are statistically independent. Thus,*

$$rank = \begin{cases} 2, & dependent \\ 1, & statistical\ independent \end{cases}$$

6.2 Independence of 3 × 3 Contingency Table

When the number of rows and columns are larger than 3, then the situation is a little changed. It is easy to see that the rank for statistical independence of a $m \times n$ contingency table is equal 1.0 as shown in Theorem 4. Also, when the rank is equal to $\min(m, n)$, two attributes are dependent.

Then, what kind of structure will a contingency matrix have when the rank is larger than 1,0 and smaller than $\min(m, n) - 1$? For illustration, let us consider the following $3times3$ contingecy table.

Example.

Let us consider the following corresponding matrix:

$$A = \begin{pmatrix} 1\,2\,3 \\ 4\,5\,6 \\ 7\,8\,9 \end{pmatrix}.$$

The determinant of A is:

$$\begin{aligned} det(A) &= 1 \times (-1)^{1+1} det \begin{pmatrix} 5\,6 \\ 8\,9 \end{pmatrix} \\ &\quad +2 \times (-1)^{1+2} det \begin{pmatrix} 4\,6 \\ 7\,9 \end{pmatrix} \\ &\quad +3 \times (-1)^{1+3} det \begin{pmatrix} 4\,5 \\ 7\,8 \end{pmatrix} \\ &= 1 \times (-3) + 2 \times 6 + 3 \times (-3) = 0 \end{aligned}$$

Thus, the rank of A is smaller than 2. On the other hand, since $(123) \neq k(456)$ and $(123) \neq k(789)$, the rank of A is not equal to 1.0 Thus, the rank of A is equal to 2.0. Actually, one of three rows can be represented by the other two rows. For example,

$$(4\,5\,6) = \frac{1}{2}\{(1\,2\,3) + (7\,8\,9)\}.$$

Therefore, in this case, we can say that two of three pairs of one attribute are dependent to the other attribute, but one pair is statistically independent of the other attribute with respect to the linear combination of two pairs. It is easy to see that this case includes the cases when two pairs are statistically independent of the other attribute, but the table becomes statistically dependent with the other attribute.

In other words, the corresponding matrix is a mixure of statistical dependence and independence. We call this case *contextual independent.* From this illustration, the following theorem is obtained:

Theorem 7. *If the rank of the corresponding matrix of a* 3×3 *contigency table is 1, then two attributes in a given contingency table are statistically independent. Thus,*

$$rank = \begin{cases} 3, & dependent \\ 2, & contextual\ independent \\ 1, & statistical\ independent \end{cases}$$

It is easy to see that this discussion can be extended into $3 \times n$ contingency tables.

6.3 Independence of $m \times n$ Contingency Table

Finally, the relation between rank and independence in a multi-way contingency table is obtained from Theorem 4.

Theorem 8. *Let the corresponding matrix of a given contingency table be a $m \times n$ matrix. If the rank of the corresponding matrix is 1, then two attributes in a given contingency table are statistically independent. If the rank of the corresponding matrix is* $\min(m, n)$*, then two attributes in a given contingency table are dependent. Otherwise, two attributes are contextual dependent, which means that several conditional probabilities can be represented by a linear combination of conditional probabilities. Thus,*

$$rank = \begin{cases} \min(m, n) & dependent \\ 2, \cdots, \min(m, n) - 1 & contextual\ independent \\ 1 & statistical\ independent \end{cases}$$

7 Conclusion

In this paper, a contingency table is interpreted from the viewpoint of granular computing and statistical independence. From the definition of statistical independence, statistical independence in a contingency table will holds when the equations of collinearity(15) are satisfied. In other words, statistical independence can be viewed as linear dependence. Then, the correspondence between contingency table and matrix, gives the theorem where the rank of the contingency matrix of a given contingency table is equal to 1 if two attributes are statistical independent. That is, all the rows of contingency table can be described by one row with the coefficient given by a marginal distribution. If the rank is maximum, then two attributes are dependent. Otherwise, some probabilistic structure can be found within attribute -value pairs in a given attribute, which we call contextual independence. Thus, matrix algebra is a key point of the analysis of a contingency table and the degree of independence, rank plays a very important role in extracting a probabilistic model.

References

1. Butz, C.J. Exploiting contextual independencies in web search and user profiling, *Proceedings of World Congress on Computational Intelligence (WCCI'2002)*, CD-ROM, 2002.
2. Coxeter, H.S.M. *Projective Geometry, 2nd Edition*, Springer, New York, 1987.
3. Polkowski, L. and Skowron, A. (Eds.) *Rough Sets and Knowledge Discovery 1*, Physica Verlag, Heidelberg, 1998.
4. Polkowski, L. and Skowron, A. (Eds.) *Rough Sets and Knowledge Discovery 2*, Physica Verlag, Heidelberg, 1998.
5. Pawlak, Z., *Rough Sets.* Kluwer Academic Publishers, Dordrecht, 1991.
6. Rao, C.R. *Linear Statistical Inference and Its Applications, 2nd Edition*, John Wiley & Sons, New York, 1973.
7. Skowron, A. and Grzymala-Busse, J. From rough set theory to evidence theory. In: Yager, R., Fedrizzi, M. and Kacprzyk, J. (eds.) *Advances in the Dempster-Shafer Theory of Evidence*, pp. 193–236, John Wiley & Sons, New York, 1994.
8. Tsumoto S and Tanaka H: Automated Discovery of Medical Expert System Rules from Clinical Databases based on Rough Sets. In: *Proceedings of the Second International Conference on Knowledge Discovery and Data Mining 96*, AAAI Press, Palo Alto CA, pp. 63–69, 1996.
9. Tsumoto, S. Knowledge discovery in clinical databases and evaluation of discovered knowledge in outpatient clinic. *Information Sciences*, **124**, 125–137, 2000.
10. Yao, Y.Y. and Wong, S.K.M., A decision theoretic framework for approximating concepts, *International Journal of Man-machine Studies*, **37**, 793–809, 1992.
11. Yao, Y.Y. and Zhong, N., An analysis of quantitative measures associated with rules, N. Zhong and L. Zhou (Eds.), *Methodologies for Knowledge Discovery and Data Mining, Proceedings of the Third Pacific-Asia Conference on Knowledge Discovery and Data Mining*, LNAI **1574**, Springer, Berlin, pp. 479–488, 1999.
12. Ziarko, W., Variable Precision Rough Set Model. *Journal of Computer and System Sciences*, 46, 39–59, 1993.

Part II

Methods of Data Mining

A Comparative Investigation on Model Selection in Binary Factor Analysis

Yujia An, Xuelei Hu and Lei Xu*

Dept. of Computer Science and Engineering, the Chinese University of Hong Kong, Shatin, N.T., Hong Kong
{yjan,xlhu,lxu}@cse.cuhk.edu.hk

Summary. Binary factor analysis has been widely used in data analysis with various applications. Most studies assume a known hidden factors number k or determine it by one of the existing model selection criteria in the literature of statistical learning. These criteria have to be implemented in two phases that first obtains a set of candidate models and then selects the "optimal" model among a family of candidates according to a model selection criterion, which incurs huge computational costs. Under the framework of Bayesian Ying-Yang (BYY) harmony learning, not only a criterion has been obtained, but also model selection can be made automatically during parameter learning without requiring a two stage implementation, with a significant saving on computational costs. This paper further investigates the BYY criterion and BYY harmony learning with automatic model selection (BYY-AUTO) in comparison with existing typical criteria, including Akaike's information criterion (AIC), the consistent Akaike's information criterion (CAIC), the Bayesian inference criterion (BIC), and the cross-validation (CV) criterion. This study is made via experiments on data sets with different sample sizes, data space dimensions, noise variances, and hidden factors numbers. Experiments have shown that in most cases BIC outperforms AIC, CAIC, and CV while the BYY criterion and BYY-AUTO are either comparable with or better than BIC. Furthermore, BYY-AUTO takes much less time than the conventional two-stage learning methods with an appropriate number k automatically determined during parameter learning. Therefore, BYY harmony learning is a more preferred tool for hidden factors number determination.

Key words: BYY harmony learning, Hidden factor, Binary factor analysis, Model selection, Automatic model selection

1 Introduction

Being different from conventional factor analysis where factors are assumed to be Gaussian, Binary Factor Analysis (BFA) regards the observable variables as

* The work described in this paper was fully supported by a grant from the Research Grant Council of the Hong Kong SAR (Project No: CUHK4225/04E).

Y. An et al.: *A Comparative Investigation on Model Selection in Binary Factor Analysis*, Studies in Computational Intelligence (SCI) **6**, 145–160 (2005)
www.springerlink.com

generated from a small number of latent binary factors. In practice, BFA has been widely used in various fields, especially the social science such as the area of political science, educational testing, psychological measurement as well as the tests of disease severity, etc [20]. Also, it can be used for data reduction, data exploration, or theory confirmation [27]. There are many researches on BFA, under the names of latent trait model (LTA), item response theory (IRT) models as well as latent class model [4, 12]. Another typical example is the multiple cause model that considers the observed samples generated from independent binary hidden factors [11, 16]. One other example is the auto-association network that is trained by back-propagation via simply copying input as the desired output [6, 7].

Determination of the hidden binary factors number is a crucial model selection problem in the implementation of BFA. In literature of statistical learning, many efforts have been made on model selection. Conventionally, it needs a two-phase style implementation that first conducts parameter learning on a set of candidate models under the maximum likelihood (ML) principle and then select the "optimal" model among the candidates according to a given model selection criterion. Popular examples of such criteria include the Akaike's information criterion (AIC) [1, 2], Bozdogan's consistent Akaike's information criterion (CAIC) [8], Schwarz's Bayesian inference criterion (BIC) [17] which formally coincides with Rissanen's minimum description length (MDL) criterion [5, 14], and cross-validation (CV) criterion [19]. However, these two-phrase featured methods consume huge computational costs.

The Bayesian Ying-Yang (BYY) harmony learning was proposed as a unified statistical learning framework firstly in 1995 [21] and systematically developed in past years. BYY harmony learning consists of a general BYY system and a fundamental harmony learning principle as a unified guide for developing new regularization techniques, a new class of criteria for model selection, and a new family of algorithms that perform parameter learning with automated model selection [23, 24, 25, 26]. By applying BYY learning to the binary factor analysis, this special case of BYY learning for BFA is obtained [27]. It can make model selection implemented either *automatically* during parameter learning or *subsequently* after parameter learning via a new class of model selection criteria [27]. This paper further investigates the BYY model selection criterion and automatic model selection ability on the implementation of BFA, in comparison with the criteria of AIC, CAIC, BIC, and CV in various situations.

This comparative study is carried out via experiments on simulated data sets of different sample sizes, data space dimensions, noise variances, and hidden factors numbers. Experiments have shown that the performance of BIC is superior to AIC, CAIC, and CV in most cases. In case of high dimensional data or large noise variance, the performance of CV is superior to AIC, CAIC, and BIC. The BYY criterion and the BYY automatic model selection learning (BYY-AUTO) are, in most cases, comparable with or even superior to the best among of BIC, AIC, CAIC, and CV. Moreover, selection of hidden factors number k by BIC, AIC, CAIC, and CV has to be made at the second stage on a set of candidate factor models obtained via parameter learning at the first stage. This two phases procedure is very time-consuming, while BYY-AUTO takes much less time than the conventional two-phase methods because an appropriate factors number k can be automatically determined during parameter learning. Thus, BYY learning is a more preferred tool for BFA.

The rest of this paper is organized as follows. Section 2 briefly describes the BFA model under the maximum likelihood learning. In Sect. 3, we further introduce not only the criteria AIC, CAIC, BIC, and CV, but also the BYY criterion and the BYY automatic model selection learning (BYY-AUTO). Comparative experiments are given in Sect. 4 and a conclusion is made in Sect. 5.

2 Binary Factor Analysis and ML Learning

The widely used factor analysis model in the literature of statistics is [3, 27]

$$x = Ay + c + e \,, \tag{1}$$

where x is a d-dimensional random vector of observable variables, e is a d-dimensional random vector of unobservable noise variables and is drawn from Gaussian, A is a $d \times k$ loading matrix, c is a d-dimensional mean vector and y is a k-dimensional random vector of unobservable latent factors. y and e are mutually independent.

Being different from conventional factor analysis where y is assumed to be Gaussian, BFA assumes y be a binary random vector [24, 27]. In this paper, we consider y is drawn from a Bernoulli distribution

$$p(y) = \prod_{j=1}^{m} [q_j \delta(y^{(j)}) + (1 - q_j)\delta(1 - y^{(j)})] \,, \tag{2}$$

where q_j is the probability that $y^{(j)}$ takes the value 1. We consider e is from a spherical Gaussian distribution, thus $p(x|y)$ has the following form:

$$p(x|y) = G(x|Ay + c, \sigma^2 I) \,, \tag{3}$$

where $G(x|Ay+c, \sigma^2 I)$ denotes a multivariate normal (Gaussian) distribution with mean $Ay + c$ and spherical covariance matrix $\sigma^2 I$ (I is a $d \times d$ identity matrix).

Given k and a set of observations $\{x_t\}_{t=1}^{n}$, one widely used method for estimating $\theta = \{A, c, \sigma^2\}$ is maximum likelihood learning. That is,

$$\hat{\theta} = \arg\max_{\theta} L(\theta) \,. \tag{4}$$

where $L(\theta)$ is the following log likelihood function

$$L(\theta) = \sum_{t=1}^{n} \ln(p(x_t))$$
$$= \sum_{t=1}^{n} \ln(\sum_{y \in D} p(x_t|y)p(y)) , \quad (5)$$

where D is the set that contains all possible values of y.

This optimization problem can be implemented by EM algorithm that iterates following steps [9, 22]:

step 1: calculate $p(y|x_t)$ by

$$p(y|x_t) = \frac{p(x_t|y)p(y)}{\sum_{y \in D} p(x_t|y)p(y)} . \quad (6)$$

step 2: update A, c and σ^2 by

$$A = \left(\sum_{t=1}^{n} \sum_{y \in D} p(y|x_t)(x_t - c)y^T \right) \left(\sum_{t=1}^{n} \sum_{y \in D} p(y|x_t)yy^T \right)^{-1} , \quad (7)$$

$$c = \frac{1}{n} \sum_{t=1}^{n} \sum_{y \in D} p(y|x_t)(x_t - Ay) , \quad (8)$$

and

$$\sigma^2 = \frac{1}{dn} \sum_{t=1}^{n} \sum_{y \in D} p(y|x_t) \|e_t\|^2 \quad (9)$$

respectively, where $e_t = x_t - Ay - c$.

3 Hidden Factors Number Determination

3.1 Typical Model Selection Criteria

Determination of hidden factors number k can be performed via several existing statistical model selection criteria. However, in practice most studies still assume a known dimension k or determine it heuristically. One main reason is that these criteria have to be implemented in a two-phase procedure that is very time-consuming. First, we need to assume a range of values of k from k_{min} to k_{max} which is assumed to contain the optimal k. At each specific k, we estimate the parameters θ under the ML learning principle. Second, we make the following selection

$$\hat{k} = \arg\min_{k} \{J(\hat{\theta}, k), k = k_{min}, \ldots, k_{max}\} , \quad (10)$$

where $J(\hat{\theta}, k)$ is a given model selection criterion.

Three typical model selection criteria are the Akaike's information criterion (AIC) [1, 2], its extension called Bozdogan's consistent Akaike's information criterion (CAIC) [8], and Schwarz's Bayesian inference criterion (BIC) [17] which coincides with Rissanen's minimum description length (MDL) criterion [5, 14]. These three model selection criteria can be summarized into the following general form [18]

$$J(\hat{\theta}, k) = -2L(\hat{\theta}) + B(n)Q(k) \tag{11}$$

where $L(\hat{\theta})$ is the log likelihood (5) based on the ML estimation $\hat{\theta}$ under a given k, and $Q(k) = d(k+1)+1$ is the number of free parameters in k-factors model. Moreover, $B(n)$ is a function with respect to the number of observations as follows:

- $B(n) = 2$ for Akaike's information criterion (AIC) [1, 2],
- $B(n) = \ln(n) + 1$ for Bozdogan's consistent Akaike's information criterion (CAIC) [8],
- $B(n) = \ln(n)$ for Schwarz's Bayesian inference criterion (BIC) [17].

Another well-known model selection technique is cross-validation (CV), by which data are repeatedly partitioned into two sets, one is used to build the model and the other is used to evaluate the statistic of interest [19]. For the ith partition, let U_i be the data subset used for testing and U_{-i} be the remainder of the data used for training, the cross-validated log-likelihood for a k-factors model is

$$J(\hat{\theta}, k) = -\frac{1}{m}\sum_{i=1}^{m} L(\hat{\theta}(U_{-i})|U_i) \tag{12}$$

where m is the number of partitions, $\hat{\theta}(U_{-i})$ denotes the ML parameter estimates of k-factors model from the ith training subset, and $L(\hat{\theta}(U_{-i})|U_i)$ is the log-likelihood evaluated on the data set U_i. Featured by m, it is usually referred as making a m-fold cross-validation or shortly m-fold CV.

3.2 BYY Harmony Learning

Determining Hidden Factors by BYY Criterion

Applying BYY harmony learning to the binary factor analysis model, the following criterion is obtained for selecting the hidden factors number k [27]

$$J(\hat{\theta}, k) = k\ln 2 + 0.5d\ln\hat{\sigma}^2 \,. \tag{13}$$

Shortly, we refer it by BYY criterion, where $\hat{\sigma}^2$ can be obtained via BYY harmony learning, i.e.,

$$\hat{\theta} = \arg\max_{\theta} H(\theta, \hat{k}) . \tag{14}$$

which is implemented by either a batch or an adaptive algorithm.

According to Sect. 4.1 in [27], especially (21), (30), (31) and (32) in [27], the specific form of $H(\theta, k)$ is given as follows

$$H(\theta, k) = \frac{d}{2} \ln \sigma^2 + \frac{1}{2n} \sum_{t=1}^{n} \frac{\|x_t - Ay_t - c\|^2}{\sigma^2}$$
$$-\frac{1}{n} \sum_{t=1}^{n} \sum_{j=1}^{k} [y_t^{(j)} lnq_j + (1 - y_t^{(j)}) ln(1 - q_j)] , \tag{15}$$

where $y_t^{(j)}$ is the j-th element in vector y_t, and

$$y_t = \arg\max_{y} H(\theta, k) . \tag{16}$$

Also, in a two-stage implementation, q_j is prefixed as 0.5.

With k fixed, (15) can be implemented via the adaptive algorithm given by Table 1 in [27]. Considering that typical model selection criteria are evaluated basing on the ML estimation via the EM algorithm made in batch, in this paper we also implement (15) in batch. Similar to the procedure given in Table 1 of [27], we iterate the following steps:
step 1: get y_t by (16),
step 2: from $\frac{\partial H(\theta,k)}{\partial \theta} = 0$, ($\theta$ include A, c, σ^2), update

$$e_t = x_t - Ay_t - c ,$$
$$\sigma^2 = \frac{\sum_{t=1}^{n} \|e_t\|^2}{dn} ,$$
$$A = \left(\sum_{t=1}^{n} (x_t - c) y_t^T \right) \left(\sum_{t=1}^{n} y_t y_t^T \right)^{-1} ,$$
$$c = \frac{1}{n} \sum_{t=1}^{n} (x_t - Ay_t) , \tag{17}$$

This iterative procedure is guaranteed to converge since it is actually the specific form of the Ying-Yang alternative procedure, see Sect. 3.2 in [26].

With k enumerated as in (10) and its corresponding parameters obtained by the above (17), we can select a best value of k by BYY criterion in (13).

Automatic Hidden Factors Determination

Furthermore, an adaptive algorithm has also been developed from implementing BYY harmony learning with appropriate hidden factors automatically determined during learning [24, 26, 27]. Instead of prefixing q_j, q_j is learned together with other parameters θ via maximizing $H(\theta, k)$, which will lead to that q_j on each extra dimension $y^{(j)}$ is pushed towards 0 or 1 and thus we can discard the corresponding dimension [24, 26, 27]. Setting k initially large enough (e.g., the dimensionality of x in this paper), we can implement the adaptive algorithm with model selection made automatically. Shortly, we refer it by BYY automatic model selection learning (BYY-AUTO).

Following the procedure given in Table 1 of [27] or the algorithm by (45) in [27], we implement BYY-AUTO by the following adaptive algorithm.

$$\begin{array}{ll}
\text{step 1:} & \text{get } y_t \text{ by (16)}, \\
\text{step 2: (a)} & \text{update} \\
& e_t = x_t - Ay_t - c, \\
& \sigma^{2\,\text{new}} = (1-\eta)\sigma^{2\,\text{old}} + \eta \text{tr}[e_t e_t^T]/d, \\
& A^{\text{new}} = A^{\text{old}} + \eta e_t y_t^T, \\
& c^{\text{new}} = c^{\text{old}} + \eta e_t, \\
& \text{where } \eta \text{ is a step constant}, \\
\text{(b)} & \text{update } q_j^{\text{new}} = b_j^{\text{new}\,2} \text{ by} \\
& b_j^{\text{new}} = b_j + \eta b_j (y_t^{(j)} - q_j)/[q_j(1-q_j)], \\
& \text{if either } q_j^{\text{new}} \to 0 \text{ or } q_j^{\text{new}} \to 1, \\
& \text{discard the } j\text{th dimension of } y. \\
\text{(c)} & \text{make eigenvalue decomposition } A = U\Lambda V'. \\
& \text{If } \Lambda \text{ is not full rank, make linear transform} \\
& \text{to } A, \text{ then delete 0 value columns of } A \text{ and} \\
& \text{the corresponding dimensions of } y \text{ to} \\
& \text{guarantee } A \text{ has full rank.}
\end{array} \tag{18}$$

The above step (c) checks the satisfaction of (40) in [27] for a full rank independent space. To save computing cost, this can also be made via updating U, Λ, V as suggested by (470) in [27].

In experiments, we have observed that the iteration by the above algorithm (18) has a considerable chance to get stuck in local maxima. We find that this problem can be considerably improved by firstly implementing the BYY harmony learning with a BI-architecture as given in Sect. 4.2 of [27] and then using the obtained parameter values as the initial parameters values for implementing the above BYY-AUTO algorithm (18). For clarity, the BI-architecture based adaptive algorithm in [27] is written below :

$$\begin{array}{ll}
\text{step 1:} & y_t = s(z_t), z_t = Wx_t + v, \\
\text{step 2: (a)} & \text{update } A,\ c,\ \sigma^2 \text{as step 1 (a) in (18)}, \\
\text{(b)} & \text{update} \\
& e_t^0 = A^{\text{old}\,T} e_t \\
& v^{\text{new}} = v^{\text{old}} + \eta D_s(z_t) e_t^0, \\
& W^{\text{new}} = W^{\text{old}} + \eta D_s(z_t) e_t^0 x_t^T,
\end{array} \tag{19}$$

where the notation $D_s(u)$ denotes a diagonal matrix with its diagonal elements $[s'(u^{(1)}), \ldots, s'(u^{(k)})]^T$, $s'(r) = ds(r)/dr$ and $s(r)$ denotes a sigmoid function $s(r) = (1 + e^{-\beta r})^{-1}$.

4 Empirical Comparative Studies

We investigate the experimental performances of the model selection criteria AIC, BIC, CAIC, 10-fold CV, BYY criterion and BYY-AUTO on four types of data sets with different sample sizes, data dimensions, noise variances, and numbers of hidden factors. In implementation, for AIC, BIC, CAIC and 10-fold CV, we use EM algorithm (6)–(9) to obtain ML estimates of A, c and σ^2. For BYY criterion we implement algorithm (17) for parameter learning. For BYY-AUTO, we first use algorithm (19) to obtain initial parameters and then implement algorithm (18) for automatic model selection during parameter learning. The observations x_t, $t = 1, \ldots, n$ are generated from $x_t = Ay_t + c + e_t$ with y_t randomly generated from a Bernoulli distribution with q_j is equal to 0.5 and e_t randomly generated from $G(x|0, \sigma^2 I)$. Experiments are repeated over 100 times to facilitate our observing on statistical behaviors. Each element of A is generated from $G(x|0, 1)$. Usually we set $k_{min} = 1$ and $k_{max} = 2k - 1$ where k is the true number of hidden factors. In addition, to clearly illustrate the curve of each criterion within a same figure we normalize the values of each curve to zero mean and unit variance.

4.1 Effects of Sample Size

We investigate the performances of every method on the data sets with different sample sizes $n = 20$, $n = 40$, and $n = 100$. In this experiment, the dimension of x is $d = 9$ and the dimension of y is $k = 3$. The noise variance σ^2 is equal to 0.5. The results are shown in Fig. 1. Table 1 illustrates the rates of underestimating, success, and overestimating of each method in 100 experiments.

When the sample size is only 20, we see that BYY and BIC select the right number 3. CAIC selects the number 2. AIC, 10-fold CV select 4. When the sample size is 100, all the criteria lead to the right number. Similar observations can be observed in Table 1. For a small sample size, CAIC tends to underestimate the number while AIC, 10-fold CV tend to overestimate the number. BYY criterion has a little risk of overestimation, and BYY-AUTO is comparable with BIC.

4.2 Effects of Data Dimension

Next we investigate the effect of data dimension on each method. The dimension of y is $k = 3$, the noise variance σ^2 is equal to 0.1, and the sample size

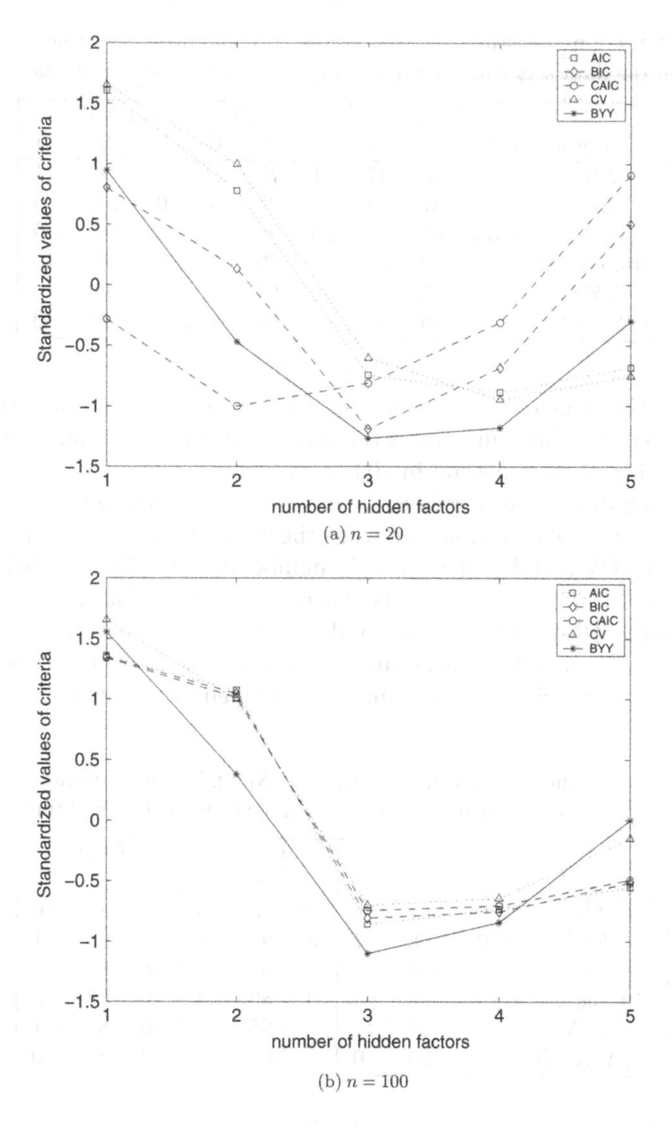

(a) $n = 20$

(b) $n = 100$

Fig. 1. The curves obtained by the criteria AIC, BIC, CAIC, 10-fold CV and BYY on the data sets of a 9-dimensional x ($d = 9$) generated from a 3-dimensional y ($k = 3$) with different sample sizes

Table 1. Rates of underestimating (U), success (S), and overestimating (O) by each criterion on the data sets with different sample sizes in 100 experiments

	$n = 20$			$n = 40$			$n = 100$		
Criteria	U	S	O	U	S	O	U	S	O
AIC	5	53	42	1	78	21	0	89	11
BIC	15	81	4	3	97	0	0	100	0
CAIC	32	67	1	10	90	0	1	99	0
10-fold CV	3	62	35	1	79	20	0	90	10
BYY	3	75	20	1	94	5	0	100	0
BYY-AUTO	4	80	16	0	93	7	1	99	0

is $n = 50$. The dimension of x is $d = 6$, $d = 15$, and $d = 25$. The results are shown in Fig. 2. Table 2 illustrates the rates of underestimating, success, and overestimating of each method in 100 experiments.

When the dimension of x is 6, we observe that all these criteria tend to select the right number 3. However, when the dimension of x is increased to 25, BYY, 10-fold CV and AIC get the right number 3, but CAIC and BIC choose the number 2. Similar observations can be obtained in Table 2. For a high dimensional x, BYY, BYY-AUTO, and 10-fold CV still have high successful rates but CAIC and BIC tend to underestimating the hidden factors number k. AIC has a slight risk to overestimate the hidden factors number.

Table 2. Rates of underestimating (U), success (S), and overestimating (O) by each criterion on the data sets with different data dimensions in 100 experiments

	$d = 6$			$d = 15$			$d = 25$		
Criteria	U	S	O	U	S	O	U	S	O
AIC	0	89	11	0	86	14	2	83	16
BIC	0	98	2	3	96	1	30	69	1
CAIC	0	100	0	7	93	0	48	52	0
10-fold CV	0	90	10	0	86	14	0	89	11
BYY	0	99	1	1	95	4	10	89	1
BYY-AUTO	1	99	0	1	93	6	13	87	0

4.3 Effects of Noise Variance

We further investigate the performance of each method on the data sets with different scales of noise added. In this example, the dimension of x is $d = 9$, the dimension of y is $k = 3$, and the sample size is $n = 50$. The noise variance σ^2 is equal to 0.05, 0.75, and 1.5. The results are shown in Fig. 3. Table 3 illustrates the rates of underestimating, success, and overestimating of each method in 100 experiments.

When the noise variance is 1.5, we see that only AIC and 10-fold CV select the right number 3, BIC, CAIC and BYY select 2 factors. When the noise

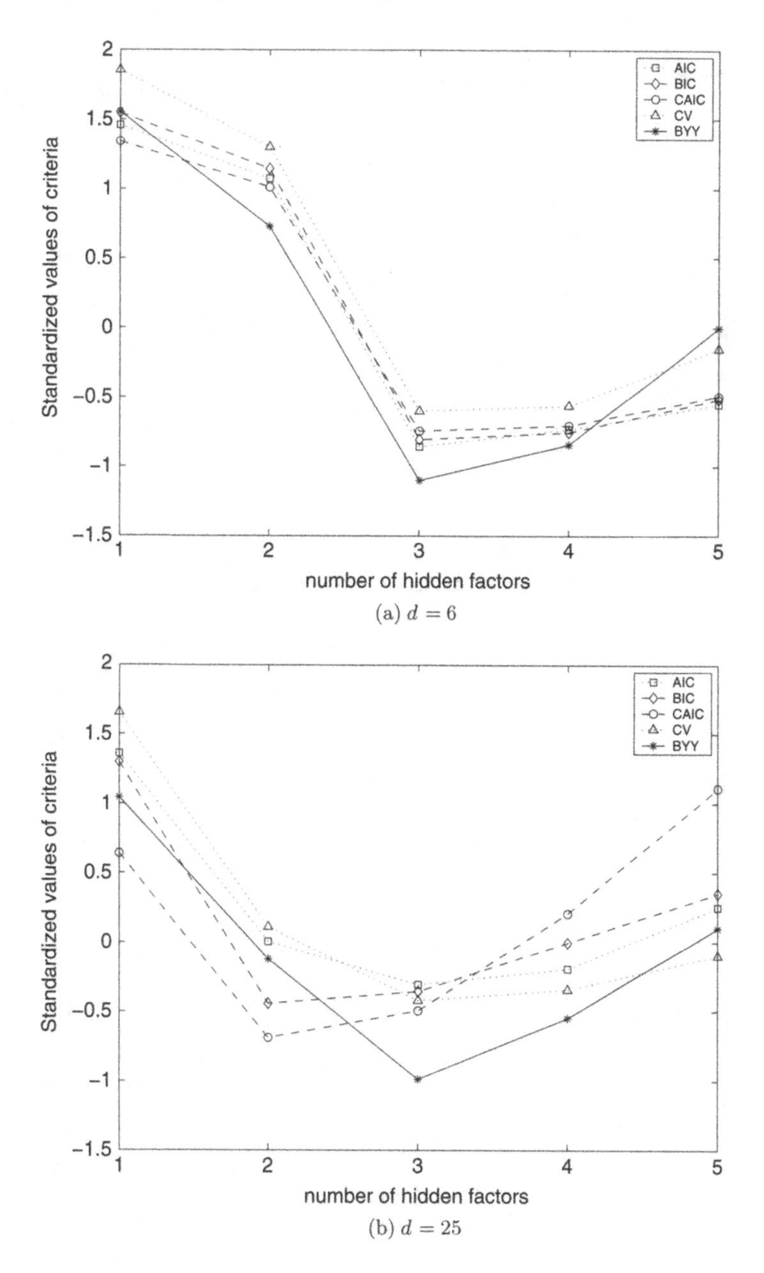

(a) $d = 6$

(b) $d = 25$

Fig. 2. The curves obtained by the criteria AIC, BIC, CAIC, 10-fold CV and BYY on the data sets of a x with different dimensions generated from a 3-dimensional y ($k = 3$)

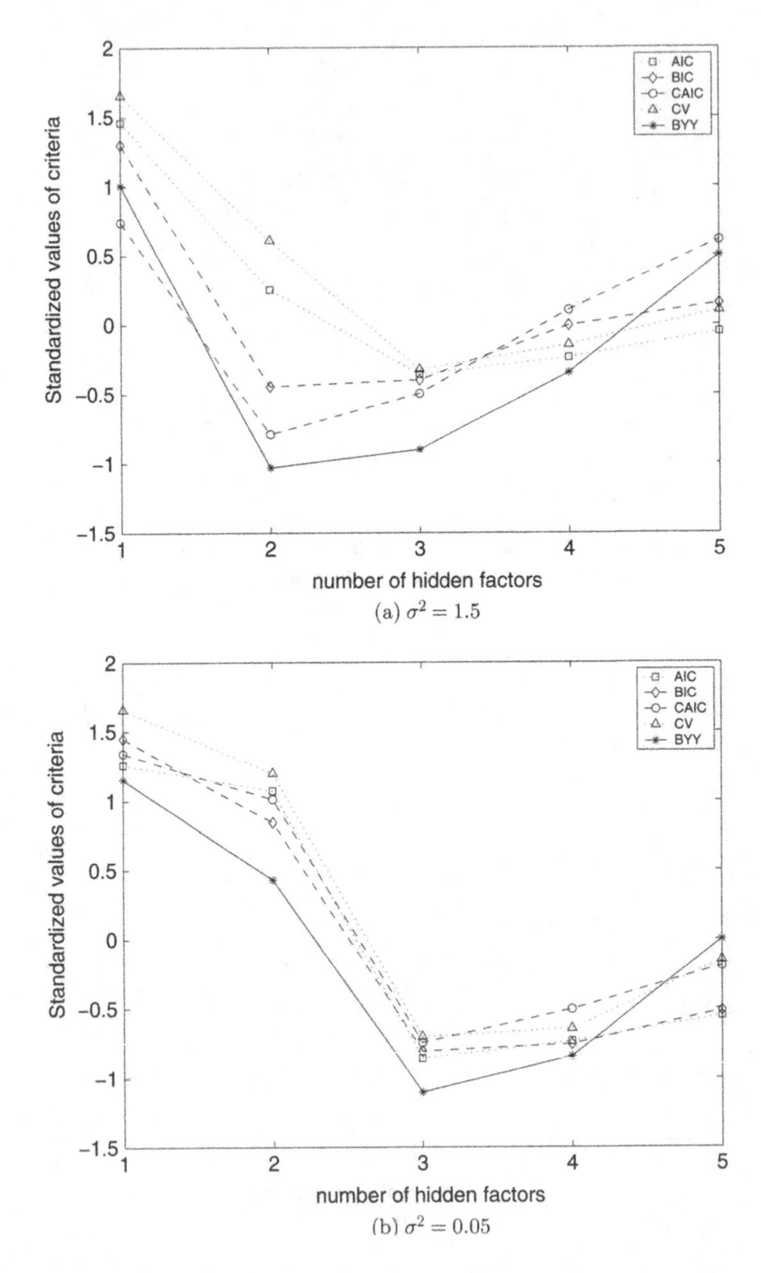

Fig. 3. The curves obtained by the criteria AIC, BIC, CAIC, 10-fold CV and BYY on the data sets of a 9-dimensional x ($d = 9$) generated from a 3-dimensional y ($k = 3$) with different noise variances

variance is 0.05 or 0.75, all the criteria lead to the right number. Similar observations can be observed in Table 3. From this table we can find, for a large noise variance, BIC, CAIC, BYY, and BYY-AUTO are high likely to underestimate the number, AIC and 10-fold CV have a slight risk of overestimate the number. For a small noise variance, CAIC, BIC, BYY and BYY-AUTO have high successful rates while AIC and 10-fold CV still have a slight risk of overestimating the number.

Table 3. Rates of underestimating (U), success (S), and overestimating (O) by each criterion on the data sets with different noise variances in 100 experiments

	$\sigma^2 = 0.05$			$\sigma^2 = 0.75$			$\sigma^2 = 1.5$		
Criteria	U	S	O	U	S	O	U	S	O
AIC	0	89	11	0	82	18	6	78	16
BIC	0	100	0	2	97	1	40	58	2
CAIC	1	99	0	6	94	0	57	43	0
10-fold CV	0	86	14	0	86	14	1	81	18
BYY	0	100	0	1	99	0	42	52	6
BYY-AUTO	0	99	1	2	98	0	46	50	4

4.4 Effects of Hidden Factor Number

Finally, we consider the effect of hidden factor number, that is, the dimension of y on each method. In this example we set $n = 50$, $d = 15$, and $\sigma^2 = 0.1$. The dimension of y is $k = 3$, $k = 6$, and $k = 10$. Table 4 illustrates the rates of underestimating, success, and overestimating of each method in 100 experiments.

As shown in Table 4, when hidden factors number is small all criteria have good performance. When hidden factors number is large AIC get a risk of overestimating.

Table 4. Rates of underestimating (U), success (S), and overestimating (O) by each criterion on simulation data sets with different hidden factors numbers in 100 experiments

	$k = 3$			$k = 6$			$k = 10$		
Criteria	U	S	O	U	S	O	U	S	O
AIC	0	86	14	0	85	15	0	72	28
BIC	2	98	0	4	96	0	9	91	0
CAIC	6	94	0	9	91	0	11	89	0
10-fold CV	0	85	15	0	85	15	0	81	19
BYY	2	98	0	4	95	1	10	86	4
BYY-AUTO	1	99	0	4	96	0	9	90	1

Table 5. CPU times on the simulation data sets with $n = 20$, $d = 9$, and $k = 3$ by using the EM algorithm for AIC, BIC, CAIC and CV, algorithm (17) for BYY criterion and (18) for BYY-AUTO in 100 experiments

Method	CPU Time (in minutes)
BYY-AUTO	1
BYY criterion	5.2
AIC, CAIC, BIC	15.3
10-fold CV	58.5

4.5 Computing Costs

All the experiments were carried out using MATLAB R12.1 v.6.1 on a P4 2GHz 512KB RAM PC. We list the computational results in Table 5 for the first example described in subsection 4.1 with a sample size $n = 20$. For AIC, BIC, CAIC and 10-fold CV, we list the total time of implementing the EM algorithm and of evaluating the criteria by $k_{max} - k_{min} + 1 = 5$ times. For BYY criterion we list the total time of implementing the algorithm (17) and of evaluating the criteria also by $k_{max} - k_{min} + 1 = 5$ times. For BYY-AUTO, we list the time of implementing the algorithm (18) as well as (19) to make initialization. It should be noted that all the values given in Table 5 are the average of 100 experiments.

The hidden factor number determination by AIC, BIC, CAIC takes a similar CPU time and takes more time than that by BYY criterion since the EM algorithm needs more iterations than the algorithm (17) needs. The m-fold cross-validation method consumed the highest computing cost because parameters have to be estimated by m times on each candidate model. The computing costs of all these criteria are much higher than that of BYY-AUTO. In a summary, the performances by BYY criterion and BYY-AUTO are either superior or comparable to those by typical statistical model selection criteria, while BYY-AUTO saves computing costs considerably. That is, BYY harmony learning is much more favorable.

5 Conclusion

We have made experimental comparisons on determining the hidden factor number in implementing BFA. The methods include four typical model selection criteria AIC, BIC, CAIC, 10-fold CV, and BYY criterion as well as BYY learning with automatic model selection. We have observed that the performances by BYY criterion and BYY-AUTO are either superior or comparable to other methods in most cases. Both BYY criterion and BYY-AUTO have high successful rates except the cases of large noise variance. BIC also got a high successful rate when the data dimension is not too high and the noise variance is not too large. CAIC has an underestimation tendency while

AIC and 10-fold CV have an overestimation tendency. Moreover, BYY-AUTO needs a much less computing time than all the considered criteria including BYY criterion.

References

1. Akaike, H.: A new look at statistical model identification. IEEE Transactions on Automatic Control, **19**, 716–723, (1974)
2. Akaike, H.: Factor analysis and AIC. Psychometrika, **52(3)**, 317–332, (1987)
3. Anderson, T.W., Rubin, H.: Statistical inference in factor analysis. Proceedings of the Third Berkeley Symposium on Mathematical Statistics and Probability, **5**, Berkeley, 111–150, (1956)
4. Bartholomew, D.J., Knott, M.: Latent variable models and factor analysis. Kendall's Library of Satistics, **7**, Oxford University Press, New York, (1999)
5. Barron, A., Rissanen, J.: The minimum description length principle in coding and modeling. IEEE Trans. Information Theory, **44**, 2743–2760, (1998)
6. Bertin, E., Arnouts, S.: Extractor: Software for source extraction. Astron. Astrophys. Suppl. Ser., **117**, 393–404, (1996)
7. Bourlard, H., Kamp, Y.: Auto-association by multilayer perceptrons and sigular value decomposition. Biol. Cybernet, **59**, 291–294, (1988)
8. Bozdogan, H.: Model selection and Akaike's information criterion (AIC): the general theory and its analytical extensions. Psychometrika, **52(3)**, 345–370, (1987)
9. Belouchrani, A., Cardoso, J.F.: Maximum likelihood source separation by the expectation-maximization technique: Deterministic and stochastic implementation. Proc. NOLTA95, 49–53, (1995)
10. Cattell, R.: The scree test for the number of factors. Multivariate Behavioral Research, **1**, 245–276, (1966)
11. Dayan, P., Zemel, R.S.: Competition and multiple cause models. Neural Computation, **7**, 565–579, (1995)
12. Heinen T.: Latent class and discrete latent trait models: Similarities and differences. Thousand Oaks, California: Sage, (1996)
13. Kaiser, H.: A second generation little jiffy. Psychometrika, **35**, 401–415, (1970)
14. Rissanen, J.: Modeling by shortest data description. Automatica, **14**, 465–471, (1978)
15. Rubin, D., Thayer. D.: EM algorithms for ML factor analysis. Psychometrika, **47(1)**, 69–76, (1982)
16. Saund, E.: A multiple cause mixture model for unsupervised learning. Neural Computation, **7**, 51–71, (1995)
17. Schwarz, G.: Estimating the dimension of a model. The Annals of Statistics, **6(2)**, 461–464, (1978)
18. Sclove, S.L.: Some aspects of model-selection criteria. Proceedings of the First US/Japan Conference on the Frontiers of Statistical Modeling: An Informational Approach., **2**. Kluwer Academic Publishers, Dordrecht, the Netherlands, 37–67, (1994)
19. Stone, M.: Use of cross-validation for the choice and assessment of a prediction function. Journal of the Royal Statistical Society, **B 36**, 111–147, (1974)

20. Treier, S., Jackman, S.: Beyond factor analysis: modern tools for social measurement. Presented at the 2002 Annual Meetings of the Western Political Science Association and the Midwest Political Science Association, (2002)
21. Xu, L.: Bayesian-Kullback coupled Ying-Yang machines: Unified learnings and new results on vector quantization. Proc. Intl. Conf. on Neural Information Processing (ICONIP95), Beijing, China, 977–988, (1995)
22. Xu, L.: Bayesian Kullback Ying-Yang Dependence Reduction Theory. Neurocomputing, **22(1–3)**, 81–112, (1998)
23. Xu, L.: Temporal BYY learning for state space approach, hidden markov model and blind source separation. IEEE Trans on Signal Processing, **48**, 2132–2144, (2000)
24. Xu, L.: BYY harmony learning, independent state space, and generalized APT financial analyses. IEEE Transactions on Neural Networks, **12(4)**, 822–849, (2001)
25. Xu, L.: BYY harmony learning, structural RPCL, and topological self-organizing on mixture models. Neural Networks, **15**, 1125–1151, (2002)
26. Xu, L.: Independent component analysis and extensions with noise and time: A Bayesian Ying-Yang learning perspective. Neural Information Processing Letters and Reviews, **1(1)**, 1–52, (2003)
27. Xu, L.: BYY learning, regularized implementation, and model selection on modular networks with one hidden layer of binary units. Neurocomputing, **51**, 277–301, (2003)
28. Xu, L.: Bi-directional BYY learning for mining structures with projected polyhedra and topological map. Proceedings of IEEE ICDM2004 Workshop on Foundations of Data Mining, Brighton, UK, Nov. 1–4, 2004, T.Y. Lin, S. Smale, T. Poggio, and C.J. Liau, eds, 2–14, (2004).

Extraction of Generalized Rules with Automated Attribute Abstraction

Yohji Shidara[1] Mineichi Kudo[2] and Atsuyoshi Nakamura[3]

[1] Division of Computer Science, Graduate School of Information Science and Technology, Hokkaido University, Sapporo 060-0814, Japan
shidara@main.ist.hokudai.ac.jp
[2] mine@main.ist.hokudai.ac.jp
[3] atsu@main.ist.hokudai.ac.jp

Summary. We propose a novel method for mining generalized rules with high support and confidence. Using our method, we can obtain generalized rules in which the abstraction of attribute values is implicitly carried out without the requirement of additional information such as information on conceptual hierarchies. Our experimental results showed that the obtained rules not only have high support and confidence but also have expressions that are conceptually meaningful.

1 Introduction

Simple generalized rules that explain a large part of a data set and have few exceptions, that is, rules with high *support* and *confidence*, are often useful in data analysis. The method of mining association rules proposed by Srikant and Agrawal [1] is often used in order to obtain such rules. However, their method does not allow multiple values to be assigned to attributes in a rule expression. This restriction sometimes prevents us from finding general rules with high support. Removing this restriction, that is, considering every subset of values as an assignment to each attribute, does not work because the removal will greatly increase computational time even if the consequence part of the rules is fixed.

This problem also arises in methods of mining by decision trees such as CART [2] and C4.5 [3]. We can generate a rule for each path from the root node to a leaf node of a decision tree but sometimes cannot find general rules that have high support when each edge represents a single-value assignment to the attribute labeled at the parent node.

One approach for overcoming this problem is to allow each attribute in a rule expression to take an abstracted value that represents a set of values, where the abstracted value must be one of the values that appear in a given hierarchy, like the one seen in some concept dictionaries [4]. In this approach, however, a hierarchy is needed, and the results depend on the given hierarchy.

Y. Shidara et al.: *Extraction of Generalized Rules with Automated Attribute Abstraction*, Studies in Computational Intelligence (SCI) **6**, 161–170 (2005)
www.springerlink.com

Our approach is to simply allow each attribute in a rule expression to take multiple values. To overcome the problem of computational time, we use an extended version of the subclass method proposed by Kudo et al. [6], which is a classification algorithm originally applied to pattern recognition problems. Unlike the association-rule and decision-tree mining mentioned above, the process of generating rules by this method is generalization starting from one instance. Thus, it is not necessary to consider all possible combinations of attribute values, a process that would involve a huge amount of computational time. In a rule generated by this method, multiple values are assigned to each attribute. If all possible values are assigned to a certain attribute, then we can eliminate the condition on the attribute from the rule, a process that simplifies the rule. For each attribute that appears in the obtained rules, the set of assigned values is possibly conceptually meaningful.

One drawback of the original subclass method is that rules found by this method do not allow exceptions. We extend this method so as to allow exceptions, thus making it possible to find more generalized rules.

In order to demonstrate the effectiveness of our method, we conducted experiments using the "Adult" data set [7] extracted from the 1994 US Census Database. The results of the experiments showed that the rules obtained by our method are ones with high support in which some attributes are removed and that a set of values that appears to be conceptually meaningful is assigned to each remaining attribute.

2 Subclass-Based Mining

Unlike association-rule mining we consider in this paper the problem of finding rules that predict values of a *fixed* class attribute. Our approach to the problem is the subclass method, which was originally proposed in the context of pattern recognition. The original subclass method is used to find a set of hyper-rectangles consistent with class labels in a multi-dimensional real space.

Here, we propose a subclass-based mining method that uses the subclass method with domains extended to categorical data.

2.1 Rule Representation

In our rule representation, the consequence part of rules is a class assignment, and their condition part is a kind of extended conjunction that assigns multiple values to each attribute. This conjunction can be seen as a special CNF (conjunctive normal form) in which only the same attribute appears in each disjunction. Among the rules of the this form, we try to find rules having the following properties:

(1) Exclusiveness: the rule does not match any instance of other classes, and
(2) Maximalness: the set S of instances explained by the rule is maximal, that is, there is no larger (more general) rule that explains all instances in S and keeps exclusiveness.

A subset of instances explained by a rule, or the rule itself, satisfying these two conditions is called a *subclass*.

2.2 Illustrative Example

A subclass can be found by starting from one instance and extending on assignment to each attribute in a rule so as to include the attribute value of each instance of the same class only when exclusiveness is preserved. We show this process by an illustrative example.

Consider the data set shown in Table 1. Here, attribute a has the value a_1, a_2 or a_3 and attribute b has the value b_1 or b_2. For simplification, we use the notion "a_i" instead of its exact expression "$a = a_i$."

Table 1. Illustrative data

Tuple	Attribute a	Attribute b	Class
1	a_1	b_1	P
2	a_1	b_2	P
3	a_2	b_2	P
4	a_3	b_2	P
5	a_2	b_1	Q
6	a_3	b_1	Q

If we think of the first tuple, to define class P, what we can have as a rule is

$$a_1 \wedge b_1 \Rightarrow P .$$

With the second tuple, we can improve this rule as

$$a_1 \wedge (b_1 \vee b_2) \Rightarrow P .$$

However, when we merge the third tuple into the rule,

$$(a_1 \vee a_2) \wedge (b_1 \vee b_2) \Rightarrow P$$

becomes unacceptable as a rule for P because tuple no. 5 with class Q breaks this rule. Thus, we should skip tuple no. 3, so should tuple no. 4. As a result, ignoring attribute b because b can be either of two possible values, b_1 and b_2, we have the rule

$$a_1 \Rightarrow P .$$

In this way, this rule gives a compact expression for tuple no. 1 and tuple no 2.

In this example, all possible rules (subclasses) are

$$\begin{aligned} a_1 &\Rightarrow P\,, \\ b_2 &\Rightarrow P\,, \text{ and} \\ (a_2 \vee a_3) \wedge b_1 &\Rightarrow Q\,. \end{aligned}$$

2.3 Subclass Method and its Application to Categorical Attributes

The subclass method is a randomized procedure for finding subclasses. In this procedure, the order σ for examination, the permutation σ for a given sequence of tuples of a class, is chosen randomly and the tuples are merged one by one according to order σ into one rule as long as the rule holds class consistency.

To obtain all possible subclasses, we must examine all permutations of instances of a class. This requires an exponential computation cost with respect to the number of instances. However, the randomized algorithm briefly described above is sufficient for finding most of the major subclasses in a reasonable computational time.

To apply this method to a categorical attribute a that can take one of n values, $a_1, a_2, \ldots, a_n$, we use the following binalization:

$$\begin{aligned} a_1 &= 011\cdots1\,, \\ a_2 &= 101\cdots1\,, \\ &\vdots \\ a_n &= 111\cdots0\,. \end{aligned}$$

With all possible combinations of these n vectors merged by a bit-AND operator, we have a lattice in which individual values correspond to the nodes immediately above the bottom node (Fig. 1).

In the subclass method, we keep a current rule corresponding to one node in the above-explained lattice that is represented by a binary vector. When we merge one vector corresponding to an individual instance into the current rule vector, the current rule vector is generalized to the least common ancestor of the two vectors. This corresponds to the operation described in the above example: for example, the binary vectors of the first two tuples, 01101 and 01110 (the first 3 bits for a and the remaining 2 bits for b), can be seen as the current rule vector and the instance to be merged. The binary vector 01100 is obtained by merging the instance using the bit-AND operation, and this produces condition $a_1 \wedge (b_1 \vee b_2)$. In the case in which all values are merged into an all-zero vector corresponding to the *root* node, such an attribute does not affect classification by the rule. Thus, we can delete the condition on the

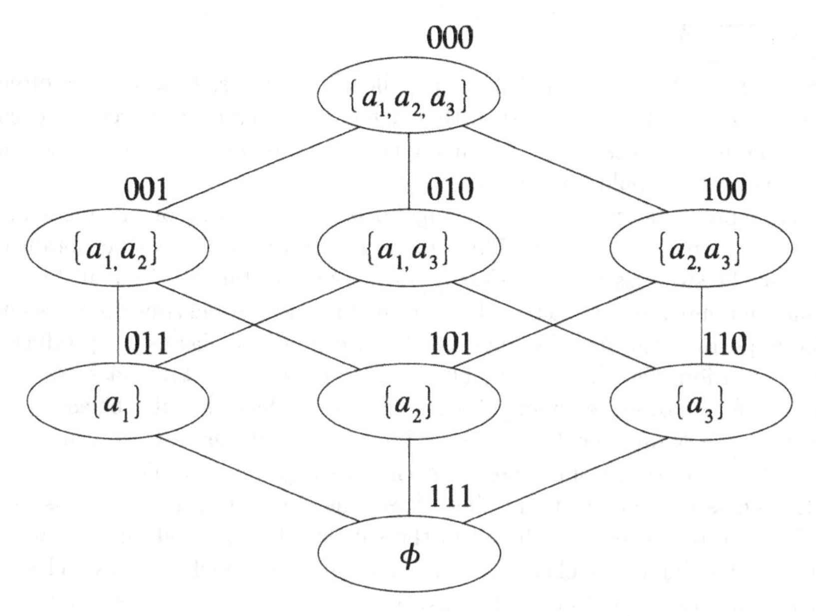

Fig. 1. A lattice and binary expressions in the case of $n = 3$

attribute. For example, condition $a_1 \wedge (b_1 \vee b_2)$ can be expressed by simple condition a_1.

For a continuous attribute, we use another binary expression in which each pair of bits shows a sub-interval in an interval prespecified on that attribute [5].

2.4 Merits

The characteristics of the subclass method that distinguish it from other methods such as decision trees are as follows.

First, the rules found by the subclass method can share the same instances; that is, an instance can be explained by more than one rule, as was shown in the case of tuple no. 2 in the example. In other words, instances are utilized maximally to find the rules. It should be noted that different paths to leaves in decision trees give different rules that explain distinct sets of instances.

Second, from a certain viewpoint, it can be considered that the subclass method carries out a kind of abstraction of attribute values. For example, in the third rule extracted from the illustrative data, $(a_2 \vee a_3) \wedge b_1 \Rightarrow Q$, we can regard the multiple values a_2 and a_3 as one abstracted value $\tilde{a}_{1,2}$ of attribute a. An abstracted value $\tilde{a}$ found by the subclass method may or may not be given an appropriate "name" from an ordinal dictionary. There is a strong possibility that such an abstraction that cannot be found in a dictionary is just the one wanted.

See [6] for other merits of the original subclass method.

2.5 Screening

The original subclass method does not allow exceptions. However, we often want a more generalized rule such that it has more compact expression even if the rule mismatches with a few instances. For this purpose, the restriction of exclusiveness should be loosened.

What prevents rules from growing up is the existence of instances belonging to non-target classes. Thus, more generalized rules can be obtained by using the subclass method after removing such instances. What instances should be removed? Instances only contained in small subclasses do not seem to be important and may be removed because small subclasses are produced in the situation where a small number of instances of a certain class are surrounded by instances belonging to other classes. To define small subclasses, we introduce the parameter θ and control the sizes of subclasses to be removed.

First, we construct subclasses of each non-target class while maintaining exclusiveness and then examine their sizes. The size of a subclass is measured by the ratio of instances included in the subclass to the total number of instances belonging to the class. We remove instances not included in subclasses whose sizes are larger than θ. Finally, we apply the subclass method to extraction of the rules for the target class using the set of instances in which some instances belonging to non-target classes are removed by the procedure described above.

We mention the selection of the θ value here. Extracted rules depend on the θ value. The lager θ value leads to general rules with higher support while the smaller θ value leads to specific rules with higher confidence. One research direction is automatic selection of the θ value by some criteria balancing support and confidence. However, it is impossible to find such a criteria that is appropriate for any data and tasks. So, in our method, we find rules interesting for us among the ones obtained by trying several different θ values.

3 Experiments

We carried out some experiments to demonstrate the effectiveness of our method. The results of those experiments are presented in this section.

3.1 Data Set

We used the "Adult" data set [7], which is a part of the 1994 US Census Database. This data set consists of training and test sets, and we used the training set only. The training set contains 32561 instances, and all of them are classified into two classes, which we will call here "rich" and "poor" classes, according to whether the person's annual income is greater or less than 50,000 dollars. The original data set has 14 attributes (6 continuous and 8 categorical), 4 of which (fnlwgt, education-num, capital-gain and capital-loss) we did not use[1]. The following attributes were used in our experiments: age (continuous), workclass (8 items), education (16 items), marital-status (7 items), occupation (14 items), relationship (6 items), race (5 items), sex (2 items), hours-per-week (continuous) and native-country (41 items). Instances having missing values were removed.

3.2 Evaluation of Rules

Let N^+ denote the number of instances in a target class and let N^- denote the number of instances in the other classes. For a certain rule of the target class, assume that $n^+ + n^-$ instances satisfy the rule condition, where n^+ instances belong to the target class and n^- instances do not. We evaluated the performance of this rule by *(class-conditioned) support*[2], *confidence* and *purity*, which are defined as follows:

$$\text{(class-conditioned) support} = \frac{n^+}{N^+},$$

$$\text{confidence} = \frac{n^+}{n^+ + n^-}, \text{ and}$$

$$\text{purity} = \frac{n^+/N^+}{n^+/N^+ + n^-/N^-}.$$

Here, we introduced *purity* to absorb the difference in populations of the two classes.

For each rule in a rule list, we also used another measure, *cumulative-support*. *Cumulative-support* of rule R_i in a rule list $(R_1, R_2, \ldots, R_m)$ is defined as the ratio of target-class instances covered by the set of rules $\{R_1, R_2, \ldots, R_i\}$.

3.3 Mining Procedure

We tried to find rule lists describing rich and poor classes. We used the randomized version of the subclass method with 1000 randomly chosen permutations. We also conducted data screening with $\theta = 0.01$. The rule lists were constructed by adding rules one-by-one that maximize *cumulative-support*.

[1] The attribute "fnlwgt" seems to be an additional attribute for some statistics and does not seem to provide significant information for this task. The attribute "education-num" appears to be another expression of "education" in continuous values. Since a person's income directly depends on the attributes "capital-gain" and "capital-loss", these attributes do not seem to be suitable for mining *interesting* rules.

[2] Generally, the definition of "support" is $n^+/(N^+ + N^-)$, but this value depends on the relative size of N^+ because $n^+ \leq N^+$. For fixed target classes, this measure does not need to depend on the relative size of N^+, so we adopted our definition. Note that our definition of "support" is equivalent to that of "recall" used in the area of information retrieval.

3.4 Results

Table 2 and Table 3 show the rules obtained. Note that "$\neg A$" means the complement of A.

The first rule for the rich class has 41.3% *support* and 60.9% *confidence*, and the first rule for the poor class has 47.9% *support* and 95.5% *confidence*. The *confidence* of the first rich rule (60.9%) seems to be rather low. However, since the rich class is smaller than the poor class (about 1/3 of the poor class), the rule should be acceptable because the number of mismatched instances of the poor class is small compared to the total number of instances of the poor class. This appears in the high *purity* value of the rule.

With the top 3 rules for each class, 56.8% of the rich class and 71.7% of the poor class are explained. These rules can explain the majority of both classes.

The first and second rules for the poor class consist of only 4 attributes and the third rule consists of 3 attributes. Since we used 10 attributes, 6 or 7 attributes were eliminated in the mining process.

Table 2. Rules for the rich class

1	**age:** ≥ 39.0 **workclass:** $\neg${Without-pay} **education:** $\neg${10th, 11th, 5th-6th, 7th-8th, 9th, Assoc-acdm, Preschool} **marital-status:** {Married-AF-spouse, Married-civ-spouse} **occupation:** $\neg${Handlers-cleaners, Other-service, Priv-house-serv} **relationship:** {Husband} **race:** $\neg${Black} **sex:** {Male} **hours-per-week:** ≥ 39.2 and < 80.4 **native-country:** {Cambodia, Canada, Cuba, England, France, Hong, Hungary, India, Iran, Ireland, Japan, Nicaragua, Scotland, United-States} [supp=0.413, conf=0.609, purity=0.825, cumulative-supp=0.413]
2	**age:** ≥ 27.0 **workclass:** $\neg${Local-gov, Without-pay} **education:** $\neg${10th, 11th, 7th-8th, 9th, HS-grad, Preschool} **marital-status:** {Married-AF-spouse, Married-civ-spouse} **occupation:** {Adm-clerical, Exec-managerial, Prof-specialty, Sales, Tech-support, Transport-moving} **relationship:** {Husband, Other-relative, Wife} **hours-per-week:** ≥ 41.2 and < 85.6 **native-country:** {Canada, China, Cuba, Ecuador, England, France, Greece, Haiti, Honduras, Hong, Iran, Ireland, Jamaica, Japan, Puerto-Rico, Scotland, Thailand, United-States, Yugoslavia} [supp=0.251, conf=0.754, purity=0.902, cumulative-supp=0.515]
3	**age:** ≥ 30.0 and < 57.0 **workclass:** $\neg${Self-emp-not-inc, State-gov, Without-pay} **education:** {12th, Bachelors, Masters} **marital-status:** {Married-AF-spouse, Married-civ-spouse} **occupation:** {Exec-managerial, Farming-fishing, Prof-specialty, Sales, Tech-support} **relationship:** {Husband, Wife} **race:** $\neg${Other} **hours-per-week:** ≥ 3.1 **native-country:** {Cambodia, Canada, Cuba, Ecuador, El-Salvador, France, Haiti, Hong, Ireland, Italy, Japan, Mexico, Philippines, Portugal, Puerto-Rico, South, United-States} [supp=0.207, conf=0.805, purity=0.926, cumulative-supp=0.568]

Table 3. Rules for the poor class

1	**education:** ¬{9th, Assoc-acdm, Bachelors, Doctorate, Masters, Prof-school} **occupation:** ¬{Exec-managerial, Prof-specialty} **relationship:** ¬{Husband} **native-country:** ¬{Canada} [supp=0.479, conf=0.955, purity=0.877, cumulative-supp=0.479]
2	**age:** < 36.0 **education:** ¬{Assoc-voc, Prof-school} **occupation:** ¬{Exec-managerial} **hours-per-week:** < 46.4 [supp=0.404, conf=0.926, purity=0.807, cumulative-supp=0.626]
3	**education:** {10th, 5th-6th, 7th-8th, HS-grad, Preschool, Some-college} **occupation:** ¬{Exec-managerial, Prof-specialty, Protective-serv, Sales, Tech-support, Transport-moving} **native-country:** ¬{Canada} [supp=0.448, conf=0.881, purity=0.711, cumulative-supp=0.717]

The attributes "education" and "occupation" seem to be important to distinguish each class from other classes. In the attribute "education" of rich rules, the values that can be labeled "higher education" are collected. On the other hand, the rules for the poor class have the opposite trend.

In the attribute "occupation" for the first poor rule, "Exec-managerial" and "Prof-specialty" are denied, but they appear in the condition of all three rules for the rich class. In general, these occupations seem to lead people to a higher income group.

The attributes "martial-status," "age" and "hours-per-week" characterize the rich class. In the "martial-status" attribute, the three rules for the rich class have both "Married-AF-spouse" and "Married-civ-spouse," which can be simply grouped as "married." As for "age" and "hours-per-week" attributes, the first rule for the rich class explains middle-aged workers, who work more than 39 hours per week. The second rule explains people who are younger and work harder.

4 Conclusions

Abstraction of attribute values is important for obtaining human-readable generalized rules, but previous methods require dictionaries of abstracted values such as conceptual hierarchies. Our method, a method based on the subclass method, does not require such additional information. This will be advantageous in applications of domains that have difficulty in building such a hierarchy.

The results of experiments using the "Adult" data set [7] showed that our method enables rules with high support to be obtained while maintaining high confidence. Moreover, the obtained rules give us the conceptual grouping of the attribute values. This is one of the parts expected for data-mining methods.

References

1. Srikant R., Agrawal R. (1996) Mining quantitative association rules in large relational tables. In Proceedings of the 1996 ACM SIGMOD International Conference on Management of Data
2. Breiman L., Friedman J.H., Olshen R.A., Stone C.J. (1984) Classification and regression trees. Technical report, Wadsworth International, Monterey, CA
3. Quinlan J.R. (1993) C4.5: Programs for machine learning. Morgan Kaufmann Publishers, 2929 Campus Drive, Suite 260, San Mateo, CA 94403
4. Kudo Y., Haraguchi M. (2000) Detecting a compact decision tree based on an appropriate abstraction. In Proceedings of the Second International Conference on Intelligent Data Engineering and Automated Learning – IDEAL2000
5. Kudo M., Shimbo M. (1993) Feature Selection Based on the Structual Indices of Categories. Pattern Recognition, 26(6):891–901
6. Kudo M., Yanagi S., Shimbo M. (1996) Construction of class regions by a randomized algorithm: A randomized subclass method. Pattern Recognition, 29(4):581–588
7. Blake C., Merz C. (1998) UCI repository of machine learning databases. http://www.ics.uci.edu/~mlearn/MLRepository.html, University of California, Irvine, Department of Information and Computer Sciences

Decision Making Based on Hybrid of Multi-Knowledge and Naïve Bayes Classifier

QingXiang Wu[1], David Bell[2], Martin McGinnity[1] and Gongde Guo[3]

[1] School of Computing and Intelligent Systems, University of Ulster at Magee, Londonderry, BT48 7JL, N. Ireland, UK
{Q.Wu, TM.McGinnity}@ulst.ac.uk

[2] Department of Computer Science, Queens University, Belfast, UK
DA.Bell@qub.ac.uk

[3] School of Computing and Mathematics, University of Ulster at Jordanstown, Newtowabbey, BT37 0QB, N. Ireland, UK
G.Guo@ulst.ac.uk

Abstract. In general, knowledge can be represented by a mapping from a hypothesis space to a decision space. Usually, multiple mappings can be obtained from an instance information system. A set of mappings, which are created based on multiple reducts in the instance information system by means of rough set theory, is defined as multi-knowledge in this paper. Uncertain rules are introduced to represent multi-knowledge. A hybrid approach of multi-knowledge and the Naïve Bayes Classifier is proposed to make decisions for unseen instances or for instances with missing attribute values. The data sets from the UCI Machine Learning Repository are applied to test this decision-making algorithm. The experimental results show that the decision accuracies for unseen instances are higher than by using other approaches in a single body of knowledge.

1 Introduction

Decision accuracy and reliability for an intelligent system depend on knowledge that extracted from the system experiences of interaction with its environments or from an instance information system. Knowledge can be represented in many forms, such as logical rules, decision trees, fuzzy rules, Baysian belief networks, and artificial neural networks. In order to extract knowledge from large databases or a data warehouse, conventional techniques in knowledge discovery or data mining always drop redundant attributes and try to obtain a compact knowledge structure. Therefore, conventional technique in knowledge discovery requires the following steps:

1. Preprocessing by discretization of the continuous attributes, optimizing the values of attributes, and eliminating useless attributes.

Q. Wu et al.: *Decision Making Based on Hybrid of Multi-Knowledge and Naïve Bayes Classifier*, Studies in Computational Intelligence (SCI) **6**, 171–184 (2005)
www.springerlink.com

2. Finding good reduct (sometimes called feature selection) from attributes of an instance information system [1, 2, 3, 9, 10, 15, 16, 24].
3. Extracting knowledge from an instance information system. Many approaches can be found in the book [11] written by Mitchell (1997) and the books [7, 8] edited by Lin & Cercone (1997) or Polkowski et al (2000).
4. Testing the rules and applying them to make decisions.

Following these steps, knowledge is obtained based on one reduct or a set of good features. The knowledge based on one reduct or a set of features is called a *single body of knowledge* in this paper. It is obvious that decision accuracy and reliability is seriously affected if some attribute values are missing in the reduct or the feature set. Therefore, we are not going to follow this convention to find a good reduct and then to extract a single body of knowledge based on this reduct. On the contrary, we propose to find multiple reducts, to create *multi-knowledge* [8, 23] and combine it with the Naïve Bayes Classifier to make decision.

In Sect. 2, after an introduction of general knowledge representation, some problems with the single body of knowledge approach are analyzed, and then multi-knowledge is defined by means of multiple reducts from the rough set theory. *Uncertain rules* are applied to represent multi-knowledge and a hybrid decision-making algorithm is proposed in Sect. 3. Experimental results are given in Sect. 4, and a conclusion is given in Sect. 5.

2 General Knowledge Representation and Multi-Knowledge

2.1 Instance Information System

Following Pawlak [2, 3] and Polkowski et al. [7], let $\boldsymbol{I} = \langle U, A \cup D \rangle$ represent an *instance information system*, where $U = \{u_1, u_2, \ldots, u_i, \ldots, u_n\}$ is a finite non-empty set, called an *instance* space or universe, where $n = |U|$ and u_i is called an *instance* in U. $A = \{a_1, a_2, a_3, \ldots, a_i, \ldots, a_m\}$, also a finite non-empty set, is a set of *attributes* of the instances, where $m = |A|$ and a_i is an attribute of a given instance. D is a non-empty set of decision attributes, and $A \cap D = 0$. In order to distinguish from information systems, an *instance information system* represents an information system with decision attributes.

For every $a \in A$ there is a domain, represented by V_a, and there is a mapping $a(u) : U \to V_a$ from U onto the domain V_a. $a(u)$ represents the value of attribute a of instance u. It is a value in set V_a.

$$V_a = a(U) = \{a(u) : u \in U\} \quad \text{for } a \in A \,. \tag{1}$$

The *vector space*, which is generated from attribute domain V_a, is denoted by

$$V_{\times A} = \mathop{\times}_{a \in A} V_a = V_{a1} \times V_{a2} \times \ldots \times V_{a|A|}$$

and

$$|V_{\times A}| = \prod_{i=1}^{|A|} |V_{a_i}| \tag{2}$$

A hypothesis space can be represented by this vector space. A hypothesis (i.e. a conjunction of attribute values for an instance) corresponds to a vector in the vector space.

$$\vec{A}(u) = (a_1(u), a_2(u), \ldots, a_{|A|}(u)) \tag{3}$$

$\vec{A}(U)$ is a set of vectors which exist in the instance information system.

$$\vec{A}(U) = \{\vec{A}(u) : u \in U\} \tag{4}$$

If $|\vec{A}(U)| = |V_{\times A}|$, the system is called a *completed instance system or completed system.* In the real world, training sets for knowledge discovery or data mining are rarely completed systems. For example, there is a training set with 4 attributes shown in Table 1 (from [11]).

Table 1. Play Tennis

U	Outlook	Temp	Humid	Wind	Play Tennis
1	Sunny	Hot	High	Weak	No
2	Sunny	Hot	High	Strong	No
3	Overcast	Hot	High	Weak	Yes
4	Rain	Mild	High	Weak	Yes
5	Rain	Cool	Normal	Weak	Yes
6	Rain	Cool	Normal	Strong	No
7	Overcast	Cool	Normal	Strong	Yes
8	Sunny	Mild	High	Weak	No
9	Sunny	Cool	Normal	Weak	Yes
10	Rain	Mild	Normal	Weak	Yes
11	Sunny	Mild	Normal	Strong	Yes
12	Overcast	Mild	High	Strong	Yes
13	Overcast	Hot	Normal	Weak	Yes
14	Rain	Mild	High	Strong	No

$|V_{\times A}| = 36$. There are 36 possible condition vectors. The number of existing condition vectors $|\vec{A}(U)| = 14$. 22 possible condition vectors did not appear in Table 1. The training set in Table 1 is not a completed instance system.

2.2 Single Body of Knowledge Representation

In general, a *classifier* or a *single body of knowledge representation* is defined as a mapping from the condition vector space to the decision space.

$$\varphi_A : V_{\times A} \rightarrow V_{\times D}$$

In fact, the mapping can be represented by many forms such as rules, decision trees, neural networks, naïve Bayes classifiers, kNN classifier, or other methods.

Let $B \subset A.V_{\times B}$ is called a subspace of $V_{\times A}$. All the vectors in $V_{\times A}$ have their components in $V_{\times B}$. The conventional approach in machine learning or data mining is to select a "best" subset B and get a mapping φ_B based on the subset B.

$$\varphi_B : V_{\times B} \rightarrow V_{\times D}$$

φ_B then replaces φ_A to make decisions or in classification. For example, the decision tree can be obtained from Table 1 by means of information entropy [11]. The decision tree can be applied to represent the mapping from $V_{\times B} \rightarrow V_{\times D}(\vec{B} \in V_{\times B}, \vec{D} \in V_{\times D})$ shown in Fig. 1. According to this mapping, a decision can be made for an unseen instance with full attribute values or only values for {Outlook, Humidity, Wind} (i.e. without temperature value) in the entire condition space.

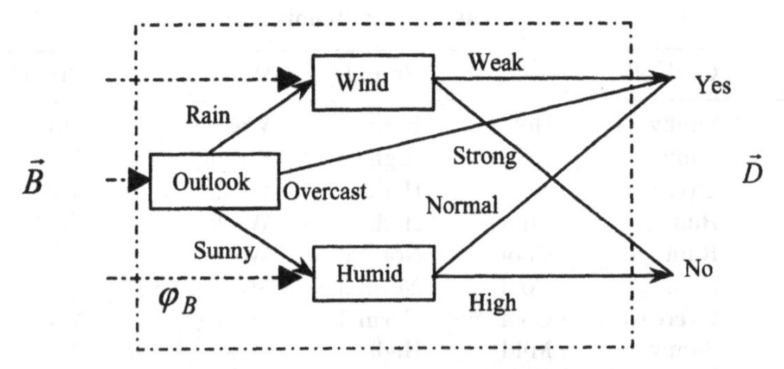

Fig. 1. Knowledge φ_B represented by $V_{\times B} \rightarrow V_{\times D}$

Alternately, the Naïve Bayes Classifier is a very practical approach in the machine learning domain. According to this approach, the knowledge from Table 1 can be represented by a mapping based on the Naïve Bayes Classifier. Suppose that attributes in an instance information system are conditionally independent, i.e.

$$P(\vec{A}\,|d_i) = P(a_1, a_2, \ldots, a_{|A|}|d_i) = \prod_j P(a_j|d_i) \tag{5}$$

The most probable decision can be expressed as follows

$$d_{MP} = \arg\max_{d_i \in V_d} P(d_i| \vec{A}) = \arg\max_{d_i \in V_d} P(\vec{A} |d_i)P(d_i)$$

$$d_{MP} = \arg\max_{d_i \in V_d} P(d_i) \prod_j P(a_j|d_i) \tag{6}$$

$$d_{MP} = \arg\max_{d_i \in V_d} Bel(d_i)$$

$$Bel(d_i) = \frac{P(d_i) \prod_j P(a_j|d_i)}{\sum_{i=1}^{|V_{\times D}|} P(d_i) \prod_j P(a_j|d_i)} \tag{7}$$

where $Bel(d_i)$, called as the belief for the decision $d_{i,}$, is applied to represent belief distribution over the decision space. Learning or knowledge extraction for the naïve Bayes classifier requires one to obtain the conditional probability table (CPT). The CPT for Table 1 is shown in Table 2. The mapping for the naïve Bayes classifier can be represented in Fig. 2. In the mapping, a condition is mapped onto the decision space with a belief distribution rather than one certain decision. This is different from decision tree representation. The final decision is made by means of the winner-take-all rule.

Table 2. Conditional Probability Table (CPT)

	P(x | d)	d = Yes	d = No
Outlook	P(Sunny | d)	2/9	3/5
	P(Oercast | d)	4/9	0/5
	P(Rain | d)	3/9	2/5
Temp	P(Hot | d)	2/9	2/5
	P(Mild | d)	4/9	2/5
	P(Cool | d)	3/9	1/5
Humid	P(High | d)	3/9	4/5
	P(Normal | d)	6/9	1/5
Wind	P(Weak | d)	6/9	2/5
	P(Strong | d)	3/9	3/5
	P(d)	9/14	5/14

One advantage of the naïve Bayes classifier is its capability of coping with missing values. For example, consider the condition vector $\vec{A}= (?, Hot, High, Weak)$ i.e. the value for Outlook is missing. This is an unseen situation for Table 1, and the value of Outlook is missing. However, the decision still can be made with three attribute values by using (7).

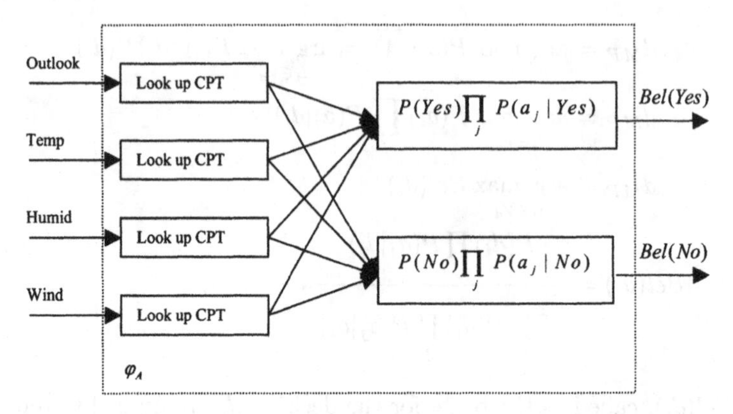

Fig. 2. Knowledge φ_A represented by the Naïve Bayes Classifier

$$\begin{aligned}
P(No)P(a_2 &= Hot|No)P(a_3 = High|No)P(a_4 = Weak|No) \\
&= (5/14)(2/5)(4/5)(2/5) = 0.046 \\
P(Yes)P(a_2 &= Hot|No)P(a_3 = High|No)P(a_4 = Weak|No) \\
&= (9/14)(2/9)(3/9)(6/9) = 0.032 \\
Bel(No) &= \frac{0.046}{0.046 + 0.032} = 0.59 \\
Bel(Yes) &= \frac{0.032}{0.046 + 0.032} = 0.41 \\
\text{Hence} \quad d_{MP} &= \arg\max_{d_i \in V_d} Bel(d_i) = \text{“}No\text{”}
\end{aligned}$$

This is a reasonable result for playing tennis under the condition vector $\vec{A} = (?, Hot, High, Weak)$. As the decision is made according to probability, this approach is good at dealing with inconsistent instance information systems.

2.3 Problems with Single Body of Knowledge Representation

Now considering the sixth instance in Table 1, we have

$$\begin{aligned}
&P(No)P(a_1 = Rain|No)P(a_2 = Cool|No)P(a_3 = Normal|No) \\
&\quad P(a_4 = Strong|No) = (5/14)(2/5)(1/5)(1/5)(3/5) = 0.0034 \\
&P(Yes)P(a_1 = Rain|Yes)P(a_2 = Cool|Yes) \\
&P(a_3 = Normal|Yes)P(a_4 = Strong|Yes) \\
&\qquad = (9/14)(3/9)(3/9)(6/9)(3/9) = 0.0158 \\
&Bel(No) = \frac{0.0034}{0.0034 + 0.0158} = 0.18 \\
&Bel(Yes) = \frac{0.0158}{0.0034 + 0.0158} = 0.82
\end{aligned}$$

The result is d_{MP} = "*Yes*". Yet the true value of the sixth instance in Table 1 is "*No*". Because the assumption of (5) is not satisfied in Table 1 and Table 1 does not provide complete instance information system, the Naïve Bayes Classifier cannot give the correct result for this instance.

On the other hand, let us look at the decision tree representation. An advantage of decision tree learning is that a very compact knowledge representation can be obtained by dropping some attributes. For example, mapping φ_B is only based on subset B of attributes {Outlook, Humidity, Wind}. The decision can be made without the Temperature value. However, there are two problems with this single body of knowledge φ_B.

1. Does the temperature not affect playing tennis?
2. How does one make a decision if a value of Outlook or Humidity or Wind is missing?

It is difficult to answer the question whether one plays tennis or not when the subset of attribute values is {Sunny, Hot, Strong} because the humidity attribute value is not in the set. These are problems of a single body of knowledge. In fact, many different mappings can be obtained from the training set. This is the idea of multi-knowledge. A simple example here can be used to demonstrate that multi-knowledge can cope better with missing values than a single body of knowledge.

In order to make a decision without value of humidity, another single body of knowledge $\varphi_{B'}$ for Table 1 can be obtained by decision tree learning. This knowledge is based on the attribute subset B' = {Outlook, Temperature, Wind} and can be represented by a mapping from condition vector space to decision space, as shown in Fig. 3.

According to the knowledge $\varphi_{B'}$, the answer is "No" for the question with {Sunny, Hot, Strong}. B and B' are two reducts of the training set in Table 1. $\Phi = \{\varphi_B, \varphi_{B'}\}$ is called *multi-knowledge.* A definition of reduct can be found in rough set theory [1, 2, 3, 4, 5, 6, 7, 8, 9, 10, 23].

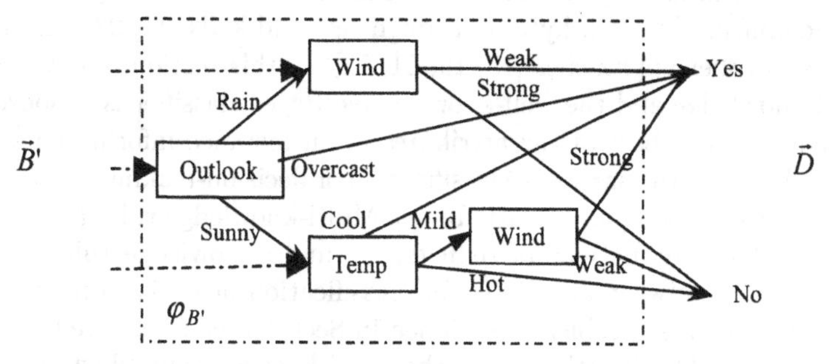

Fig. 3. Knowledge $\varphi_{B'}$ represented by $V_{\times B'} \rightarrow V_{\times D}$

2.4 Definition of Multi-Knowledge

Definition Given an instance information system $I = \langle U, A \cup D \rangle$. *Multi-knowledge* is defined as follows

$$\Phi = \{\varphi_B | B \in RED\} \tag{8}$$

where φ_B is a mapping with respect to attribute subset B. RED is a set of good reducts from the training set. Different approaches to find reducts in an instance information system can be found in the literature e.g. [2, 3, 4, 5, 6, 7, 8, 9, 10, 12, 13, 14, 23, 24]. The algorithm for finding multiple good reducts for multi-knowledge can be found in [7, 9, 12, 20].

Outlook and Wind are *indispensable attributes* for making decision in this example (*indispensable attributes* are also called core attributes). The algorithms for finding *indispensable attributes* can be found in literature e.g. [3, 13, 14, 20]. It is difficult to make a decision by means of multi-knowledge φ_B and $\varphi_{B'}$ if the Outlook or Wind attribute value is missing. On the other hand, the Naïve Bayes Classifier is good at dealing with this situation. This implies that multi-knowledge and the Naïve classifier have their own merits and shortcomings. The combination of the Naïve Bayes Classifier and multi-knowledge gives an opportunity to improve decision accuracy and reliability for instance information systems with unseen instance and missing values.

3 Hybrid of Multi-Knowledge and the Naïve Bayes Classifier

Many successful applications are achieved by various intelligent techniques. As each intelligent technique (e.g. machine learning or data mining approach) has its own merits and shortcomings, many cases are very difficult to handle using a single technique or approach. Hybrid approaches are widely applied to solve problems in intelligent systems: for example, neuro-fuzzy systems, rough neural computing [27, 28], hybrid of rough sets and kNN [17, 23], and hybrid of kNN and other approaches [18, 19, 21, 22]. In this section, a novel hybrid of multi-knowledge and the well-known naïve Bayes classifier is proposed.

A reduct is a subset of the attributes in an instance information system such that by using the subset in classification or decision-making, the accuracy is the same as by using all the attributes. Multi-knowledge rules are based on attribute value conjunctions of reducts, i.e. multi-knowledge rules can take in relationships between attributes for classification or decision-making. The naïve Bayes classifier problem mentioned in Sect. 2.3 can be solved by multi-knowledge rules. On the other hand, the multi-knowledge problem for missing values of indispensable attributes can be dealt with using the naïve Bayes classifier. Therefore, a hybrid of multi-knowledge and the naïve Bayes classifier can, in principle, reach high decision accuracy.

Let B be a reduct in an instance information system. To introduce an uncertain rule to represent knowledge φ_B, decision is represented by a belief distribution over the decision space instead of a given value; this is different from the multi-knowledge in [12]. The format of the rule is defined as follows

$$\varphi_B(\vec{B}) \to \vec{M} \quad \text{for } \vec{B} \in V_{\times B} \tag{9}$$

where $\vec{M}$ is the belief distribution over the decision space, $\vec{M}= (m_1, m_2, \ldots, m_{|M|})$. From the Bayes Classifier (7), the most probable decision can be obtained by

$$d_{MP} = \arg\max_{d_i \in V_d} P(\vec{B}\,|d_i)P(d_i)$$

Let $m_i = P(\vec{B}\,|d_i)P(d_i)/\sum_i P(\vec{B}\,|d_i)P(d_i)$, and $\sum_i^{|M|} m_i = 1$. The most probable decision is obtained by

$$d_{MP} = \arg\max_{d_i \in V_d} m_i \tag{10}$$

For Table 1, two reducts are found by the algorithm in [20].

$$RED_1 = \{\text{Outlook, Humid, Wind}\}$$
$$RED_2 = \{\text{Outlook, Temp, Wind}\}$$

According to the format of the rule in (9), the rule group for RED_1 is represented as follows.

Rule(RED_1)	Outlook	Temp	Humid	Wind	m_{no}	m_{yes}
G1r1	Sunny	*	High	*	1	0
G1r2	Sunny	*	Normal	*	0	1
G1r3	Overcast	*	*	*	0	1
G1r4	Rain	*	*	Weak	0	1
G1r5	Rain	*	*	Strong	1	0

where * represents arbitrary values in the attribute domain. The rule group for RED_2 can be obtained by analogy. If conditions of an unseen instance are satisfied in multiple rule groups at the same time, m_i is replaced with its product with respect to the groups. If the instance information system is not a consistent system, $m_i \in [0, 1]$ is a real number. For example, consider an attribute value conjunction {Rain, Mild}; the belief distribution can be calculated by

$$m_i = P(\vec{B}\,|d_i)P(d_i) \Big/ \sum_i P(\vec{B}\,|d_i)P(d_i)$$

There are three objects with the conjunction (Rain, Mild). One object supports decision "No". Two objects support "Yes".

$$P((Rain, Mild)|\text{“}No\text{”})P(\text{“}No\text{”}) = (1/5)(5/14) = 0.071$$
$$P((Rain, Mild)|\text{“}Yes\text{”})P(\text{“}Yes\text{”}) = (2/5)(5/14) = 0.142$$
$$\sum_i P(\vec{B}\,|d_i)P(d_i) = 0.071 + 0.142 = 0.213$$
$$m_1 = 0.071/0.213 = 0.333$$
$$m_2 = 0.142/0.213 = 0.667$$

So if (*Rain, Mild*) then we have (0.667 *Yes*, 0.333 *No*). This is called an uncertain rule. Knowledge for inconsistent systems can be represented by such rules. For example, the data set for the data mining competition in KDD 2000 is an inconsistent system.

There are two belief distributions in the hybrid system. One is from the uncertain rules of multi-knowledge, another is from the naïve Bayes classifier. Applying these two belief distributions, the final decision is made by

$$d_{MP} = \underset{d_i \in V_d}{\arg\max}\ m_i Bel(d_i) \tag{11}$$

where m_i is from multi-knowledge, and $Bel(d_i)$ is from the Naïve Bayes Classifier. For the problem with sixth instance in Table 1, the condition vector is {*Rain, Cool, Normal, Strong*}. The Bayes belief distribution for this case was obtained in Sect. 2.3

$$Bel(\text{no} = 0.18,\ \text{yes} = 0.82)$$

Multi-knowledge decision belief distribution can be obtained from rule G1r5 in rule group (RED_1).

$$m_{no} = 1, m_{yes} = 0$$

Therefore,

$$d_{MP} = \underset{d_i \in V_d}{\arg\max}\ m_i Bel(d_i) = no$$

The formal decision making algorithm is as follows.

A3.1 *Decision Algorithm Based on Multi-knowledge and Bayes Classifier*

1. Input training set
2. Find multi-reducts *RED*.
3. Create rule groups $\Phi = \{\varphi_B | B \in RED\}$
4. Create probability distribution $P(d_i)$ and $P(a_i|d_i)$
5. Input an unseen instance u
6. If condition vector $\vec{B}$ of instance u is matched in rule groups //if multiple rules are matched, $\vec{M}$ is the product of belief distributions
7. $\vec{M} \leftarrow \varphi_B(\vec{B})$
8. $d_{MP} = \underset{d_i \in V_d}{\arg\max}\ m_i Bel(d_i)$

9. else // for *indispensable attribute* value missing
10. $d_{MP} = \arg\max_{d_i \in V_d} P(d_i) \prod_j P(a_j|d_i)$
11. End if

This algorithm includes two sections. The multi-knowledge rule group and the probability are calculated in Sect. 1 from steps 1 to 4 of the algorithm. For some training sets, there are not any redundant attributes i.e. if any attribute is removed, the decision accuracy will be seriously affected. All attributes are regarded as one reduct to generate multi-knowledge rule group. Decision for a new instance is made in Sect. 2 from steps 5 to 10.

4 Experimental Results

In order to test the algorithms, benchmark data sets from the UCI Machine Learning Repository were applied. The experimental results are shown in Table 3. The results are calculated by using the ten-fold cross validation standard.

Table 3. Comparable results

Data Set	At	Ins	IA	MR	C5.0	Bays	MKB
BreastCancer ♣	10	699	1	19*	95.1	96.8	97.6
Crx_data	15	690	1	60*	85.1	81.0	85.1
Heart	13	270	0	109*	78.1	76.3	83.3
Lung	56	32	0	50	46.7	57.1	96.7
Iris_data	4	150	0	4*	94.0	94.0	96.7
Promoter	57	106	0	57	78.8	89.7	98.1
Splice	60	3190	0	60	94.1	93.5	94.3
Hepatitis ♣	19	155	0	16	62.0	69.6	76.1
Hungarian ♣	13	294	0	9	79.2	83.3	85.4
Ionosphere	34	351	0	34	88.3	82.6	90.6
Bupa	6	345	0	9*	67.2	64.4	65.5
Bridges(V2) ♣	11	108	2	7*	58.2	63.9	65.7

Column ***At*** represents the number of attributes in the data set. Column ***Ins*** represents the number of instances. Column ***IA*** denotes the number of ***I****ndispensable* ***A****ttributes* which are found by the algorithm in [20]. Column ***MR*** is the number of ***M****ultple* ***R****educts* which are also found by the algorithm in [20]. "♣" indicates the data set with missing values. "*" indicates that all the reducts in a data set have been found by exhaustive search. Exhaustive search could not be done for data sets such as Lung and Splice because the numbers of attributes and instances is too large. The decision accuracy by the Naïve Bayes Classifier is in the column labeled ***Bays***. ***MKB*** denotes the decision accuracy based on the hybrid approach of ***M**ulti-**K**nowledge and the Naïve **B**ayes Classifier*. The accuracies in column ***C5.0*** are calculated by the C5.0 tree in the commercial software Clementine 5.2. It can be seen that the more reducts in a data set, the more the decision accuracy is improved for the hybrid approach of multi-knowledge and the naïve Bayes classifier. These results are comparable with other approaches such as combinations of kNN and rough sets [17] and other complex classifiers [18, 19, 21].

5 Conclusion

The Naïve Bayes Classifier is a highly practical machine learning method. In some domains its performance has been shown to be comparable to that of neural networks and decision tree learning. For example, it gives a high decision accuracy for the Breast-cancer and Iris_data sets. However, the Naïve Bayes Classifier is based on the assumption that the attribute values are conditionally independent for the given decision value. If this assumption is not satisfied, the decision accuracy will degrade. The multi-knowledge rule set supplies a complement for the Naïve Bayes Classifier. Using a hybrid Naïve Bayes Classifier and multi-knowledge approach, high decision accuracies are reached in many data sets, as shown in Table 3. The concept of multi-knowledge φ and uncertain rule have been applied to represent knowledge. A decision can be made with incomplete attributes using the proposed algorithm. Therefore, there are wide applications, for example, a robot with multi-knowledge is able to identify a changing environment [20].

References

1. Zhong Ning and Dong Junzhen, "Using rough sets with heuristics for feature selection", Journal of Intelligent Information Systems, vol. 16, Kluwer Academic Publishers. Manufactured in The Netherlands, (2001) 199–214.
2. Pawlak, Z. "Rough Sets", International Journal of Computer and Information Sciences, 11, (1982) 341–356.
3. Pawlak Z. Rough sets: theoretical aspects data analysis, Kluwer Academic Publishers, Dordrecht, (1991).
4. Yao Y.Y. Wong S.K.M. and Lin T.Y. A Review of Rough Set Models, in: Lin T.Y. and Cercone N. (eds) Rough Sets and Data Mining: Analysis for Imprecise Data, Kluwer Academic, Boston, Mass, London, (1997) 25–60.
5. Skowron A. and Rauszer C. The Discernibility Matrices and Functions in Information Systems. In: Slowinski R. (ed), Intelligent Decision Support, Boston, MA: Kluwer Academic Publishers, (1992) 331–362.
6. Yao, Y.Y. and Lin, T.Y., Generalization of rough sets using modal logic, Intelligent Automation and Soft Computing, An International Journal, Vol. 2, No. 2, (1996) 103–120.

7. Polkowski L., Tsumoto S., Lin T.Y., "Rough Set Methods and Applications", New Developments in Knowledge Discovery in Information Systems, Physica-Verlag, A Springer-Verlag Company, (2000).
8. Dietterich T.G. Machine Learning Research: Four Current Directions AI Magazine. 18 (4), (1997) 97–136.
9. Wróblewski J. Genetic algorithms in decomposition and classification problem. In: Polkowski L, Skowron A (eds): Rough Sets in Knowledge Discovery. Physica-Verlag, Heidelberg, 2: (1998) 471–487.
10. Bazan J., Skowron A., Synak P. Dynamic Reducts as a Tool for Extracting Laws from Decision Tables, Proc. Symp. on Methodologies for Intelligent Systems, Charlotte, NC, USA, Oct. 16–19, Lecture Notes in Artificial Intelligence, Vol. 869, Springer Verlag, (1994) 346–355.
11. Mitchell T.M., "Machine Learning", Co-published by the MIT Press and TheMcGraw-Hill Companies, Inc., (1997).
12. Hu X., Cecone N. and Ziarko W. Generation of multiple knowledge from databases based on rough set theory, in: Lin TY and Cercone N (eds) Rough Set and Data Mining, Kluwer Academic Publishers, (1997) 109–121.
13. Bell D.A., Guan J.W. "Computational methods for rough classification and discovery", Journal of the American Society for Information Science, Special Topic Issue on Data Mining, (1997).
14. Guan J.W., and Bell D.A. "Rough Computational Methods for Information Systems", Artificial Intelligence, vol. 105 (1998) 77–103.
15. Kohavi R., Frasca B., "Useful feature subsets and rough set reducts", In the International Workshop on Rough Sets and Soft Computing (RSSC), (1994).
16. Bell D. and Wang H. "A Formalism for Relevance and Its Application in Feature Subset Selection", Machine learning, vol. 41, Kluwer Academic Publishers. Manufactured in The Netherlands, (2000) 175–195.
17. Bao Y., Du X. and Ishii N. "Improving Performance of the K-Nearest Neighbor Classification by GA and Tolerant Rough Sets", International Journal of Knowledge-Based Intelligent Engineering Systems, 7(2): (2003) 54–61.
18. Guo G.D., Wang H., Bell D. "Asymmetric neighbourhood selection and support aggregation for effective classification", 8th Ibero-American Conference on Artifical Intelligence (IBERAMIA 02), LECT NOTES COMPUT SC 2888: (2003) 986–996.
19. Guo G.D., Wang H., Bell D., et al. "KNN model-based approach in classification", OTM Confederated International Conference CoopIS, DOA and ODBASE, LECT NOTES ARTIF INT 2527: (2002) 21–31.
20. Wu Q.X. and Bell D.A. "Multi-Knowledge Extraction and Application", In proceedings of 9th international conference on Rough Sets, Fuzzy Sets, Data Mining, and Granular Computing, (eds) Wang G.Y., Liu Q., Yao Y.Y. and Skowron A., LNAI 2639, Springer, Berlin, (2003) 274–279.
21. Wilson D.R. and Martinez T.R. "An Integrated Instance-Based Learning Algorithm", Computational Intelligence, 16(1), (2000) pp. 1–28.
22. Wilson D.R. and Martinez T.R. "Reduction Techniques for Instance-Based Learning Algorithms", Machine Learning, 381, (2000) pp. 257–28.
23. Wróblewski J. Adaptive aspects of combining approximation spaces. In: Pal S.K., Polkowski L., Skowron A. (eds.): Rough-Neural Computing. Techniques for computing with words. Springer-Verlag (Cognitive Technologies), Berlin, Heidelberg, (2003) pp. 139–156.

24. Bazan J., Nguyen H.S., Nguyen S.H., Synak P., Wróblewski J. Rough Set Algorithms in Classification Problem. In: L. Polkowski, S. Tsumoto and T.Y. Lin (eds.): Rough Set Methods and Applications. Physica-Verlag, Heidelberg, New York, (2000) pp. 49–88.
25. Krzysztof Dembczynski and Roman Pindur and Robert Susmaga, Generation of Exhaustive Set of Rules within Dominance-based Rough Set Approach, Electronic Notes in Theoretical Computer Science, Vol. 82 (4) (2003).
26. Izak D., Wróblewski J. 2002. Approximate bayesian network classifiers. In: Alpigini J.J., Peters J.F., Skowron A., Zhong N. (eds.): Rough Sets and Current Trends in Computing. Proc. of 3rd International Conference RSCTC 2002, Malvern, PA, USA. Springer-Verlag (LNAI 2475), Berlin, Heidelberg, (2002) pp. 365–372.
27. Pal S.K., Polkowski L., Skowron A. (eds.): Rough-Neural Computing. Techniques for computing with words. Springer-Verlag (Cognitive Technologies), Berlin, Heidelberg 2003.
28. Pawlak, Z. Combining Rough Sets and Bayes' Rule, Computational Intelligence, volume 17, issue 3, (2001) pp. 401–409.

First-Order Logic Based Formalism for Temporal Data Mining*

Paul Cotofrei[1] and Kilian Stoffel[2]

[1] University of Neuchâtel, Pierre-à-Mazel, 7, 2000, Neuchâtel, Switzerland
paul.cotofrei@unine.ch
[2] University of Neuchâtel, Pierre-à-Mazel, 7, 2000, Neuchâtel, Switzerland
kilian.stoffel@unine.ch

Summary. In this article we define a formalism for a methodology that has as purpose the discovery of knowledge, represented in the form of general Horn clauses, inferred from databases with a temporal dimension. To obtain what we called *temporal rules*, a discretisation phase that extracts *events* from raw data is applied first, followed by an induction phase, which constructs classification trees from these events. The theoretical framework we proposed, based on first-order temporal logic, permits us to define the main notions (event, temporal rule, constraint) in a formal way. The concept of *consistent linear time structure* allows us to introduce the notions of *general interpretation* and of *confidence*. These notions open the possibility to use statistical approaches in the design of algorithms for inferring higher order temporal rules, denoted temporal meta-rules.

1 Introduction

Data mining is the process of discovering interesting knowledge, such as patterns, associations, changes, anomalies and significant structures, in large amounts of data stored in databases, data warehouses, or other data repositories. Due to the wide availability of huge amounts of data in digital form, and the need for turning such data into useful information, data mining has attracted a great deal of attention in information industry in recent years.

In many applications, the data of interest comprise multiple sequences that evolve over time. Examples include financial market data, currency exchange rates, network traffic data, signals from biomedical sources, etc. Although traditional time series techniques can sometimes produce accurate results, few can provide easy understandable results. However, a drastically increasing number of users with a limited statistical background would like to use these tools. In the same time, we have a number of tools developed by researchers in

* Supported by the Swiss National Science Foundation (grant N° 2100-063 730).

P. Cotofrei and K. Stoffel: *First-Order Logic Based Formalism for Temporal Data Mining*, Studies in Computational Intelligence (SCI) **6**, 185–210 (2005)
www.springerlink.com

the field of artificial intelligence, which produce understandable rules. However, they have to use ad-hoc, domain-specific techniques for transforming the time series to a "learner-friendly" representation. These techniques fail to take into account both the special problems and special heuristics applicable to temporal data and therefore often results in unreadable concept description.

As a possible solution to overcome these problems, we proposed in [9] a methodology that integrates techniques developed both in the field of machine learning and in the field of statistics. The machine learning approach is used to extract symbolic knowledge and the statistical approach is used to perform numerical analysis of the raw data. The overall goal consists in developing a series of methods able to extract/generate temporal rules, having the following characteristics:

- Contain explicitly a temporal (or at least a sequential) dimension.
- Capture the correlation between time series.
- Predict/forecast values/shapes/behavior of sequences (denoted events).

1.1 State of Art

The domain of temporal data mining focuses on the discovery of causal relationships among events that may be ordered in time and may be causally related. The contributions in this domain encompass the discovery of temporal rule, of sequences and of patterns. However, in many respects this is just a terminological heterogeneity among researchers that are, nevertheless, addressing the same problem, albeit from different starting points and domains.

The main tasks concerning the information extraction from time series database and on which the researchers concentrated their efforts over the last years may be divided in several directions. *Similarity/Pattern Querying* concerns the measure of similarity between two sequences or sub-sequences respectively. Different methods were developed, such as window stitching [1] or dynamic time warping based matching [23, 24]. *Clustering/Classification* concentrates on optimal algorithms for clustering/classifying sub-sequences of time series into groups/classes of similar sub-sequences. Different techniques were proposed: Hidden Markov Models (HMM) [25], Dynamic Bayes Networks (DBNs) [13], Recurrent Neural Networks [14], supervised classification using piecewise polynomial modelling [30] or agglomerative clustering based on enhancing the time series with a line segment representation [22]. *Pattern finding/Prediction* methods concern the search for periodicity patterns (fully or partially periodic) in time series databases. For full periodicity search there is a rich collection of statistic methods, like FFT [26]. For partial periodicity search, different algorithms were developed, which explore properties related to partial periodicity such as the max-subpattern-hit-set property [16] or point-wise periodicity [15]. *Temporal Rules' extraction* approach concentrated to the extraction of explicit rules from time series, like temporal association rules [6] or cyclic association rules [31]. Adaptive methods for finding rules whose conditions refer to patterns in time series were described in [11, 17, 37] and a general methodology for classification and extraction of temporal rules was proposed in [9].

Although there is a rich bibliography concerning formalism for temporal databases, there are very few articles on this topic for temporal data mining. In [2, 5, 29] general frameworks for temporal mining are proposed, but usually the researches on causal and temporal rules are more concentrated on the methodological/algorithmic aspect, and less on the theoretical aspect. In this article, we extend our methodology with an innovative formalism based on first-order temporal logic, which permits an abstract view on temporal rules. The formalism allows also the application of an inference phase in which higher order temporal rules (called temporal meta-rules) are inferred from local temporal rules, the lasts being extracted from different sequences of data. Using this strategy, known in the literature as higher order mining [36], we can guarantee the scalability (the capacity to handle huge databases) of our methodological approach, by applying statistical and machine learning tools.

It is important to mention that there are two distinct approaches concerning the time structure in a framework for temporal data mining. The first conceives the time as an ordered sequence of points and it is usually employed in temporal database applications. The second is based on intervals and it is predominant in AI applications [3, 8, 19, 35]. The difference induced at the temporal logic level, by the two approaches, is expressed in the set of temporal predicates: they are unary in the first case and binary for the second.

The rest of the paper is structured as follows. In the next section, the main steps of the methodology (the discretisation phase and the inference phase) are presented. Section 3 contains an extensively description of the first-order temporal logic formalism (definitions of the main terms – *event, temporal rules, confidence* – and concepts – *consistent linear time structure, general interpretation*). Section 4 reviews the methodology for temporal rules extraction, in the frame of the proposed formalism. The notion of temporal meta-rules and the algorithms for inferring such high order rules are described in Sect. 5. Finally, the last section summarizes our work and emphasize what we consider an important and still open problem of our formalism.

2 The Methodology

The approaches concerning the information extraction from time series, described above, have mainly two shortcomings, which we tried to overcome.

The first problem is the type of knowledge inferred by the systems, which most often is difficult to be understood by a human user. In a wide range of applications, (e.g. almost all decision making processes) it is unacceptable to produce rules that are not understandable for an end user. Therefore, we decided to develop inference methods that produce knowledge represented in the form of general Horn clauses, which are at least comprehensible for a moderately sophisticated user. In the fourth approach, (*Temporal Rules' extraction*), a similar representation is used. However, the rules inferred by these systems have a more restricted form than the rules we propose.

The second problem consists in the number of time series investigated during the inference process. Almost all methods mentioned above are based on one-dimensional data, i.e. they are restricted to one time series. The methods we propose are able to handle multi-dimensional data.

Two of the most important scientific communities which brought relevant contributions to data analysis (the statisticians and database researchers) chose two different ways: statisticians concentrated on the continuous aspect of the data and the large majority of statistical models are continuous models, whereas the database community concentrated much more on the discrete aspects, and in consequence, on discrete models. For our methodology, we adopt a mixture of these two approaches, which represents a better description of the reality of data and which generally allows us to benefit from the advantages of both approaches.

The two main steps of the methodology for temporal rules extraction are structured in the following way:

1. Transforming sequential raw data into sequences of events: Roughly speaking, an event can be seen as a labelled sequence of points extracted from the raw data and characterized by a finite set of predefined features. The features describing the different events are extracted using statistical methods.
2. Inferring temporal rules: We apply a first induction process, using sets of events as training sets, to obtain several classification trees. Local temporal rules are then extracted from these classification trees and a final inference process will generate the set of temporal meta-rules.

2.1 Phase One

The procedure that creates a database of events from the initial raw data can be divided into two steps: time series discretisation, which extracts the discrete aspect, and global feature calculation, which captures the continuous aspect.

Time Series Discretisation During this step, the sequence of raw data is "translated" into a sequence of discrete symbols. By an abuse of language, an event means a subsequence having a particular shape. In the literature, different methods were proposed for the problem of discretizing times series using a finite alphabet (window's clustering method [11], ideal prototype template [22], scale-space filtering [18]). In window's clustering method, all contiguous subsequences of fixed length w are classified in clusters using a similarity measure and these clusters receive a name (a symbol or a string of symbols). If we set $w = 2$ and classify not the sequences $x_i x_{i+1}$, but the normalized sequence of the differences $x_{i+1} - x_i$, using the quantile of the normal distribution, we obtain the discretisation algorithm proposed by [21]. But the biggest weakness of these methods which use a fixed length window is the sensibility to the noise. Therefore, the scale-space filtering method, which finds the boundaries of the subsequences having a persistent behavior over multiple degree of smoothing, seems to be more appropriate and must be considered as a first compulsory pre-processing phase.

Global Feature Calculation During this step, one extracts various features from each sub-sequence as a whole. Typical global features include global maxima, global minima, means and standard deviation of the values of the sequence as well as the value of some specific point of the sequence, such as the value of the first or of the last point. Of course, it is possible that specific events will demand specific features, necessary for their description (e.g. the slope of the best-fitting line or the second real Fourier coefficient). The optimal set of global features is hard to be defined in advance, but as long as these features are simple descriptive statistics, they can be easily added or removed form the process.

Example 1. Consider a database containing daily price variations of a given stock. After the application of the first phase we obtain an ordered sequence of events. Each event has the form (*name*, v_1, v_2), where the name is one of the strings {*peak, flat, valley*} – we are interested only in three kinds of shapes – and v_1, v_2 represents the mean, respectively, the standard error – we chose only two features as determinant for the event. The statistics are calculated using daily prices, supposed to be subsequences of length $w = 12$.

2.2 Phase Two

During the second phase, we create a set of temporal rules inferred from the database of events, obtained in phase one. Two important steps can be defined here:

- Application of a first induction process, using the event database as training database, to obtain a set of *classification trees*. From each classification tree, the corresponding set of temporal rules is extracted.
- Application of a second inference process using the previously inferred temporal rules sets to obtain the final set of temporal meta-rules.

First Induction Process There are different approaches for extracting rules from a set of events. Association Rules [6], Inductive Logic Programming [35], Classification Trees [20] are the most popular ones. For our methodology we selected the *classification tree approach*. It is a powerful tool used to predict memberships of cases or objects in the classes of a categorical dependent variable from their measurements on one or more predictor variables (or attributes). A classification tree is constructed by recursively partitioning a learning sample of data in which the class and the values of the predictor variables for each case are known. Each partition is represented by a node in the tree. A variety of classification tree programs has been developed and we may mention QUEST [27], CART [4], FACT [28] and last, but not least, C4.5 [32]. Our option was the C4.5 like approach. The tree resulting by applying the C4.5 algorithm is constructed to minimize the observed error rate, using equal priors. This criterion seems to be satisfactory in the frame of sequential data and has the advantage to not favor certain events. To successively create the partitions, the C4.5 algorithm uses two forms of tests in each node: a standard test for discrete attributes, with one outcome ($A = x$) for each possible value x of the attribute A, and a binary test, for continuous attributes, with outcomes $A \leq z$ and $A > z$, where z is a threshold value. Finally, each path from the root to a leave corresponds to a rule representing a conjunction of tests outcomes (see the example from Fig. 1).

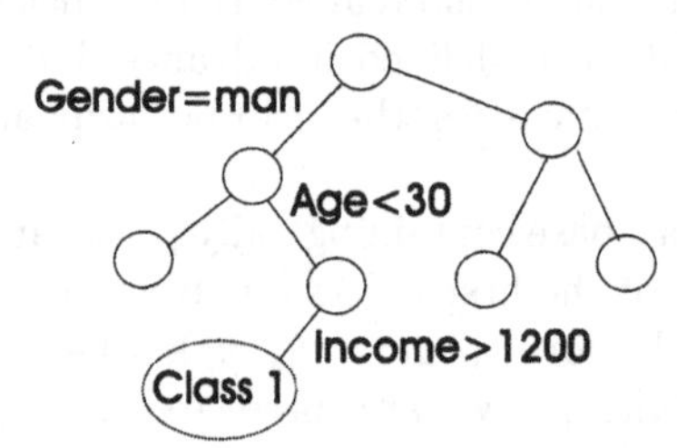

Fig. 1. Rule corresponding to a path from the root to the leaf "Class 1", expressed as a conjunction of three outcome tests implying each a different attribute

Before to apply the decision tree algorithm to a database of events, an important problem has to be solved first: establishing the training set. An n-tuple in the training set contains $n - 1$ values of the predictor variables (or attributes) and one value of the categorical dependent variable, which represents the class. There are two different approaches on how the sequence that represents the classification (the values of the categorical dependent variable) is obtained. In a supervised methodology, an expert gives this sequence. The situation becomes more difficult when there is no prior knowledge about the possible classifications. Suppose that, following the example we gave, we are interested in seeing if a given stock value depends on other stock values. As the dependent variable (the stock price) is not categorical, it cannot represent a valid classification used to create a classification tree. The solution is to use the sequence of the names of events extracted from the continuous time series as sequence of classes.

Let us suppose that we have k sequences representing the predictor variables $q_1, q_2, \ldots, q_k$. Each q_{ij}, $i = 1, \ldots, k$, $j = 1, \ldots, n$ is the name of an event (*Remark:* we consider a simplified case, with no feature as predictor variable, but without influence on the following rationing). We have also a

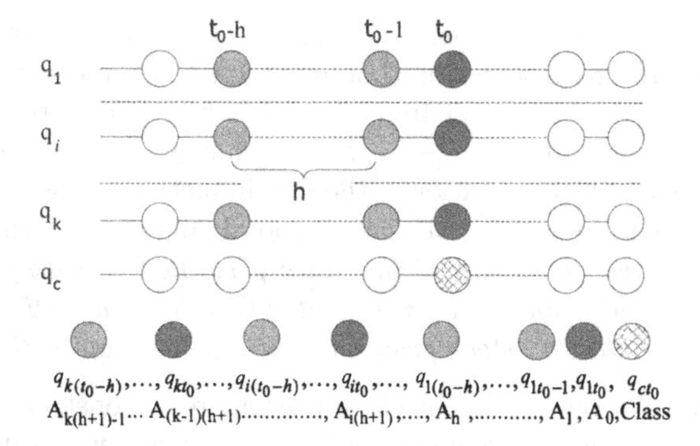

Fig. 2. Graphical representation of the first tuple and the list of corresponding attributes

sequence $q_c = q_{c1}, \ldots, q_{cn}$ representing the classification. The training set will be constructed using a procedure depending on three parameters. The first, t_0, represents a time instant considered as *present time*. Practically, the first tuple contains the class q_{ct_0} and there is no tuple in the training set containing an event that starts after time t_0. The second, t_p, represents a time interval and controls the further back in time class $q_{c(t_0-t_p)}$ included in the training set. Consequently, the number of tuples in the training set is $t_p + 1$. The third parameter, h, controls the influence of the past events $q_{i(t-1)}, \ldots, q_{i(t-h)}$ on the actual event q_{it}. This parameter ($h$istory) reflects the idea that the class q_{ct} depends not only on the events at time t, but also on the events occurred before time t. Finally, each tuple contains $k(h+1)$ events (or values for $k(h+1)$ attributes, in the terminology of classification trees) and one class value (see Fig. 2). The first tuple is $q_{ct_0}, q_{1t_0}, \ldots, q_{1(t_0-h)}, \ldots, q_{k(t_0-h)}$ and the last $q_{c(t_0-t_p)}, q_{1(t_0-t_p)}, \ldots, q_{k(t_0-t_p-h)}$. To adopt this particular strategy for the construction of the training set, we made an assumption: the events q_{ij}, $i = 1, \ldots, k$, j a fixed value, occur all at the same time instant. The same assumption allows us to solve another implementation problem: the time information is not processed during the classification tree construction, (time is not a predictor variable), but the temporal dimension must be captured by the temporal rules. The solution we chose to encode the temporal information is to create a map between the index of the attributes (or predictor variables) and the order in time of the events. The $k(h+1)$ attributes are indexed as $\{A_0, A_1, \ldots, A_h, \ldots, A_{2h}, \ldots A_{k(h+1)-1}\}$. As we can see in Fig. 2, in each tuple the values of the attributes from the set $\{A_0, A_{h+1}, \ldots, A_{(k-1)(h+1)}\}$ represent events which occur at the same time moment as the class event, those of the set $\{A_1, A_{h+2}, \ldots, A_{(k-1)(h+1)+1}\}$ represent events which occur one time moment before the same class event, and so on. Let be $\{i_0, \ldots, i_m\}$ the set of indexes of the attributes that appear in the body of the rule (i.e. the rule has the form

$$\textit{If } (A_{i_0} = e_0) \textit{ and } (A_{i_1} = e_1) \textit{ and } \ldots \textit{ and } (A_{i_m} = e_m) \textit{ Then Class } e,$$

where e_{i_j} are events from the sequences $\{q_1, \ldots, q_k\}$ and e is an event from the sequence q_c. If t represents the time instant when the event in the head of the rule occurs, then an event from the rule's body, corresponding to the attribute A_{i_j}, occurred at time $t - \bar{i}_j$, where $\bar{i}_j$ means i *modulo* $(h+1)$.

Example 2. After the application of the first induction process, on the same stock prices database, we may obtain a temporal rule of the form: *If, during two consecutive days, the shape of the stock price had a* **peak** *form, with a mean less than* ***a*** *and a standard error greater than* ***b****, then, with a confidence of* ***c%****, three days later the stock price variation will present a* ***flat****.*

Second Inference Process Different classification trees, constructed from different training sets, generate finally different sets of temporal rules. The process that tries to infer temporal meta-rules from these sets of local rules is derived from the rules pruning strategy used by C4.5 system. Because this strategy may be theoretically applied not only to the rules generated by C4.5 algorithm, but to all rules having the form of a general Horn clause, a modelling process of our methodology, at an abstract level, looks not only feasible, but also necessary. To obtain an abstract view of temporal rules we propose a formalism based on first-order temporal logic. This formalism allows not only to model the main concepts used by the algorithms applied during the different steps of the methodology, but also to give a common frame to many of temporal rules extraction techniques, mentioned in the literature.

3 Formalism of Temporal Rules

Time is ubiquitous in information systems, but the mode of representation/perception varies in function of the purpose of the analysis [7, 12]. Firstly, there is a choice of a *temporal ontology*, which can be based either on *time points* (instants) or on *intervals* (periods). Secondly, time may have a *discrete* or a *continuous* structure. Finally, there is a choice of *linear* vs. *nonlinear* time (e.g. acyclic graph). For our methodology, we chose a temporal domain represented by linearly ordered discrete instants. This is imposed by the discrete representation of all databases.

Definition 1 *A single-dimensional linearly ordered temporal domain is a structure* $T_P = (T, <)$*, where* T *is a set of time instants and "<" a linear order on* T*.*

Databases being first-order structures, the first-order logic represents a natural formalism for their description. Consequently, the first-order temporal logic expresses the formalism of temporal databases. For the purposes of our methodology we consider a restricted first-order temporal language L which contains only constant symbols $\{c, d, ..\}$, n-ary ($n \geq 1$) function symbols $\{f, g, ..\}$, variable symbols $\{y_1, y_2, ...\}$, n-ary predicate symbols ($n \geq 1$, so no proposition symbols), the set of relational symbols $\{=, <, \leq, >, \geq\}$, the logical connective $\{\wedge\}$ and a temporal connective of the form X_k, $k \in \boldsymbol{Z}$, where k strictly positive means *next k times*, k strictly negative means *last k times* and $k = 0$ means *now*.

The syntax of L defines terms, atomic formulae and compound formulae. The *terms* of L are defined inductively by the following rules:

T1. Each constant is a term.
T2. Each variable is a term.
T3. If $t_1, t_2, \ldots, t_n$ are terms and f is an n-ary function symbol then $f(t_1, \ldots, t_n)$ is a term.

The atomic formulae (or atoms) of L are defined by the following rules:

A1. If $t_1, \ldots, t_n$ are terms and P is an n-ary predicate symbol then $P(t_1, \ldots, t_n)$ is an atom.
A2. If t_1, t_2 are terms and ρ is a relational symbol then $t_1 \rho t_2$ is an atom (also called relational atom).

Finally, the (compound) formulae of L are defined inductively as follow:

F1. Each atomic formula is a formula.
F2. If p, q are formulae then $(p \wedge q)$, $X_k p$ are formulae.

A Horn clause is a formula of the form $B_1 \wedge \cdots \wedge B_m \rightarrow B_{m+1}$ where each B_i is a positive (non-negated) atom. The atoms B_i, $i = 1, \ldots, m$ are called implication clauses, whereas B_{m+1} is known as the implicated clause. Syntactically, we cannot express Horn clauses in our language L because the logical connective $\rightarrow$ is not defined. However, to allow the description of rules, which formally look like a Horn clause, we introduce a new logical connective, $\mapsto$, which practically will represent a rewrite of the connective $\wedge$. Therefore, a formula in L of the form $p \mapsto q$ is syntactically equivalent to the formula $p \wedge q$. When and under what conditions we may use the new connective, is explained in the next definitions.

Definition 2 *An event (or temporal atom) is an atom formed by the predicate symbol E followed by a bracketed n-tuple of terms ($n \geq 1$) $E(t_1, t_2, \ldots, t_n)$. The first term of the tuple, t_1, is a constant symbol representing the name of the event and all others terms are expressed according to the rule T3 ($t_i = f(t_{i1}, \ldots, t_{ik_i})$). A short temporal atom (or the event's head) is the atom $E(t_1)$.*

For each constant symbol t used as an event name, two other constant symbols, *start_t* and *stop_t*, are included in our language L. Consequently, for each temporal atom $E(t_1, t_2, \ldots, t_n)$, two temporal atoms, $E(start\_t_1, t_2, \ldots, t_n)$ and $E(stop\_t_1, t_2, \ldots, t_n)$, are defined.

Definition 3 *A constraint formula for the event $E(t_1, t_2, \ldots t_n)$ is a conjunctive compound formula, $E(t_1, t_2, \ldots t_n) \wedge C_1 \wedge C_2 \wedge \cdots \wedge C_k$, where each C_j is a relational atom. The first term of C_j is one of the terms t_i, $i = 1 \ldots n$ and the second term is a constant symbol.*

For a short temporal atom $E(t_1)$, the only constraint formula that is permitted is $E(t_1) \wedge (t_1 = c)$. We denote such constraint formula as *short constraint formula.*

Definition 4 *A temporal rule is a formula of the form $H_1 \wedge \cdots \wedge H_m \mapsto H_{m+1}$, where H_{m+1} is a short constraint formula and H_i are constraint formulae, prefixed by the temporal connectives X_{-k}, $k \geq 0$. The maximum value of the index k is called the time window of the temporal rule.*

As a consequence of the Definition 4, a conjunction of constraint formulae $H_1 \wedge H_2 \wedge \cdots \wedge H_n$, each formula prefixed by temporal connectives X_{-k}, $k \geq 0$, may be rewritten as $H_{\sigma(1)} \wedge \cdots \wedge H_{\sigma(n-1)} \mapsto H_{\sigma(n)}$, — σ being a permutation of $\{1..n\}$ — only if there is a short constraint formula $H_{\sigma(n)}$ prefixed by X_0.

Remark. The reason for which we did not permit the expression of the implication connective in our language is related on the truth table for a formula $p \to q$: even if p is false, the formula is still true, which is unacceptable for a temporal rationing of the form *cause→ effect.*

If we change in Definition 2 the conditions imposed to the terms $t_i, i = 1...n$ to "each term t_i is a variable symbol", we obtain the definition of a temporal atom template. We denote a such template as $E(y_1, \ldots, y_n)$. Following the same rationing, a constraint formula template for $E(y_1, \ldots, y_n)$ is a conjunctive compound formula, $C_1 \wedge C_2 \wedge \cdots \wedge C_k$, where the first term of each relational atom C_j is one of the variables y_i, $i = 1 \ldots n$. Consequently, a short constraint formula template is the relational atom $y_1 = c$. Finally, by replacing in Definition 4 the notion "constraint formula" with "constraint formula template" we obtain the definition of a temporal rule template. Practically, the only formulae constructed in L are temporal atoms, constraint formulae, temporal rules and the corresponding templates.

The semantics of L is provided by an interpretation I over a domain D (in our formalism, D is always a linearly ordered domain). The interpretation assigns an appropriate meaning over D to the (non-logical) symbols of L. More precisely I assigns a meaning to the symbols of L as follows:

- for an n-ary predicate symbol P, $n \geq 1$, the meaning $I(P)$ is a function $D^n \to B$, where B is the set $\{$*true, false*$\}$,
- for an n-ary function symbol, f, $n \geq 1$, the meaning $I(f)$ is a function $D^n \to D$,
- for a constant symbol c the meaning $I(c)$ is an element of D,
- for a variable symbol y the meaning $I(y)$ is a element of D.

The interpretation I is extended to arbitrary terms as $I(f(t_1, \ldots, t_n)) = I(f)(I(t_1), \ldots, I(t_n))$. For atomic formulae and compound formulae, we have:

- If $t_1, \ldots, t_n$ are terms and P is an n-ary predicate symbol then $I(P(t_1, \ldots, t_n)) = true$ iff $I(P)(I(t_1), \ldots, I(t_n)) = true$.
- If t_1, t_2 are terms and ρ is a relational symbol then $I(t_1 \rho t_2) = true$ iff $I(t_1) \rho I(t_2)$.

- If p, q are formulae then $I(p \wedge q) = true$ iff $I(p) = true$ and $I(q) = true$.

Usually, the domain D is imposed during the discretisation phase, which is a pre-processing phase used in almost all knowledge extraction methodologies. Based on Definition 2, an event can be seen as a labelled (constant symbol t_1) sequence of points extracted from raw data and characterized by a finite set of features (terms $t_2, \cdots, t_n$). Let be D_e the set containing all the strings used as event names. We will extend this set by adding, for each $e \in D_e$, the strings *start_e* and *stop_e*. Finally, the domain D is the union $D_e \cup D_f$, where D_e is the extended set of strings and D_f represents the union of all sub-domains corresponding to chosen features.

To define a first-order linear temporal logic based on L, we need a structure having a temporal dimension and capable to capture the relationship between a time moment and the interpretation I at this moment.

Definition 5 *Given L and a domain D, a (first order) linear time structure is a triple* $M = (S, x, \boldsymbol{I})$, *where S is a set of states,* $x : \mathbf{N} \rightarrow S$ *is an infinite sequence of states* $(s_0, s_1, \ldots, s_n, \ldots)$ *and* $\boldsymbol{I}$ *is a function that associates with each state s an interpretation* $\boldsymbol{I}_s$ *of all symbols from L.*

In the framework of linear temporal logic, the set of symbols is divided into two classes, the class of global symbols and the class of local symbols. Intuitively, a global symbol w has the same interpretation in each state, i.e. $\boldsymbol{I}_s(w) = \boldsymbol{I}_{s'}(w) = \boldsymbol{I}(w)$, for all $s, s' \in S$; the interpretation of a local symbol may vary, depending on the state at which is evaluated. The formalism of temporal rules assumes that all function symbols (including constants) and all relational symbols are global, whereas the predicate symbols and variable symbols are local. Consequently, as the temporal atoms, constraint formulae, temporal rules and the corresponding templates are expressed using the predicate symbol E or the variable symbols y_i, the meaning of truth for these formulae depend on the state at which are evaluated. Given a first order time structure M and a formula p, we denote the instant i (or equivalently, the state s_i) for which $\boldsymbol{I}_{s_i}(p) = true$ by $i \models p$, i.e. at time instant i the formula p is true. Therefore, $i \models E(t_1, \ldots, t_n)$ means that at time i an event with the name $\boldsymbol{I}(t_1)$ and characterized by the global features $\boldsymbol{I}(t_2), \ldots, \boldsymbol{I}(t_n)$ occurs. Using this definition, we can also define:

- $i \models E(start\_t_1, \ldots, t_n)$ iff $i \models E(t_1, \ldots, t_n)$ and $(i-1) \not\models E(t_1, \ldots, t_n)$,
- $i \models E(stop\_t_1, \ldots, t_n)$ iff $i \models E(t_1, \ldots, t_n)$ and $(i+1) \not\models E(t_1, \ldots, t_n)$

Concerning the event template $E(y_1, \ldots, y_n)$, the interpretation of the variable symbols y_j at the state s_i, $\boldsymbol{I}_{s_i}(y_j)$, is chosen such that $i \models E(y_1, \ldots, y_n)$ for every time moment i. Because

- $i \models p \wedge q$ if and only if $i \models q$ and $i \models q$, and
- $i \models X_k p$ if and only if $i + k \models p$,

a constraint formula (template) is true at time i if and only if all relational atoms are true at time i and $i \models E(t_1, \ldots, t_n)$, whereas a temporal rule (template) is true at time i if and only if $i \models H_{m+1}$ and $i \models (H_1 \wedge \cdots \wedge H_m)$.

Now suppose that the following assumptions are true:

A. For each formula p in L, there is an algorithm that calculates the value of the interpretation $\boldsymbol{I}_s(p)$, for each state s, in a finite number of steps.
B. There are states (called incomplete states) that do not contain enough information to calculate the interpretation for all formulae defined at these states.
C. It is possible to establish a measure, (called *general interpretation*) about the degree of truth of a compound formula along the entire sequence of states $(s_0, s_1, \ldots, s_n, \ldots)$.

The first assumption express the calculability of the interpretation $\boldsymbol{I}$. The second assumption express the situation when only the body of a temporal rule can be evaluated at time moment i, but not the head of the rule. Therefore, for the state s_i, we cannot calculate the interpretation of the temporal rule and the only solution is to estimate it using a general interpretation. This solution is expressed by the third assumption. (*Remark:* The second assumption violates the condition about the existence of an interpretation in each state s_i, as defined in Definition 5. But it is well known that in data mining sometimes data are incomplete or are missing. Therefore, we must modify this condition as "***I*** *is a function that associates with* **almost** *each state s an interpretation* $\boldsymbol{I}_s$ *of all symbols from L*").

However, to ensure that this general interpretation is well defined, the linear time structure must present some property of consistency. Practically, this means that if we take any sufficiently large subset of time instants, the conclusions we may infer from this subset are sufficiently close from those inferred from the entire set of time instants. Therefore,

Definition 6 *Given L and a linear time structure M, we say that M is a consistent time structure for L if, for every atomic formula p, the limit* $supp(p) = \lim_{n \to \infty} \frac{\#A}{n}$ *exists, where* $A = \{i \in \{0, \ldots, n\} | i \models p\}$ *and # means "cardinality". The notation supp(p) denotes the support (of truth) of p.*

Now we define the general interpretation for an n-ary predicate symbol P as:

Definition 7 *Given L and a consistent linear time structure M for L, the general interpretation* I_G *for an n-ary predicate P is a function* $D^n \to \{true\} \times [0,1]$*, such that, for each n-tuple of terms* $\{t_1, \ldots, t_n\}$*,* $I_G(P(t_1, \ldots, t_n)) = (true, supp(P(t_1, \ldots, t_n))$.

The general interpretation is naturally extended to constraint formulae, temporal rules and the corresponding templates. There is another useful measure, called *confidence*, but available only for temporal rules (templates). This measure is calculated as a limit ratio between the number of certain applications (time instants where both the body and the head of the rule are true) and the number of potential applications (time instants where only the body of the rule is true). The reason for this choice is related to the presence of incomplete states, where the interpretation for the implicated clause cannot be calculated.

Definition 8 *The confidence of a temporal rule (template)* $H_1 \wedge \cdots \wedge H_m \mapsto H_{m+1}$ *is the limit* $\lim_{n \to \infty} \frac{\#A}{\#B}$, *where* $A = \{i \in \{0, \ldots, n\} | i \models H_1 \wedge \cdots \wedge H_m \wedge H_{m+1}\}$ *and* $B = \{i \in \{0, \ldots, n\} | i \models H_1 \wedge \cdots \wedge H_m\}$.

For different reasons, (the user has not access to the entire sequence of states, or the states he has access to are incomplete), the general interpretation cannot be calculated. A solution is to estimate I_G using a finite linear time structure, i.e. a model.

Definition 9 *Given L and a consistent time structure* $M = (S, x, \boldsymbol{I})$, *a model for M is a structure* $\tilde{M} = (\tilde{T}, \tilde{x})$ *where* $\tilde{T}$ *is a finite temporal domain* $\{i_1, \ldots, i_n\}$, $\tilde{x}$ *is the subsequence of states* $\{x_{i_1}, \ldots, x_{i_n}\}$ *(the restriction of x to the temporal domain* $\tilde{T}$*) and for each* $i_j, j = 1, \ldots, n$, *the state* x_{i_j} *is a complete state.*

Now we may define the estimator for a general interpretation:

Definition 10 *Given L and a model* $\tilde{M}$ *for M, an estimator of the general interpretation for an n-ary predicate P,* $\tilde{I}_G(P)$, *is a function* $D^n \to \{true\} \times [0, 1]$, *assigning to each atomic formula* $p = P(t_1, \ldots, t_n)$ *the value* true *with a support equal to the ratio* $\frac{\#A}{\#\tilde{T}}$, *where* $A = \{i \in \tilde{T} | i \models p\}$. *The notation* $supp(p, \tilde{M})$ *will denote the estimated support of p, given* $\tilde{M}$.

Once again, the estimation of the confidence for a temporal rule (template) is defined as:

Definition 11 *Given a model* $\tilde{M} = (\tilde{T}, \tilde{x})$ *for M, the estimation of the confidence for the temporal rule (template)* $H_1 \wedge \cdots \wedge H_m \mapsto H_{m+1}$ *is the ratio* $\frac{\#A}{\#B}$, *where* $A = \{i \in \tilde{T} | i \models H_1 \wedge \cdots \wedge H_m \wedge H_{m+1}\}$ *and* $B = \{i \in \tilde{T} | i \models H_1 \wedge \cdots \wedge H_m\}$.

4 Methodology Versus Formalism

As it was extensively presented in Sect. 2, the methodology for temporal rules extraction may be structured in two phases. During the first phase one transform sequential raw data into sequences of events. Practically, this means to establish the set of events, identified by names, and the set of features, common for all events.

In the frame of our formalism, during this phase we establish the set of temporal atoms which can be defined syntactically in L. For this we start by defining the first-order temporal language L. Considering as raw data the database described in Example 1, we include in L a 3-ary predicate symbol E, three variable symbols $y_i, i = 1..3$, two 12-ary function symbols f and g, two sets of constant symbols – $\{d_1, \ldots, d_6\}$ and $\{c_1, \ldots, c_n\}$ – and the usual set of relational symbols and logical (temporal) connectives. As we showed in the above example and according to the syntactic rules of L, an event is defined as $E(d_i, f(c_{j_1}, \ldots, c_{j_{12}}), g(c_{k_1}, \ldots, c_{k_{12}}))$, whereas an event template is defined as $E(y_1, y_2, y_3)$. Also provided during this phase is the semantics of L. Firstly, the domain $D = D_e \cup D_f$ (see Sect. 3) is defined. According to the results of the discretisation algorithm applied on raw data from the cited example, the domain D_e is defined as {*peak, start_peak, stop_peak, flat, start_flat, stop_flat, valley, start_valley, stop_valley*}. During the step *global features calculation*, two features – the mean and the standard error – were selected to capture the continuous aspect of the events. Consequently, the domain $D_f = \Re^+$, as the stock prices are positives real numbers and the features are statistical functions.

Secondly, a linear time structure M, i.e. a triple $(S, x, \boldsymbol{I})$ (see Def. 5) is specified. The database of events, obtained after the first phase of the methodology, contains tuples with three values, (v_1, v_2, v_3). For a tuple with a recording index i, the first value expresses the name of the event – *peak, flat, valley* – which occurs at time moment i and the two other values express the result of the two features. Therefore, we define a state s as a triple (v_1, v_2, v_3), the set S as the set of all tuples from database and the sequence x as the ordered sequence of tuples in database (see Table 1).

Table 1. The first nine states of the linear time structure M (example)

Index	State	Index	State	Index	State
1	$(peak, 10, 1.5)$	4	$(flat, 1, 0.5)$	7	$(valley, 15, 1.9)$
2	$(peak, 10, 1.5)$	5	$(flat, 1, 0.5)$	8	$(flat, 3, 1.1)$
3	$(peak, 14, 2.2)$	6	$(flat, 1, 0.5)$	9	$(peak, 12, 1.2)$

At this stage the interpretation of all symbols (global and local symbols) can be defined. For the global symbols (function symbols and relational symbols), the interpretation is quite intuitive. Therefore, the meaning $\boldsymbol{I}(d_j)$ is an element of D_e, the meaning $\boldsymbol{I}(c_j)$, $j = 1..n$, is a positive real number, whereas the meaning $\boldsymbol{I}(f)$, respectively $\boldsymbol{I}(g)$, is the function $f : D_f^{12} \rightarrow D_f$, $f(\boldsymbol{x}) = \overline{\boldsymbol{x}}$, respectively the function $g : D_f^{12} \rightarrow D_f$, $g(\boldsymbol{x}) = \text{se}(\boldsymbol{x})$ – we used the standard notations in statistics for the mean and standard error estimators.

The interpretation of a local symbol (the variable symbols y_i and the predicate symbol E) depends on the state at which is evaluated. According

Table 2. The temporal atoms evaluated *true* at the first nine states of M (example)

State	Temporal Atom
1	$E(peak, 10, 1.5)$, $E(start\_peak, 10, 1.5)$
2	$E(peak, 10, 1.5)$, $E(stop\_peak, 10, 1.5)$
3	$E(peak, 14, 2.2)$, $E(start\_peak, 14, 2.2)$, $E(stop\_peak, 14, 2.2)$
4	$E(flat, 1, 0.5)$, $E(start\_flat, 1, 0.5)$
5	$E(flat, 1, 0.5)$
6	$E(flat, 1, 0.5)$, $E(stop\_flat, 1, 0.5)$
7	$E(valley, 15, 1.9)$, $E(start\_valley, 15, 1.9)$, $E(stop\_valley, 15, 1.9)$
8	$E(flat, 3, 1.1)$, $E(start\_flat, 3, 1.1)$, $E(stop\_flat, 3, 1.1)$
9	$E(peak, 12, 1.2)$, $E(start\_peak, 12, 1.2)$, $E(stop\_peak, 12, 1.2)$

to the assumption A (see Sect. 3), the function $\boldsymbol{I}_{s_i}(E)$ defined on D^3 with values in $B = \{true, false\}$ is provided by a finite algorithm. This algorithm will receive at input at least the state s_i and will provide at output one of the values from B. Therefore, the interpretation of $E(t_1, t_2, t_3)$ evaluated at s_i is defined as:

Algorithm 1 ***Temporal atom evaluation***

Consider the state $s_i = (v_1, v_2, v_3)$
If $(\boldsymbol{I}_{s_i}(t_1) = v_1)$ *and* $(\boldsymbol{I}_{s_i}(t_2) = v_2)$ *and* $(\boldsymbol{I}_{s_i}(t_3) = v_3)$
  Then $\boldsymbol{I}_{s_i}(E(t_1, t_2, t_3)) = true$
  Else $\boldsymbol{I}_{s_i}(E(t_1, t_2, t_3)) = false$

Finally, the interpretation of the variable symbol y_j at the state s_i is given by $\boldsymbol{I}_{s_i}(y_j) = v_j, j = 1..3$, which satisfies the condition imposed to the interpretation of temporal atom template (see Sect. 3), which is $\boldsymbol{I}_{s_i}E(y_1, y_2, y_3) = true$ for each state s_i. Having well-defined the language L, the syntax and the semantics of L, as well as the linear time structure M, we can construct the temporal atoms evaluated as *true* at time moment i (see Table 2).

During the second phase of the methodology, we generate a set of temporal rules inferred from the database of events, obtained in phase one. The first induction process consists in creating classification trees, each based on a different training set. In the frame of our formalism, choosing a training set is equivalent to choose a model $\tilde{M}$ for the linear time structure M. All the states from these models are complete states, because the algorithm which construct the tree must know, for each time moment, the set of predictor events and the corresponding dependent event. Once the classification tree constructed, the test contained in each node becomes a relational atom and the set of all relational atoms situated on a path from root to a leaf become a constraint formula template. The variable symbols y_i included in the template are generated by the following rule:

- if the attribute concerned by the test is related to the event name, it is replaced by y_1
- if the attribute concerned by the test is related to the feature mean (respectively standard error), it is replaced by y_2 (respectively y_3).

The constraint formula template becomes a temporal rule template by adding temporal connectives according to the procedure which links a temporal dimension to a rule generated by C4.5 algorithm. Finally, the confidence of temporal rule template is calculated according to Definition 11.

Remark: The values of the categorical dependent variable, or the classes, may be obtained either in a supervised mode (e.g. given by an expert) or in an unsupervised mode (e.g. the names of the events). In the last case, we may restrict the possible values to the set $\{start\_event_1, stop\_event_1, \ldots, start\_event_n, stop\_event_n\}$).

To exemplify the induction process, from training set to temporal rule templates, consider the following model $\tilde{M} = \{s_1, \ldots, s_{100}\}$. According to the procedure for training set selection (see Sect. 2), in a first step we must indicate the sequences q_i, $i = 1..k$, of predictor variables and the sequence q_c of class values. The information on sequences q_i is extracted from the structure of the states from the model $\tilde{M}$: given the state $s_i = (v_1, v_2, v_3)$, $1 \leq i \leq 100$, we define $q_{ji} = v_j$, $j = 1..3$. Therefore, q_1 is the sequence of the event names, q_2 is the sequence of the mean values and q_3 is the sequence of the standard error values. As there is no predefined classification (unsupervised mode), the sequence q_c is defined as $q_{ci} = q_{1i}, i = 1..100$. The next step consists in defining the parameters t_0, t_p and h, which are set as $t_0 = 100$, $t_p = 96$ and $h = 3$ (the methodology for finding the optimal value of the parameter h is based on the analysis of classification errors and is described in [10]). Concerning the tuples from the training set, there is a minor difference compared with the procedure from Sect. 2 – the sequence of class values (q_c) being the same

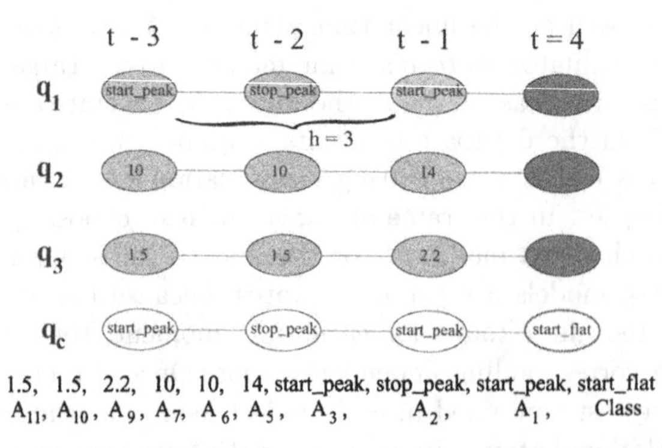

Fig. 3. Graphical representation of the last tuple of the training set based on states from Table 1 and defined by the parameters $t_0 = 100$, $t_p = 96$ and $h = 3$ (including the list of corresponding attributes)

as one of the sequence of predictor variables (q_1), we can not include in a tuple which contains the class value q_{ct} the predictor values q_{1t}, q_{2t} and q_{3t}. In other words, we can not accept that the same event occurred at time t to be simultaneously on the left and on the right side of a temporal rule. As we can see in Fig. 3, a tuple contains now $k \cdot h = 9$ predictor values instead of $k(h+1) = 12$, and there is no attribute having an index i such that i *modulo* $(h+1)$ to be 0. Suppose now that one of the rule generated by C4.5 algorithm using the previous defined training set has the form

If (A_3=*start_peak*) *and* ($A_7 < 11$) *and* (A_1=*start_peak*) *Then Class start_valley*

By convention, the event in the head of the rule occurs always at time moment $t = 0$, so an event from the body of the rule, corresponding to the attribute A_i, occurs at time moment $-(i \text{ modulo } 4)$. By applying this observation and the convention on how to use symbol variables in a temporal rule, we obtain the following temporal rule template

$$X_{-3}(y_1 = start\_peak) \wedge X_{-3}(y_2 < 11) \\ \wedge X_{-1}(y_1 = start\_peak) \mapsto X_0(y_1 = start\_valley)$$

The induction process is repeatedly applied on different models $\tilde{M}_i$ of the time structure M, which will generate in the end different sets of temporal rule templates (see Table 3). It is possible to obtain the same template in two different sets, but with different confidence, or templates with the same head, but with different bodies.

The second inference process is designed to obtain temporal meta-rules, which are temporal rule templates in accordance with Definition 4, but supposed to have a small variability of the estimated confidence among different models. Therefore, a temporal meta-rule may be applied with the same confidence in any state, complete or incomplete. To obtain such temporal rules, we

Table 3. Different temporal rule templates extracted from two models $\tilde{M}$ using the induction process (example)

Model	Temporal Rule Templates
$s_1 \ldots s_{100}$	$X_{-3}(y_1 = start\_peak) \wedge X_{-3}(y_2 < 11) \wedge$ $\wedge X_{-1}(y_1 = start\_peak) \mapsto X_0(y_1 = start\_valley)$ $X_{-2}(y_1 = start\_peak) \wedge X_{-2}(y_3 < 1.1) \wedge$ $\wedge X_{-1}(y_1 = stop\_flat) \mapsto X_0(y_1 = start\_valley)$ ....
$s_{300} \ldots s_{399}$	$X_{-2}(y_1 = peak) \wedge X_{-2}(y_3 < 1.1) \wedge$ $\wedge X_{-2}(y_2 \geq 12.3) \mapsto X_0(y_1 = start\_valley)$ $X_{-4}(y_1 = stop\_peak) \wedge X_{-3}(y_1 = start\_flat) \wedge X_{-3}(y_2 >= 3.2) \wedge$ $\wedge X_{-3}(y_3 < 0.4) \wedge X_{-1}(y_1 = stop\_flat) \mapsto X_0(y_1 = start\_peak)$ ....

apply strategies which cut irrelevant relational atoms, according with some criterions, from the implication clauses of temporal rules templates obtained during the first induction process. The process of inferring temporal meta-rules is related to a new approach in data mining, called *higher order mining*, i.e. mining from the results of previous mining runs. According to this approach, the rules generated by the first induction process are first order rules and those generated by the second inference process (i.e. temporal meta-rules) are higher order rules. The formalism we proposed does not impose what methodology to use to discover first order temporal rules. As long as these rules satisfy the syntactic form described in Definition 4, the strategy (including algorithms, criterions, statistical methods) developed to infer temporal meta-rules might be applied.

5 Temporal Meta-Rules

Suppose that for a given model $\tilde{M}$ we dispose of a set of temporal rules templates, extracted from the corresponding classification tree. It is very likely that some temporal rules templates contain implication clauses that are irrelevant, i.e. after their deletion, the general interpretation of the templates remain unchanged (*Remark*: in the following, by the notion "implication clause" we consider a relational atom prefixed by the temporal connective X_{-k}). In the frame of a consistent time structure M, it is obviously that we cannot delete an implication clause from a temporal rule template (denoted TR) if the resulting template (noted TR^-) has a lower confidence. But for a given model $\tilde{M}$, we calculate an estimate (denoted $co(TR, \tilde{M})$) of the true confidence (denoted $co(TR)$). Because this estimator has a binomial distribution, we may establish a confidence interval for $co(TR)$ and, consequently, we accept to delete an implication clause from TR if and only if the lower confidence limit of $co(TR^-, \tilde{M})$ is greater than the lower confidence limit of $co(TR, \tilde{M})$.

The estimator $co(TR, \tilde{M})$ being the ratio $\#A/\#B$, (see Def. 11), a confidence interval for this value is constructed using a normal distribution depending on $\#A$ and $\#B$ (more precisely, the normal distribution has mean $\pi = \#A/\#B$ and variance $\sigma^2 = \pi(1-\pi)/\#B$). The lower limit of the interval is $L_\alpha(A, B) = \pi - z_\alpha \sigma$, where z_α is a quantile of the normal distribution for a given confidence level α. The algorithm which generalize a single temporal rule template TR, by deleting a single implication clause, is presented in the following:

Algorithm 2 ***Generalization 1-delete***

Step 1 Let $TR = H_1 \wedge \cdots \wedge H_m \mapsto H_{m+1}$ *be a temporal rule template. Let* $\aleph = \cup\{C_j\}$, *where* C_j *are all the implication clauses that appear in the body of the template. Rewrite TR, by an abuse of notation, as* $\aleph \mapsto H_{m+1}$. *If* $n = \#\aleph$, *denote by* $C_1, \ldots, C_n$ *the list of all implication clauses from* $\aleph$.

Step 2 For each $i = 1, \ldots, n$ *do*

$$\aleph^- = \aleph - C_i, \quad TR_i^- = \aleph^- \mapsto H_{m+1}$$
$$A = \{i \in \tilde{T} | i \models \aleph \wedge H_{m+1}\}, B = \{i \in \tilde{T} | i \models \aleph\}$$
$$A^- = \{i \in \tilde{T} | i \models \aleph^- \wedge H_{m+1}\}, B^- = \{i \in \tilde{T} | i \models \aleph^-\}$$
$$co(TR, \tilde{M}) = \#A / \#B, \; co(TR_i^-, \tilde{M}) = \#A^- / \#B^-$$
If $L_\alpha(A, B) \leq L_\alpha(A^-, B^-)$ *then store* TR_i^-

Step 3 Keep only the generalized temporal rule template TR_i^- *for which* $L_\alpha(A^-, B^-)$ *is maximal.*

The core of the algorithm is the **Step 2**, where the sets used to estimate the confidence of the initial template, TR, and of the generalized template, TR^-, i.e. A, B, A^- and B^-, are calculated. The complexity of this algorithm is linear in n. Using the criterion of lower confidence limit, (or LCL), we define the temporal meta-rule inferred from TR as the temporal rule template with a maximum set of implication clauses deleted from $\aleph$ and having the maximum lower confidence limit greater than $L_\alpha(A, B)$. An algorithm designed to find the largest subset of implication clauses that can be deleted will have an exponential complexity. A possible solution is to use the Algorithm 2 in successive steps until no more deletion is possible, but without having the guarantee that we will get the global maximum.

As example, consider the first temporal rule template from Table 3 and suppose that $\#A = 20$ and $\#B = 40$ (giving an estimate $co(TR, \tilde{M}) = 0.5$, with $L_{0.95}(0.5) = 0.345$). Looking at Table 4, we find two implication clauses which could be deleted (the first and the second) with a maximum $L_\alpha(A^-, B^-)$ given by the second clause. As a remark, by deleting the first implication clause, the resulting template has an estimate of the confidence (0.489) less than of the original template (0.5), but a lower confidence limit 0.349 greater than $L_{0.95}(0.5)$, which allows the deletion operation.
If we apply again the Algorithm 2 on the template

$$X_{-3}(y_1 = start\_peak) \wedge X_{-1}(y_1 = start\_peak) \mapsto X_0(y_1 = start\_valley)$$

(denoted TR^-), we find that no other implication clause can be deleted, i.e. TR^- is the temporal meta rule according to the criterion LCL inferred from

Table 4. Parameters calculated in Step 2 of the Algorithm 2 by deleting one implication clause from the template $X_{-3}(y_1 = start\_peak) \wedge X_{-3}(y_2 < 11) \wedge X_{-1}(y_1 = start\_peak) \mapsto X_0(y_1 = start\_valley)$

Deleted Implication Clause	$\#A^-$	$\#B^-$	$co(TR_i^-, \tilde{M})$	$L_\alpha(A^-, B^-)$
$X_{-3}(y_1 = start\_peak)$	24	49	0.489	0.349
$X_{-3}(y_2 < 11)$	30	50	0.60	0.464
$X_{-1}(y_1 = start\_peak)$	22	48	0.458	0.317

the temporal rule template

$$X_{-3}(y_1 = start\_peak) \wedge X_{-3}(y_2 < 11) \wedge$$
$$X_{-1}(y_1 = start\_peak) \mapsto X_0(y_1 = start\_valley) .$$

Suppose now that we dispose of two models, $\tilde{M}_1 = (\tilde{T}_1, \tilde{x}_1)$ and $\tilde{M}_2 = (\tilde{T}_2, \tilde{x}_2)$, and for each model we have a set of temporal rule templates with the same implicated clause H (sets denoted S_1, respectively S_2). Let S be a subset of the union $S_1 \cup S_2$. If $TR_j \in S$, $j = 1, \ldots, n$, $TR_j = H_1 \wedge \cdots \wedge H_{m_j} \mapsto H$, then consider the sets

$$A_j = \{i \in \tilde{T}_1 \cup \tilde{T}_2 | i \models H_1 \wedge \ldots \wedge H_{m_j} \wedge H\}, \mathbf{A} = \cup A_j ,$$
$$B_j = \{i \in \tilde{T}_1 \cup \tilde{T}_2 | i \models H_1 \wedge \ldots \wedge H_{m_j}\}, \mathbf{B} = \cup B_j ,$$
$$\mathbf{C} = \{i \in \tilde{T}_1 \cup \tilde{T}_2 | i \models H\} .$$

The performance of the subset S can be summarized by the number of *false positives* (time instants where the implication clauses of each template from S are true, but not the clause H) and the number of *false negatives* (time instants where the clause H is true, but none of the implication clauses of the templates from S). Practically, the number of *false positives* is $fp = \#(\mathbf{B} - \mathbf{A})$ and the number of *false negatives* is $fn = \#(\mathbf{C} - \mathbf{B})$. The worth of the subset S of temporal rule templates is assessed using the Minimum Description Length Principle (MDLP) [33, 34]. This provides a basis for offsetting the accuracy of a theory (here, a subset of templates) against its complexity. The principle is simple: a Sender and a Receiver have both the same models $\tilde{M}_1$ and $\tilde{M}_2$, but the states from the models of the Receiver are incomplete states (the interpretation of the implicated clause cannot be calculated). The sender must communicate the missing information to the Receiver by transmitting a theory together with the exceptions to this theory. He may choose either a simple theory with a great number of exceptions or a complex theory with fewer exceptions. The MLD Principle states that the best theory will minimize the number of bits required to encode the total message consisting of the theory together with its associated exceptions. This is a particular instantiation of the MLDP, called two-part code version, which states that, among the set of candidate hypotheses $\mathcal{H}$, the best hypothesis to explain a set of data is one which minimizes the sum of the length, in bits, of the description of the hypothesis, and the length, in bits, of the description of the data encoded with the help of the hypothesis (which usually amounts to specifying the errors the hypothesis makes on the data). In the case where there are different hypotheses for which the sum attains its minimum, we select those with a minimum description length.

To encode a temporal rule template from S, we must specify its implication clauses (the implicated clause being the same for all rules, there is no need to encoded it). Because the order of the implication clauses is not important, the number of required bits is reduced by $\kappa \log_2(m!)$, where m is the number of implication clauses and κ is a constant depending on encoding procedure. The number of bits required to encode the set S is the sum of encoding length for each template from S reduced by $\kappa \log_2(n!)$ (the order of the n templates from S is not important). The exceptions are encoded by indicating the sets *false positive* and *false negative.* If $b = \#\mathbf{B}$ and $N = \#(\tilde{T}_1 + \tilde{T}_2)$ then the number of bits required is $\kappa \log_2 \left(\binom{b}{fp} \right) + \kappa \log_2 \left(\binom{N-b}{fn} \right)$, because we have $\binom{b}{fp}$ possibilities to choose the *false positives* among the cases covered by the rules and $\binom{N-b}{fn}$ possibilities to indicate the *false negatives* among the uncovered cases. The total number of bits required to encode the message is then equal to *theory bits + exceptions bits.*

Using the criterion of MDLP, we define as temporal meta-rules inferred from a set of temporal rule templates (implying the same clause and extracted from at least two different models), the subset S that minimizes the total encoding length. An algorithm designed to extensively search this subset S has an exponential complexity, but in practice (and especially when $\#(S_1 + S_2) > 10$) we may use different non-optimal strategies (hill-climbing, genetic algorithms, simulated annealing), having a polynomial complexity.

For a practical implementation of an encoding procedure in the frame of the formalism, one use a notion from the theory of probability, i.e. the entropy. Given a finite set S, the entropy of S is defined as $\mathcal{I}(S) = -\sum_{v \in S} freq(v) \cdot \log_2(freq(v))$, where $freq(v)$ means the frequency of the element v in S. This measure attains its maximum when all frequencies are equal. Consider now a model $\tilde{M}$, characterized by the states $s_1, \ldots, s_n$, where each state s_i is defined by a m-tuple $(v_{i_1}, \ldots, v_{i_m})$. Based on these states consider the sets $A_j, j = 1..m$, where $A_j = \bigcup_{i=1..n}\{v_{i_j}\}$. Let be a temporal rule template inducted from the model $\tilde{M}$ and let be $X_{-k}(y_j \ \rho \ c)$ an implication clause from this template, with $j \in \{1 \ldots m\}$ and ρ a relational symbol. We define the encoding length for $X_{-k}(y_j \ \rho \ c)$ to be $\mathcal{I}(A_j)$. The encoding length of a temporal rule template having k implication clauses is thence equal with $\log_2(k)$ plus the sum of encoding length for each clause, reduced by $\log_2(k!)$ (order is not important), but augmented with $\log_2(m \cdot h_{max})$, where h_{max} is the time window of the template. The last quantity expresses the encoding length for the temporal dimension of the rule. Finally, the encoding length of q temporal rule templates is $\log_2(q)$ plus the sum of encoding length for each template, reduced by $\log_2(q!)$ (order is not important), whereas the encoding length of the exceptions is given by $\log_2 \left(\binom{b}{fp} \right) + \log_2 \left(\binom{N-b}{fn} \right)$.

As example, consider the set of temporal rule templates from Table 3 having as implicated clause $X_0(y_1 = start\_valley)$. To facilitate the notation, we denote with TR_1, TR_2 and TR_3 the three concerned templates, written in this order in the mentioned Table. Therefore $S_1 = \{TR_1, TR_2\}$, $S_2 = \{TR_3\}$ and states used to calculate the entropy of the sets $A_j, j = 1..3$ are $\{s_1, \ldots, s_{100}, s_{300}, \ldots, s_{399}\}$. The encoding length for each subset $S \subseteq S_1 \cup S_2$

is calculated in the last column of Table 5, value which is the sum of the templates encoding length (second column) and the exceptions encoding length (third column). As an observation, even if the set $\{TR_1, TR_2\}$ has more templates than the set $\{TR_3\}$, the encoding length for the two templates (14.34) is less than the encoding length of the last template (17.94). The conclusion to be drawn by looking at the last column of Table 5 is that the temporal meta rules, according to the MDLP criterion and inferred from the set $\{TR_1, TR_2, TR_3\}$ (based on the states $\{s_1, \ldots, s_{100}\}$, $\{s_{300}, \ldots, s_{399}\}$) is the subset $S = \{TR_1, TR_2\}$.

Table 5. The encoding length of different subsets of temporal rule templates having as implicated clause $X_0(y_1 = start\_valley)$, based on states $\{s_1, \ldots, s_{100}\}$ and $\{s_{300}, \ldots, s_{399}\}$

Subset S	Templates Length	Exceptions Length	Total Length
$\{TR_1\}$	8.88	70.36	79.24
$\{TR_2\}$	7.48	66.64	74.12
$\{TR_3\}$	17.94	67.43	85.37
$\{TR_1, TR_2\}$	14.34	46.15	60.49
$\{TR_1, TR_3\}$	24.82	41.2	66.02
$\{TR_2, TR_3\}$	23.42	38.00	61.42
$\{TR_1, TR_2, TR_3\}$	31.72	30.43	62.15

Because the two definitions of temporal meta-rules differ not only in criterion (LCL, respectively MLDP), but also in the number of initial models (one, respectively at least two), the second inference process is applied in two steps. During the first step, temporal meta-rules are inferred from each set of temporal rule templates based on a single model. During the second step, temporal meta-rules are inferred from each set of temporal rules created during the step one and having the same implicated clause (see Fig. 4).

There is another reason to apply firstly the LCL criterion: the resulting temporal meta-rules are less redundant concerning the set of implication clauses and so the encoding procedures, used by MLDP criterion, don't need an adjustment against this effect.

6 Conclusions

In this article we constructed a theoretical framework for a methodology introduced in [9], which has as finality the discovery of knowledge, represented in the form of general Horn clauses, inferred from databases with a temporal dimension. To obtain what we called *temporal rules*, a discretisation phase that extracts *events* from raw data is applied first, followed by an inductive

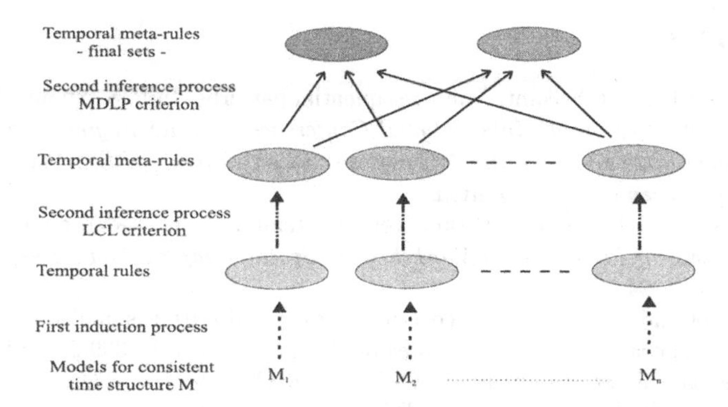

Fig. 4. Graphical representation of the second inference process

phase, which constructs classification trees from these events. The discrete and continuous characteristics of an *event*, according to its definition, allow us to use statistical tools as well as techniques from artificial intelligence on the same data.

The theoretical framework we proposed, based on first-order temporal logic, permits to define the main notions (event, temporal rule, constraint) in a formal way. The concept of consistent linear time structure allows us to introduce the notions of *general interpretation*, of *support* and of *confidence*, the lasts two measure being the expression of the two similar concepts used in data mining.

Also included in the proposed framework, the process of inferring temporal meta-rules is related to a new approach in data mining, called *higher order mining*, i.e. mining from the results of previous mining runs. According to this approach, the rules generated by the first induction process are first order rules and those generated by the second inference process (i.e temporal meta-rules) are higher order rules. Our formalism do not impose which methodology must be used to discover first order rules. As long as these rules may be expressed according to Definition 4, the strategy (here including algorithms, criterions, statistical methods), developed to infer temporal meta-rules may be applied.

It is important to mention that the condition of the existence of the limit, in the definition of consistent linear time structure, is a fundamental one: it express the fact that the structure M represents a homogenous model and therefore the conclusions (or inferences) based on a finite model $\tilde{M}$ for M are consistent. However, at this moment, we do not know methods which may certified that a given model is consistent. In our opinion, the only feasible approach to this problem is the development of methods and procedure for detecting the change points in the model and, in this direction, the analysis of the evolution of temporal meta-rules seems a very promising starting point.

References

1. R. Agrawal and R. Srikant. Mining sequential patterns. In P. S. Yu and A. S. P. Chen, editors, *Eleventh International Conference on Data Engineering*, pp. 3–14, Taipei, Taiwan, 1995. IEEE Computer Society Press. URL `citeseer.ist.psu.edu/agrawal95mining.html`.
2. S. Al-Naemi. A theoretical framework for temporal knowledge discovery. In *Proceedings of International Workshop on Spatio-Temporal Databases*, pp. 23–33, Spain, 1994.
3. G. Berger and A. Tuzhilin. Discovering Unexpected Patterns in Temporal Data using Temporal Logic. *Lecture Notes in Computer Science*, 1399:281–309, 1998.
4. L. Breiman, J. Friedman, R. A. Olshen, and C. J. Stone. *Classification and Regression Trees.* Wadsworth & Brooks/ Cole Advanced Books & Software, 1984.
5. X. Chen and I. Petrounias. A Framework for Temporal Data Mining. *Lecture Notes in Computer Science*, 1460:796–805, 1998.
6. X. Chen and I. Petrounias. Discovering Temporal Association Rules: Algorithms, Language and System. In *Proceedings of the 6th International Conference on Data Engineering*, p. 306, San Diego, USA, 2000.
7. J. Chomicki and D. Toman. Temporal Logic in Information Systems. *BRICS Lecture Series*, LS-97-1:1–42, 1997.
8. P. Cohen. Fluent Learning: Elucidating the Structure of Episodes. In *Advances in Intelligent Data Analysis*, pp. 268–277. Springer Verlang, 2001.
9. P. Cotofrei and K. Stoffel. Classification Rules + Time = Temporal Rules. In *Lecture Notes in Computer Science, vol 2329*, pp. 572–581. Springer Verlang, 2002a.
10. P. Cotofrei and K. Stoffel. Rule Extraction from Time Series Databases using Classification Trees. In *Proceedings of IASTED International Conference*, pp. 572–581, Insbruck, Austria, 2002b.
11. G. Das, K. Lin, H. Mannila, G. Renganathan, and P. Smyth. Rule Discovery from Time Series. In *Proceedings of the 4th Conference on Knowledge Discovery and Data Mining*, pp. 16–22, 1998.
12. E. A. Emerson. Temporal and Modal Logic. *Handbook of Theoretical Computer Science*, pp. 995–1072, 1990.
13. N. Friedman, K. Murphy, and S. Russel. Learning the structure of dynamic probabilistic networks. In *Proceedings of the 14th Conference on Uncertainty in Artificial Intelligence*, pp. 139–147. AAAI Press, 1998.
14. G. Guimares. Temporal knowledge discovery for multivariate time series with enhanced self-organizing maps. In *Proceedings of the IEEE-INNS-ENNS Int. Joint Conference on Neural Networks*, pp. 165–170. IEEE Computer Society, 2000.
15. J. Han, G. Dong, and Y. Yin. Efficient Mining of Partial Periodic Patterns in Time Series Database. In *Proceedings of International Conference on Data Engineering*, pp. 106–115, Sydeny, Australia, 1999.
16. J. Han, W. Gong, and Y. Yin. Mining Segment-Wise Periodic Patterns in Time-Related Databases. In *Proceedings of the 4th Conference on Knowledge Discovery and Data Mining*, pp. 214–218, 1998.
17. F. Hoppner. Learning Temporal Rules from State Sequences. In *IJCAI Workshop on Learning from Temporal and Spatial Data*, pp. 25–31, Seattle, USA, 2001.

18. F. Hoppner. Discovery of core episodes from sequences. In *Pattern Detection and Discovery*, pp. 199–213, 2002.
19. P. Kam and A. W. Fu. Discovering Temporal Patterns for Interval-based Events. *Lecture Notes in Computer Science*, 1874:317–326, 2000.
20. K. Karimi and H. Hamilton. Finding Temporal Relations: Causal Bayesian Networks vs. C4.5. In *Proceedings of the 12th International Symposium on Methodologies for Intelligent Systems*, Charlotte, USA, 2000.
21. E. Keogh, S. Lonardi, and B. Chiu. Finding Surprising Patterns in a Time Series Database in Linear Time and Space. In *Proceedings of 8th ACM SIKDD International Conference on Knowledge Discovery and Data Mining*, pp. 550–556, Edmonton, Canada, 2002a.
22. E. Keogh and M. J. Pazzani. An Enhanced Representation of Time Series which Allows Fast and Accurate Classification, Clustering and Relevance Feedback. In *Proceedings of the 4th Conference on Knowledge Discovery and Data Mining*, pp. 239–243, 1998.
23. E. J. Keogh, S. Chu, D. Hart, and M. J. Pazzani. Iterative Deepening Dynamic Time Warping for Time Series. In *Proceedings of Second SIAM International Conference on Data Mining*, 2002b.
24. E. J. Keogh and M. J. Pazzani. Scalling up Dynamic Type Warping to Massive Datasets. In *Proceedings of the 3rd European Conference PKDD*, pp. 1–11, 1999.
25. W. Lin, M. A. Orgun, and G. J. Williams. Temporal Data Mining using Hidden Markov-Local Polynomial Models. In *Proceedings of the 5th International Conference PAKDD, Lecture Notes in Computer Science*, volume 2035, pp. 324–335, 2001.
26. H. Loether and D. McTavish. *Descriptive and Inferential Statistics: An introduction.* Allyn and Bacon, 1993.
27. W. Loh and Y. Shih. Split Selection Methods for Classification Trees. *Statistica Sinica*, 7:815–840, 1997.
28. W. Loh and N. Vanichsetakul. Tree-structured classification via generalized discriminant analysis. *Journal of the American Statistical Association*, 83(403): 715–725, September 1988.
29. D. Malerba, F. Esposito, and F. Lisi. A logical framework for frequent pattern discovery in spatial data. In *Proceedings of 5th Conference Knowledge Discovery in Data*, 2001.
30. S. Mangaranis. *Supervised Classification with Temporal Data.* PhD thesis, Computer Science Department, School of Engineering, Vanderbilt University, 1997.
31. B. Ozden, S. Ramaswamy, and A. Silberschatz. Cyclic Association Rules. In *Proceedings of International Conference on Data Engineering*, pp. 412–421, Orlando, USA, 1998.
32. J. R. Quinlan. *C4.5: Programa for Machine Learning.* Morgan Kauffmann, San Mateo, California, 1993.
33. J. R. Quinlan and R. L. Rivest. Inferring decision trees using Minimum Description Length Principle. *Information and Computation*, 3:227–248, 1989.
34. J. Rissanen. Modelling by Shortest Data Description. *Automatica*, 14:465–471, 1978.
35. J. Rodriguez, C. Alonso, and H. Boström. Learning first order logic time series classifiers: Rules and boosting. In *Proceedings of 4th European Conference on Principles of Data Mining and Knowledge Discovery*, pp. 299–308, 2000.

36. M. Spiliopoulou and J. Roddick. Higher order mining: modelling and mining the results of knowledge discovery. In *Proceedings of the 2nd International Conference on Data Mining, Methods and Databases*, pp. 309–320, UK, 2000.
37. S. Tsumoto. Rule Discovery in Large Time-Series Medical Databases. In *Proceedings of the 3rd Conference PKDD*, pp. 23–31. Lecture Notes in Computer Science, 1074, 1999.

An Alternative Approach to Mining Association Rules

Jan Rauch and Milan Šimůnek

Faculty of Informatics and Statistics, University of Economics, Prague, nám. W. Churchilla 4, 130 67 Praha 3, Czech Republic
rauch@vse.cz, simunek@vse.cz

Summary. An alternative approach to mining association rules is presented. It is based on representation of analysed data by suitable strings of bits. This approach was developed for the GUHA method of mechanising hypothesis formation more than 30 years ago. The procedure 4ft-Miner that is contemporary application of this approach is described. It mines for various types of association rules including conditional association rules. The 4ft-Miner procedure is a part of the academic system LISp-Miner for KDD research and teaching.

Key words: Data mining, association rules, procedure 4ft-Miner, system LISp-Miner, bit string approach to data mining

1 Introduction

An association rule is commonly understood to be an expression of the form $X \rightarrow Y$, where X and Y are sets of items. The association rule $X \rightarrow Y$ means that transactions containing items of set X tend to contain items of set Y. There are two measures of intensity of an association rule – *confidence* and *support*.

An association rule discovery task is a task to find all the association rules of the form $X \rightarrow Y$ such that the support and confidence of $X \rightarrow Y$ are above the user–defined thresholds *minconf* and *minsup*. The conventional a-priori algorithm of association rules discovery proceeds in two steps. All frequent itemsets are found in the first step. A frequent itemset is a set of items that is included in at least minsup transactions. Association rules with a confidence of at least minconf are generated in the second step [1].

Particular items can be represented by Boolean attributes and a Boolean data matrix can represent the whole set of transactions. The algorithm can be modified to deal with attributes with more than two values. Thus, the association rules for example of the form $A(a_1) \wedge B(b_3) \rightarrow C(c_7)$ can be mined. We assume that the attribute A has k particular values $a_1, \ldots, a_k$.

J. Rauch and M. Šimůnek: *An Alternative Approach to Mining Association Rules*, Studies in Computational Intelligence (SCI) **6**, 211–231 (2005)
www.springerlink.com

The expression $A(a_1)$ denotes the Boolean attribute that is true if the value of attribute A is a_1 etc.

The goal of this chapter is to draw attention to an alternative approach to mining association rules. This approach is based on representing each possible value of an attribute by a single string of bits [10, 11]. In this way it is possible to mine for association rules of the form, for example, $A(\alpha) \wedge B(\beta) \rightarrow C(\gamma)$ where α is not a single value but a subset of all the possible values of the attribute A. The expression $A(\alpha)$ denotes the Boolean attribute that is true for a particular row of data matrix if the value of A in this row belongs to α, and the same is true for $B(\beta)$ and $C(\gamma)$.

The bit string approach means that it is easy to compute all the necessary frequencies. Then we can mine not only for association rules based on confidence and support but also for rules corresponding to various additional relations of Boolean attributes including relations described by statistical hypotheses tests. Association rules of this type are introduced in Sect. 2. The bit string approach also makes it easy to mine for conditional association rules that are mentioned also in Sect. 2.

The bit string approach was used several times in the implementation of the GUHA method of mechanising hypotheses formation [3, 4, 5, 7]. We present the GUHA procedure 4ft-Miner that is a contemporary implementation of this approach (see Sect. 2). The 4ft-Miner procedure has very fine tools to tune the set of potentially interesting association rules to be automatically generated and verified and also tools to filter and to sort the set of found association rules. Some of these tools were invented on the basis of more than 30 years experience with the GUHA method. There are also some new tools never implemented in former implementations of the GUHA method. These tools come from 8 years experience with the procedure 4ft-Miner.

The principles of the bit string approach are described in Sect. 4. The final algorithm is very fast and it is approximately linearly dependent on the number of rows of the analysed data matrix. Time and memory complexity are discussed in Sect. 5. The 4ft-Miner procedure is a part of the academic system *LISp-Miner* for KDD research and teaching. Main features of the LISp-Miner system are mentioned in Sect. 6.

2 Association Rules

The association rule here is understood to be an expression $\varphi \approx \psi$ where φ and ψ are Boolean attributes. The association rule $\varphi \approx \psi$ means that the Boolean attributes φ and ψ are associated in the way given by the symbol $\approx$.

The symbol $\approx$ is called the *4ft-quantifier*. It corresponds to a condition concerning a four-fold contingency table of φ and ψ. Various types of dependencies of φ and ψ can be expressed by 4ft-quantifiers.

The association rule $\varphi \approx \psi$ concerns the analysed data matrix $\mathcal{M}$. The rule $\varphi \approx \psi$ is *true in data matrix* $\mathcal{M}$ if the condition corresponding to the

4ft-quantifier is satisfied in the four-fold contingency table of φ and ψ in $\mathcal{M}$, otherwise $\varphi \approx \psi$ is *false in data matrix* $\mathcal{M}$.

We define the value $Val(\varphi \approx \psi, \mathcal{M})$ of $\varphi \approx \psi$ in $\mathcal{M}$ as $Val(\varphi \approx \psi, \mathcal{M}) = 1$ if $\varphi \approx \psi$ is true in $\mathcal{M}$, and as $Val(\varphi \approx \psi, \mathcal{M}) = 0$ if $\varphi \approx \psi$ is false in $\mathcal{M}$.

The four-fold contingency table of φ and ψ in data matrix $\mathcal{M}$ is a quadruple $\langle a, b, c, d \rangle$ of natural numbers such that

- a is the number of rows of $\mathcal{M}$ satisfying both φ and ψ
- b is the number of rows of $\mathcal{M}$ satisfying φ and not satisfying ψ
- c is the number of rows of $\mathcal{M}$ not satisfying φ and satisfying ψ
- d is the number of rows of $\mathcal{M}$ satisfying neither φ nor ψ.

The four-fold contingency table (the *4ft table*) of φ and ψ in $\mathcal{M}$ is denoted by $4ft(\varphi, \psi, \mathcal{M})$. In addition we sometimes use $r = a + b$, $s = c + d$, $k = a + c$, $l = b + d$ and $n = a + b + c + d$, see Table 1.

Table 1. 4ft table $4ft(\varphi, \psi, \mathcal{M})$ of φ and ψ in $\mathcal{M}$

$\mathcal{M}$	ψ	$\neg\psi$	
φ	a	b	r
$\neg\varphi$	c	d	s
	k	l	n

The Boolean attributes φ and ψ are derived from the columns of data matrix $\mathcal{M}$. We assume there is a finite number of possible values for each column of $\mathcal{M}$. *Basic Boolean attributes* are created first. The basic Boolean attribute is an expression of the form $A(\alpha)$ where $\alpha \subset \{a_1, \ldots a_k\}$ and $\{a_1, \ldots a_k\}$ is the set of all possible values of the column A. The basic Boolean attribute $A(\alpha)$ is true in row o of $\mathcal{M}$ if it is $a \in \alpha$ where a is the value of the attribute A in row o. Boolean attributes φ and ψ are derived from basic Boolean attributes using propositional connectives $\vee$, $\wedge$ and $\neg$ in the usual way.

The value of a Boolean attribute φ in row o of $\mathcal{M}$ is denoted by $\varphi(o, \mathcal{M})$. It is $\varphi(o, \mathcal{M}) = 1$ if φ is true in row o of $\mathcal{M}$, if φ is false then $\varphi(o, \mathcal{M}) = 0$.

We usually call the columns of the data matrix $\mathcal{M}$ *attributes*. The possible values of attributes are called *categories*. We assume that the rows of $\mathcal{M}$ correspond to the observed objects see Fig. 1. There are also some examples of basic Boolean attributes and Boolean attributes in Fig. 1. We write only $A_1(1)$ instead of $A_1(\{1\})$, $A_1(1, 2)$ instead of $A_1(\{1, 2\})$ etc.

There are various 4ft-quantifiers. Some examples are given below. Further 4ft-quantifiers are defined in [3, 6, 16] and also at `http://lispminer.vse.cz/`.

The 4ft-quantifier $\Rightarrow_{p,Base}$ of *founded implication* [3] is defined for $0 < p \leq 1$ and $Base > 0$ by the condition $\frac{a}{a+b} \geq p \wedge a \geq Base$. The association rule $\varphi \Rightarrow_{p,Base} \psi$ means that at least $100p$ per cent of rows of $\mathcal{M}$ satisfying

Object i.e. Row of $\mathcal{M}$	Columns of $\mathcal{M}$ i.e. Attributes				Examples of Boolean Attributes			
					Basic Boolean Attributes		Boolean Attributes	
	A_1	A_2	...	A_K	$A_1(1)$	$A_2(1,4,5)$	$A_1(1) \wedge A_2(1,4,5)$	$\neg A_K(6)$
o_1	1	9	...	4	1	0	0	1
o_2	1	4	...	6	1	1	1	0
o_3	3	5	...	7	0	1	0	1
$\vdots$	$\vdots$	$\vdots$	$\ddots$	$\vdots$	$\vdots$	$\vdots$	$\vdots$	$\vdots$
o_n	2	2	...	6	0	0	0	0

Fig. 1. Data matrix $\mathcal{M}$ and examples of Boolean attributes

φ satisfy also ψ and that there are at least *Base* rows of $\mathcal{M}$ satisfying both φ and ψ.

The 4ft-quantifier $\Leftrightarrow_{p,Base}$ of *founded double implication* [6] is defined for $0 < p \leq 1$ and $Base > 0$ by the condition $\frac{a}{a+b+c} \geq p \wedge a \geq Base$. The association rule $\varphi \Leftrightarrow_{p,Base} \psi$ means that at least $100p$ per cent of rows of $\mathcal{M}$ satisfying φ or ψ satisfy both φ and ψ and that there are at least *Base* rows of $\mathcal{M}$ satisfying both φ and ψ.

Fisher's quantifier $\sim_{\alpha,Base}$ [3] is defined for $0 < \alpha < 0.5$ and $Base > 0$ by the condition $\sum_{i=a}^{\min(r,k)} \frac{\binom{k}{i}\binom{n-k}{r-i}}{\binom{r}{n}} \leq \alpha \wedge ad > bc \wedge a \geq Base$. This quantifier corresponds to the statistical test (on the level α) of the null hypothesis of independence of φ and ψ against the alternative one of the positive dependence.

The 4ft-quantifier $\rightarrow_{conf,sup}$ is for $0 < conf < 1$ and $0 < sup < 1$ defined by the condition $\frac{a}{a+b} \geq conf \wedge \frac{a}{n} \geq sup$. It corresponds to the "classical" association rule with confidence *conf* and support *sup*.

Each 4ft-quantifier $\approx$ can be understood as a $\{0,1\}$ – valued function defined for all 4ft tables $\langle a,b,c,d \rangle$ such that $\approx (a,b,c,d) = 1$ if and only if the condition corresponding to $\approx$ is satisfied for $\langle a,b,c,d \rangle$. For example

- $\Rightarrow_{p,Base} (a,b,c,d) = 1$ if and only if $\frac{a}{a+b} \geq p \wedge a \geq Base$
- $\Leftrightarrow_{p,Base} (a,b,c,d) = 1$ if and only if $\frac{a}{a+b+c} \geq p \wedge a \geq Base$.

If the condition depends only on a and b then we write only $\approx (a,b)$ instead of $\approx (a,b,c,d)$, e.g. we write $\Rightarrow_{p,Base} (a,b)$ instead of $\Rightarrow_{p,Base} (a,b,c,d)$. Similarly we write only $\Leftrightarrow_{p,Base} (a,b,c)$ instead of $\Leftrightarrow_{p,Base} (a,b,c,d)$ etc.

The 4ft-Miner procedure mines also for *conditional association rules* of the form $\varphi \approx \psi/\chi$ where φ, ψ and χ are Boolean attributes. The intuitive meaning of $\varphi \approx \psi/\chi$ is that φ and ψ are in the relation given by 4ft-quantifier $\approx$ when the condition χ is satisfied. Both the association rule $\varphi \approx \psi$ and the conditional association rule $\varphi \approx \psi/\chi$ concern the analysed data matrix $\mathcal{M}$. The Boolean attributes φ, ψ and χ are derived from the columns of the analysed data matrix in the same way as described for φ and ψ above.

The conditional association rule $\varphi \approx \psi/\chi$ is *true in data matrix* $\mathcal{M}$ if there is both a row of $\mathcal{M}$ satisfying χ and if the association rule $\varphi \approx \psi$ is true in data matrix $\mathcal{M}/\chi$, otherwise $\varphi \approx \psi$ is *false in data matrix* $\mathcal{M}$. The data matrix $\mathcal{M}/\chi$ consists of all rows of data matrix $\mathcal{M}$ satisfying χ. We define the value $Val(\varphi \approx \psi/\chi, \mathcal{M})$ of $\varphi \approx \psi/\chi$ in $\mathcal{M}$ as $Val(\varphi \approx \psi/\chi, \mathcal{M}) = 1$ if $\varphi \approx \psi$ is true in $\mathcal{M}/\chi$ and as $Val(\varphi \approx \psi/\chi, \mathcal{M}) = 0$ if $\varphi \approx \psi$ is false in $\mathcal{M}/\chi$.

3 4ft-Miner Procedure

The procedure 4ft-Miner mines for association rules $\varphi \approx \psi$ and for conditional association rules $\varphi \approx \psi/\chi$. We would like to point out that 4ft-Miner is a GUHA procedure in the sense of [3]. We also use the terminology introduced in [3] e.g. the notions *antecedent, succedent, relevant question*, etc.

The input of 4ft-Miner consists of

- the analysed data matrix
- several parameters defining the set of association rules to be automatically generated and tested.

The analysed data matrix is created from a database table. Any database accessible by the ODBC can be used. The columns of the database table are transformed into attributes (i.e. columns) of the analysed data matrix. There is a special module *DataSource* in the LISP-Miner system intended for these transformations. It is, for example, possible to use the original values from the given column of the database table as the categories of the defined attribute. It is also possible to define new categories as intervals of the given length. Moreover the DataSource module can generate the given number of equifrequency intervals as new categories.

The association rules to be automatically generated and verified are called *relevant questions* [3]. There are very fine tools to tune a definition of a set of relevant questions. The parameters defining the set of relevant questions are described in Sect. 3.1.

The output of 4ft-Miner procedure consists of all *prime association rules.* The association rule is prime if it is both true in the analysed data matrix and if it does not immediately follow from other simpler output association rules. The definition of the prime association rule depends on the properties of the used 4ft-quantifier.

Let us e.g. consider the 4ft-quantifier $\Rightarrow_{p,Base}$ of founded implication (see Sect. 2). If the association rule $A(a_1) \Rightarrow_{p,Base} B(b_1)$ is true then the association rule $A(a_1) \Rightarrow_{p,Base} B(b_1, b_2)$ is always also true. Thus if the association rule $A(a_1) \Rightarrow_{p,Base} B(b_1)$ is part of an output then the association rule $A(a_1) \Rightarrow_{p,Base} B(b_1, b_2)$ is not prime and thus it is not listed in the output.

There are interesting and important results concerning deduction rules among association rules that are related to the definition of the prime association rule see [3, 16]. The precise definition of the prime association rule does not fall with the scope of this chapter. Please note that the 4ft-Miner procedure deals also with missing information [3, 13]. The details concerning dealing with missing information again does not fall with the scope of this chapter.

There are also great possibilities of filtering and sorting the output set rules. An example of 4ft-Miner application is given in Sect. 3.2.

3.1 Input of 4ft-Miner

The 4ft-Miner procedure mines for association rules of the form $\varphi \approx \psi$ and for conditional association rules of the form $\varphi \approx \psi/\chi$. The Boolean attribute φ is called *antecedent*, ψ is called *succedent* [3] and χ is called *condition.*

Definition of the set of relevant questions consists of

- definition of a *set of relevant antecedents*
- definition of a *set of relevant succedents*
- definition of a *set of relevant conditions*
- definition of the 4ft-quantifier $\approx$.

Antecedent, succedent and condition are conjunctions of *literals.* Literal is a basic Boolean attribute $A(\alpha)$ or a negation $\neg A(\alpha)$ of a basic Boolean attribute. The set α is a *coefficient* of the literals $A(\alpha)$ and $\neg A(\alpha)$. The expression

$$A(a_1, a_7) \wedge B(b_2, b_5, b_9) \Rightarrow_{p,Base} C(c_4) \wedge \neg D(d_3)$$

is an example of an association rule. Here $A(a_1, a_7)$, $B(b_2, b_5, b_9)$, $C(c_4)$ and $\neg D(d_3)$ are literals. Moreover a_1 and a_7 are categories of A and $\{a_1, a_7\}$ is a coefficient of $A(a_1, a_7)$ (we write $A(a_1, a_7)$ instead of $A(\{a_1, a_7\})$, the same is true for additional literals).

To increase the possibilities of defining the set of relevant questions is antecedent φ defined as a conjunction

$$\varphi = \varphi_1 \wedge \varphi_2 \wedge \cdots \wedge \varphi_k$$

where $\varphi_1, \varphi_2, \ldots, \varphi_k$ are *partial antecedents.* Each φ_i is chosen from one *set of relevant partial antecedents.* A partial antecedent is thus a conjunction of literals. The *length of the partial antecedent* is the number of literals in this conjunction.

The definition of the set of relevant antecedents is given by at least one definition of the set of relevant partial antecedents. The set of partial antecedents is given in the following manner:

- the minimum and maximum length of the partial antecedent is defined
- a set of attributes from which literals will be generated is given
- some attributes can be marked as *basic*, each partial antecedent then must contain at least one basic attribute
- a simple definition of the set of all literals to be generated is given for each attribute.

- *classes of equivalence* can be defined, each attribute belongs to a maximum one class of equivalence; no partial antecedent can contain two or more attributes from one class of equivalence.

We would like to point out that the minimum length of the partial antecedent can be 0 and this results in some conjunctions of length 0. The value of the conjunction of length 0 is identical to 1 thus the partial antecedent φ_i of the length 0 can be omitted from the antecedent.

The *length of the literal* is the number of categories in its coefficient. The set of all literals to be generated for a particular attribute is given by:

- the type of coefficient; there are seven types of coefficients: *subsets, intervals, cyclic intervals, left cuts, right cuts, cuts, one particular category*
- the minimum and the maximum length of the literal
- positive/negative literal option:
  - generate only positive literals
  - generate only negative literals
  - generate both positive and negative literals

We use the attribute A with categories {1, 2, 3, 4, 5} to give examples of particular types of coefficients:

- *subsets*: definition of subsets of length 2–3 gives literals A(1,2), A(1,3), A(1,4), A(1,5), A(2,3), ..., A(4,5), A(1,2,3), A(1,2,4), A(1,2,5), A(2,3,4), ..., A(3,4,5)
- *intervals*: definition of intervals of length 2–3 gives literals A(1,2), A(2,3), A(3,4), A(4,5), A(1,2,3), A(2,3,4) and A(3,4,5)
- *cyclic intervals*: definition of intervals of length 2–3 gives literals A(1,2), A(2,3), A(3,4), A(4,5), A(5,1) A(1,2,3), A(2,3,4), A(3,4,5), A(4,5,1) and A(5,1,2)
- *left cuts*: definition of left cuts with a maximum length of 3 defines literals A(1), A(1,2) and A(1,2,3)
- *right cuts*: definition of right cuts with a maximum length of 4 defines literals A(5), A(5,4), A(5,4,3) and A(5,4,3,2)
- *cuts* means both left cuts and right cuts
- *one particular value* means one literal with one chosen category, e.g. A(2).

Let us emphasize that even if the type *subsets* covers all the other types of literals it still has sense to use the particular subtypes. There are e.g. more than 10^{13} literals – subsets with length 10 for the attribute *Age* with 100 categories. At the same time, there are only 91 literals of the type *intervals* with length 10.

Definitions of the set of relevant succedents and of the set of relevant conditions are analogous. The set of relevant antecedents, the set of relevant succedent and the set of relevant conditions can overlap. However, association rules with more than one literal created from the same attribute are not generated.

There are 16 various types of 4ft-quantifiers. Examples of types of the 4ft-quantifier are founded implication, lower critical implication, founded double implication etc. (see Sect. 2). To define the 4ft-quantifier means to choose the type of quantifier and to define the values of its parameters (e.g p and *Base* for the founded implication $\Rightarrow_{p,Base}$).

3.2 4ft-Miner Application Example

An example of the analysis of financial data is given in this section. The analysed data matrix *LoanDetails* is derived from data on a fictitious bank see `http://lisp.vse.cz/pkdd99/Challenge/`. There are 6181 rows in the *LoanDetails* data matrix. Each row corresponds to a particular loan.

There are 12 attributes derived from the columns of the *LoanDetails* data matrix. The attributes can be divided into four groups.

The group *Client* has four attributes

- *Sex* with 2 categories *female* and *male*
- *Age* with 5 categories – intervals $\langle 20, 30), \ldots, \langle 60, 70)$
- *Salary* with 5 categories *very low*, *low*, *average*, *high* and *very high*
- *District* with 77 categories i.e. districts where the clients live.

The group *Loan* has four attributes

- *Amount* i.e. the amount of borrowed money in thousands of Czech crowns divided into 6 categories – intervals $\langle 0;20)$, $\langle 20;50)$, $\langle 50;100)$, $\langle 100;250)$, $\langle 250;500)$ and ≥ 500.
- *Repayment* i.e. repayment in thousands of Czech crowns with 10 categories – intervals $(0,1\rangle$, ..., $(8,9\rangle$ and > 9.
- *Months* i.e. the number of months to repay the whole amount with values divided into 5 categories – intervals $1-12$, $13-24$, ..., $49-60$.
- *Quality* with 2 categories *good* (i.e. the loan is already paid up or it is repaid without problems) and *bad* (i.e. the loan finished and it is not paid up or it is repaid with problems).

The group *Crimes* has two attributes

- *Crimes95* i.e. the number of crimes in the District of the client in the year 1995 divided into 6 categories – intervals $\langle 0;2000)$, $\langle 2000;4000)$, ..., $\langle 8000;10000)$ and ≥ 10000.
- *Crimes96* i.e. the number of crimes in the District of the client in the year 1996 divided into the same 6 categories as for the attribute *Crimes95*.

The group *Unemployment* has two attributes

- *Unemployment95* i.e. unemployment in the District of the client in the year 1995 (measured in percentage) divided into 8 categories – intervals $\langle 0;1), \ldots, \langle 7;8)$.

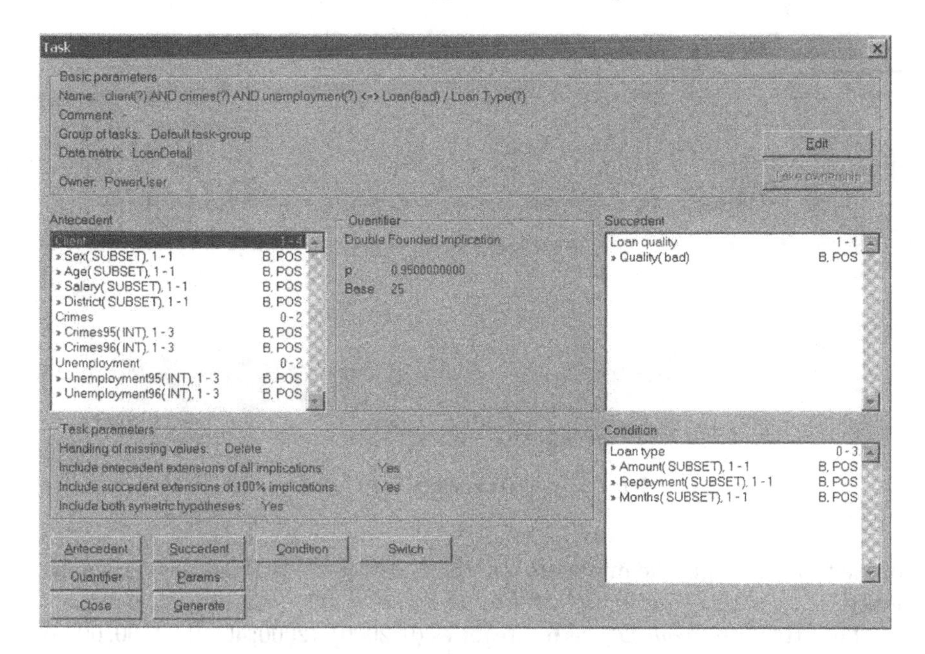

Fig. 2. 4ft-Miner input parameters example

- *Unemployment96* i.e. unemployment in the District of the client in the year 1996 (measured in percentage) divided into 10 categories – intervals $\langle 0;1),\ldots,\langle 9;10)$.

We solve the task *What combinations of characteristics of a client and of his district are (almost) equivalent to a bad loan and for what type of loan?*. It can be written in a slightly formal way like an association rule:

$$\text{Client}(?) \wedge \text{Crimes}(?) \wedge \text{Unemployment}(?) \Leftrightarrow^{*} \text{Quality}(\text{bad}) \;/\; \text{LoanType}(?)$$

This task can be solved by the procedure 4ft-Miner with the parameters described in Fig. 2. There are definitions of three partial antecedents. Partial antecedent *Client* has a minimum length of 1 and a maximum length of 4, no class of equivalence and four attributes see Fig. 3. All four attributes are marked as basic and positive literals with a coefficient of the type *subsets* are automatically generated for them. Both the minimum and the maximum length of the coefficient is 1 for all attributes. This means that there are 2 literals for attribute *Sex*, 5 literals for attribute *Age*, 5 literals for attribute *Salary* and 77 literals for attribute *District.* Thus there are 8423 partial antecedents *Client* corresponding to the expression "Client(?)" given this way.

The partial antecedent *Crimes* has a minimum length of 0 and a maximum length of 2, no class of equivalence and two attributes *Crimes95* and *Crimes96.* Both attributes are marked as basic and have positive literals with coefficients

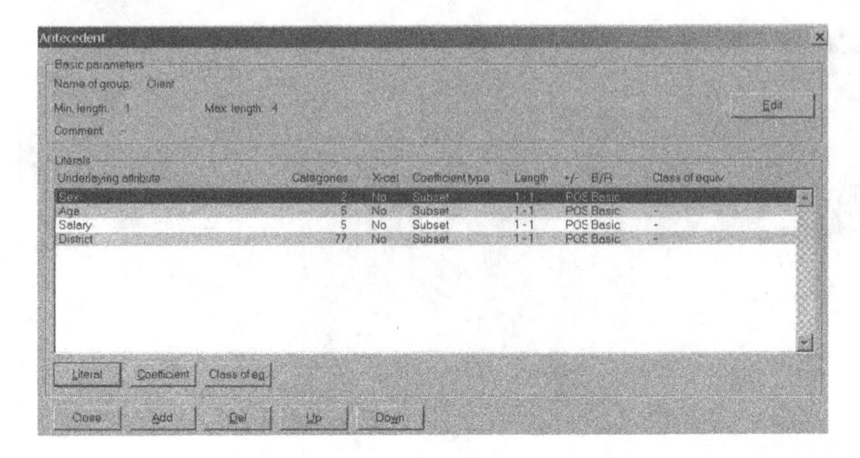

Fig. 3. Partial antecedent *Client*

of the type *interval.* The minimum length of coefficients is 1 and the maximum length is 3 for both attributes see Fig. 2.

The attribute *Crimes95* has 6 categories $\langle 0; 2000)$, $\langle 2000; 4000)$, $\langle 4000; 6000)$, $\langle 6000; 8000)$, $\langle 8000; 10000)$ and ≥ 10000 see above. Let us denote them 1,2,3,4,5, and 6 respectively to show more clearly the corresponding literals. There are the following literals with coefficients–intervals of length 1–3:

Crimes95(1) i.e. *Crimes95*($\langle 0; 2000)$)
Crimes95(1,2) i.e. *Crimes95*($\langle 0; 2000), \langle 2000; 4000)$)
Crimes95(1,2,3) i.e. *Crimes95*($\langle 0; 2000), \langle 2000; 4000), \langle 4000; 6000)$)
Crimes95(2) i.e. *Crimes95*($\langle 2000; 4000)$)
...
Crimes95(5,6) i.e. *Crimes95*($\langle 8000; 10000), \geq 10000$)
Crimes95(6) i.e. *Crimes95*(≥ 10000).

There are 15 such literals together. The attribute *Crimes96* also has 6 categories and thus there are also 15 literals for it. It means that there are 256 partial antecedents *Crimes* corresponding to the expression "Crimes(?)" defined this way. There is one partial antecedent *Crimes* of the length 0, 30 partial antecedents *Crimes* of the length 1 and 225 partial antecedents *Crimes* of the length 2. The partial antecedent *Crimes* of the length 0 is identically true but we have to consider it when counting the total number of antecedents.

We would like to point out that the literal *Crimes95*($\langle 0; 2000), \langle 2000; 4000)$) corresponds to the basic Boolean attribute *Crimes95*($\langle 0; 4000)$) with one category equivalent to the interval $\langle 0; 4000)$ etc.

Partial antecedent *Unemployment* has two attributes *Unemployment95* with 8 categories and *Unemployment96* with 10 categories. Further parameters are the same as for the partial antecedent *Crimes.* This means that there

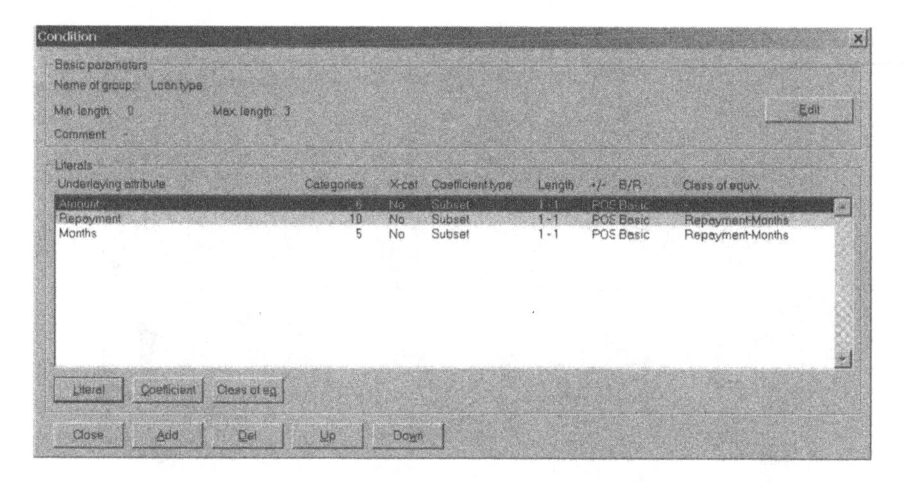

Fig. 4. Partial condition *LoanType*

are 614 partial antecedents *Unemployment* correspondindg to the expression "Unemployment(?)".

Note that there are 1 323 960 832 (i.e. 8 423 * 256 * 614) antecedents of the form Client(?) $\wedge$ Crimes(?) $\wedge$ Unemployment(?).

There is one partial succedent *Loan Quality* that has one literal *Quality(bad)* of the type *one particular category.*

There is one partial condition *LoanType* with a minimum length of 0 and a maximum length of 3 and with three attributes see Fig. 4. All three attributes are marked as basic and have positive literals with a coefficient of the type *subset.* Both the minimum and maximum length of the coefficients is 1 for all attributes.

This means that there are 6 literals for attribute *Amount*, 10 literals for attribute *Repayment* and 5 literals for attribute *Months.* There is one class of equivalence called *Repayment-Months* that contains attributes *Repayment* and *Months.* This means that no partial condition *LoanType* can contain both *Repayment* and *Months.* We define this class of equivalence because the attributes *Repayment* and *Months* are strongly dependent. This means that there are 92 partial conditions *LoanType.*

We use the 4ft-quantifier $\Leftrightarrow_{0.95,25}$ of founded double implication with parameters $p = 0.95$ and $Base = 25$ (see Sect. 2) instead of the symbol $\Leftrightarrow^*$. This means that we search for characteristics Client(?) of clients and characteristics Crimes(?) and Unemployment(?) of districts such that their combination is almost equivalent to bad loans of type LoanType(?).

There are more than $12 * 10^{10}$ relevant questions of the form

$$\text{Client}(?) \wedge \text{Crimes}(?) \wedge \text{Unemployment}(?) \Leftrightarrow^* \text{Quality}(\text{bad}) \ / \ \text{LoanType}(?)$$

defined this way. The procedure 4ft-Miner solves the task of finding all the prime association rules in 8 minutes and 6 seconds (on PC with Intel Pentium4

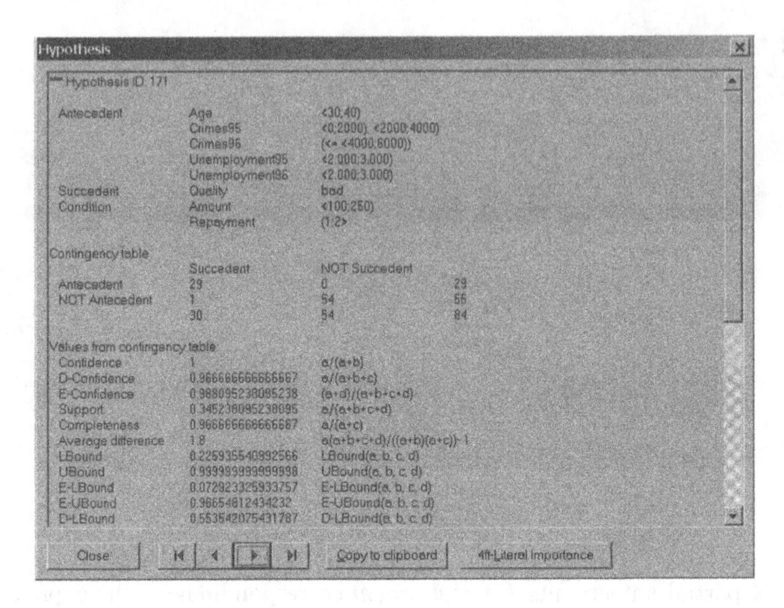

Fig. 5. 4ft-Miner output association rule example

on 3 GHz and 512MB RAM). Due to various optimizations less than $5.4 * 10^6$ of the relevant questions were actually verified. 290 prime association rules were found.

An example of an output rule is given in Fig. 5. The rule shown in Fig. 5 we can write in the form

$$Age\langle 30, 40) \ \wedge \ DISTRICT \ \Leftrightarrow_{0.967,29} \ Quality(bad) \ / \ TYPE$$

where
$DISTRICT \ = \ CRIME \ \wedge \ UNEMPLOYMENT$
$CRIME \ = \ Crimes95\langle 0, 4000) \ \wedge \ Crimes96\langle 0, 6000)$
$UNEMPLOYMENT \ = \ Unemployment95\langle 2.0, 3.0) \ \wedge \ Unemployment96\langle 2.0, 3.0)$
$TYPE \ = \ Amount\langle 100, 250) \ \wedge \ Repayment(1, 2\rangle$
Here we write $Age\langle 30, 40)$ instead of $Age(\langle 30, 40))$, $Crimes95\langle 0, 4000)$ instead of $Crimes95(\langle 0, 2000), \langle 2000, 4000))$. We also write $Crimes96\langle 0, 6000)$ instead of $Crimes96(<= \langle 4000, 6000))$ that is the 4ft-Miner abbreviation for the literal $Crimes96(\langle 0, 2000), \langle 2000, 4000), \langle 4000, 6000))$ and analogously for further literals. The 4ft table for this rule is given in Table 2.

Data matrix *LoanDetails* / *TYPE* consists of 84 rows (i.e. loans) of data matrix *LoanDetails* satisfying the Boolean attribute *TYPE*. There are 30 loans satisfying $Age\langle 30, 40) \wedge DISTRICT$ or $Quality(bad)$ and 29 of them (i.e. $96.7 = 100 * \frac{29}{29+1+0}$ per cent) satisfy both $Age\langle 30, 40) \wedge DISTRICT$ and $Quality(bad)$. This means that the combination $Age\langle 30, 40) \ \wedge DISTRICT$ of the characteristics of the client and of his district is almost (at the level 96.7 per cent) equivalent to bad loan for the type of loan $Amount\langle 100, 250) \wedge Repayment\langle 1, 2)$.

Table 2. 4ft table of the rule given in Fig. 5

LoanDetails / TYPE	*Quality(bad)*	$\neg$ *Quality(bad)*	
$Age\langle 30,40) \wedge DISTRICT$	29	0	29
$\neg$ ($Age\langle 30,40) \wedge DISTRICT$)	1	54	55
	30	54	84

There are lot of possibilities of sorting and filtering the found prime association rules. We can, for example, ask 4ft-Miner to show only association rules that satisfy these conditions:

- only attributes *District*, *Crimes95* and *Unemployment95* are contained in antecedent
- there are at least 30 clients satisfying both antecedent and succedent.

This requirement can be expressed in the window *Filter* shown in Fig. 6. Our requirements are marked by dotted arrows. Four association rules satisfy these conditions and 4ft-Miner lists them in the form shown in Fig. 7.

Let us emphasize that these examples are not a serious analysis of the problem of bad loans. They only demonstrate some of the main features of the 4ft-Miner procedure that implements the bit string approach to mining association rules introduced in Sect. 4. Some considerations of time and space complexity are in Sect. 5.

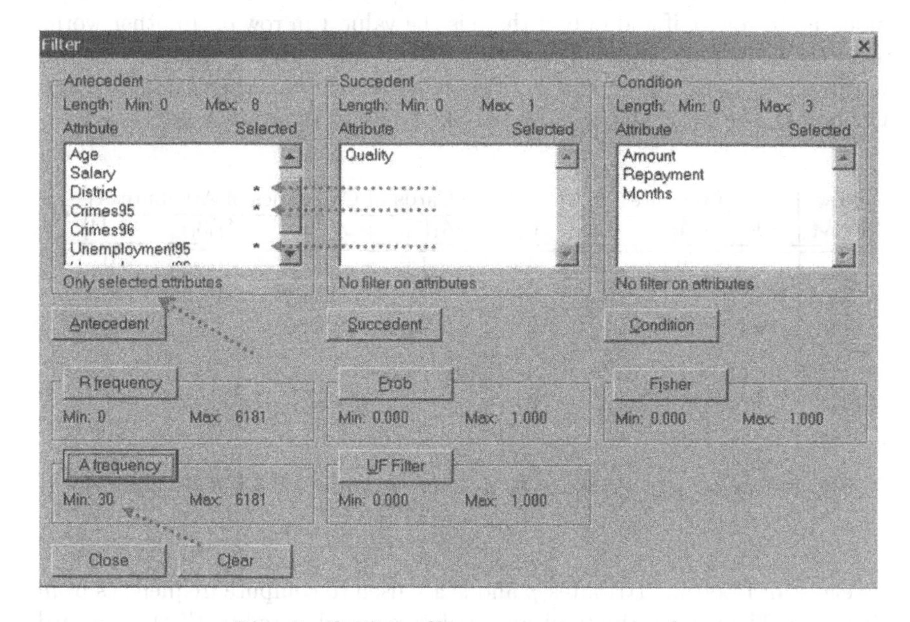

Fig. 6. 4ft-Miner filter parameters example

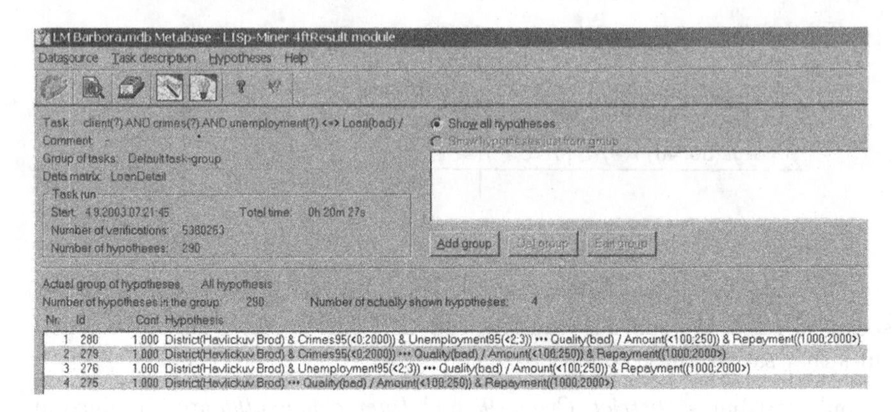

Fig. 7. 4ft-Miner output rules list example

4 Bit String Approach

We use the data matrix $\mathcal{M}$ from Fig. 8 to present the principles of the bit string approach. The basic principle is to present each attribute (i.e. columns of $\mathcal{M}$) by cards of its categories. Let $\{1,2,3,4\}$ be the set of all categories of the attribute A_1. Then the attribute A_1 is represented by cards $A_1[1]$, $A_1[2]$, $A_1[3]$, $A_1[4]$ of categories 1,2,3,4 respectively.

The card $A_1[1]$ of category 1 is the string of bits. Each row of data matrix $\mathcal{M}$ corresponds to one bit of the card $A_1[1]$. There is "1" in the bit corresponding to row o_i if and only if there is the value 1 in row o_i. In other words there is "1" in the i-th bit of the card $A_1[1]$ if and only if the basic Boolean attribute $A_1(1)$ is true in row o_i. The same is true for other categories and attributes. The cards $A_1[1]$, $A_1[2]$, $A_1[3]$, $A_1[4]$ are shown in Fig. 8.

Row	Attributes of $\mathcal{M}$				Cards of Categories of Attribute A_1			
of $\mathcal{M}$	A_1	A_2	...	A_K	$A_1[1]$	$A_1[2]$	$A_1[3]$	$A_1[4]$
o_1	1	9	...	4	1	0	0	0
o_2	1	4	...	6	1	0	0	0
o_3	3	5	...	7	0	0	1	0
$\vdots$	$\vdots$	$\vdots$	$\ddots$	$\vdots$	$\vdots$	$\vdots$	$\vdots$	$\vdots$
o_{n-1}	4	1	...	8	0	0	0	1
o_n	2	2	...	6	0	1	0	0

Fig. 8. Cards $A_1[1]$, $A_1[2]$, $A_1[3]$, $A_1[4]$ of categories 1,2,3,4 of attribute A_1

Cards of Boolean attributes φ and ψ are used to compute frequencies from 4ft tables. The card of the Boolean attributes φ is denoted by $\mathcal{C}(\varphi)$. The card $\mathcal{C}(\varphi)$ is a string of bits that is analogous to the card of the category. Each row of the data matrix corresponds to one bit of $\mathcal{C}(\varphi)$ and there is "1" in the i-th bit if and only if φ is true in row o_i.

It is evident that $\mathcal{C}(\varphi \wedge \psi) = \mathcal{C}(\varphi) \dot{\wedge} \mathcal{C}(\psi)$, $\mathcal{C}(\varphi \vee \psi) = \mathcal{C}(\varphi) \dot{\vee} \mathcal{C}(\psi)$, $\mathcal{C}(\neg\varphi) = \dot{\neg}\, \mathcal{C}(\varphi)$. Here $\mathcal{C}(\varphi) \dot{\wedge} \mathcal{C}(\psi)$ is a bit-wise conjunction of bit strings $\mathcal{C}(\varphi)$ and $\mathcal{C}(\psi)$, analogously for $\dot{\vee}$ and $\dot{\neg}$. Moreover it is $\mathcal{C}(A_1(1,2)) = A_1[1] \dot{\vee} A_1[2]$ for the basic Boolean attribute $A_1(1,2)$ etc.

It is important that the bit-wise Boolean operations $\dot{\wedge}$, $\dot{\vee}$ and $\dot{\neg}$ are carried out by very fast computer instructions. Very fast computer instructions are also used to carry out a bit string function $Count(\xi)$ returning the number of values "1" in the bit string ξ. This function is used to compute frequencies a, b, c, d from 4ft table $4ft(\varphi, \psi, \mathcal{M})$ see Table 3.

Table 3. 4ft table $4ft(\varphi, \psi, \mathcal{M})$

$\mathcal{M}$	ψ	$\neg\psi$
φ	a	b
$\neg\varphi$	c	d

It is $a = Count(\mathcal{C}(\varphi) \dot{\wedge} \mathcal{C}(\psi))$, $b = Count(\mathcal{C}[\varphi]) - a$, $c = Count(\mathcal{C}[\psi]) - a$, $d = n - a - b - c$ where n is the total number of rows in the data matrix $\mathcal{M}$.

The algorithm of the 4ft-Miner is sketched in Fig. 9. This algorithm is for

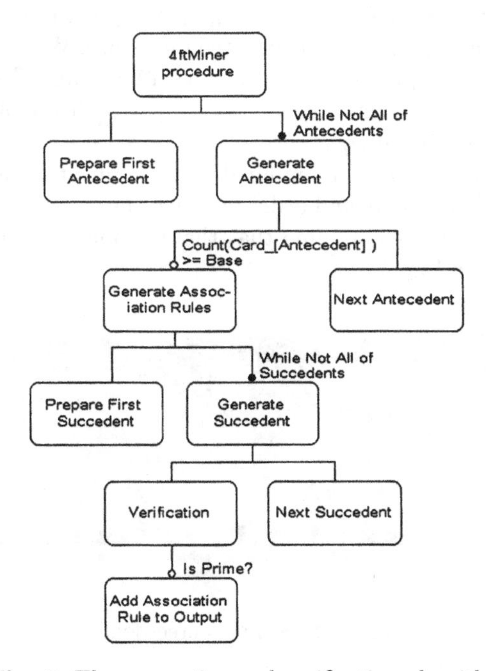

Fig. 9. The generation and verification algorithm

unconditional association rules $\varphi \approx \psi$. The algorithm for conditional association rules is similar. The algorithm maintains special data structures *traces of literals*, *trace of antecedent* φ and *trace of succedent* ψ.

The trace TA of antecedent $\varphi = \alpha_1 \wedge \alpha_2 \wedge \ldots \wedge \alpha_K$ with literals α_1, $\alpha_2, \ldots, \alpha_K$ consists of K steps – $\mathtt{TA}_1, \mathtt{TA}_2, \ldots, \mathtt{TA}_K$. The i-th step $\mathtt{TA}_i$ is the card of the conjunction $\alpha_1 \wedge \alpha_2 \wedge \ldots \wedge \alpha_i$. The last step is the card $\mathcal{C}(\varphi)$ of the whole antecedent φ, see Fig. 10. The trace TS of succedent ψ is analogous to the trace TA of antecedent φ.

Step	Computation of the Step
$\mathtt{TA}_1$	$\mathcal{C}(\alpha_1)$
$\mathtt{TA}_2$	$\mathtt{TA}_1 \dot{\wedge} \mathcal{C}(\alpha_2)$
$\vdots$	$\vdots$
$\mathtt{TA}_i$	$\mathtt{TA}_{i-1} \dot{\wedge} \mathcal{C}(\alpha_i)$
$\vdots$	$\vdots$
$\mathtt{TA}_{K-1}$	$\mathtt{TA}_{K-2} \dot{\wedge} \mathcal{C}(\alpha_{K-1})$
$\mathtt{TA}_K$	$\mathtt{TA}_{K-1} \dot{\wedge} \mathcal{C}(\alpha_K) = \mathcal{C}(\alpha_1 \wedge \ldots \wedge \alpha_i \wedge \ldots \wedge \alpha_K)$

Fig. 10. Trace of antecedent $\alpha_1 \wedge \alpha_2 \wedge \ldots \wedge \alpha_{K-1} \wedge \alpha_K$

The antecedents are generated in the depth-first way. This means that for all possible antecedents $\alpha_1 \wedge \cdots \wedge \alpha_{K-1} \wedge \alpha_K$ with the same beginning $\alpha_1 \wedge \cdots \wedge \alpha_{K-1}$ the step $\mathtt{TA}_{K-1}$ is computed only once. We need one operation only – the bit-wise conjunction of bit strings $\mathtt{TA}_{K-1}$ and $\mathcal{C}(\alpha'_K)$ to compute the card of the new antecedent $\alpha_1 \wedge \cdots \wedge \alpha_{K-1} \wedge \alpha'_K$ if the old-one literal α_K is replaced by a new literal α'_K.

The trace of the literal is a similar data structure that is maintained for each literal. The antecedents and order in which the algorithm generates them are outlined in Fig. 12. We assume that there is only one partial antecedent defined according to Fig. 11.

Antecedent

Basic parameters

Name of group: Sample antecedet

Min. length: 1 Max. length: 3

Comment: -

Literals

Underlaying attribute	Categories	X-cat	Coefficient type	Length	+/-	B/R	Class of equiv.
Sex	2	No	Subset	1-1	pos	Basic	-
Salary	5	No	Subset	1-1	pos	Basic	-
District	77	No	Subset	1-3	pos	Basic	-

Fig. 11. Sample antecedent

We further assume that there are five categories *very low*, *low*, *average*, *high* and *very high* of the attribute *Salary*, that the first five districts are *Benesov*, *Beroun*, *Blansko*, *Breclav* and *Brno* and that the last three districts are *Zdar*, *Zlin*, *Znojmo*, see also Sect. 3.2.

$Sex(female)$
$Sex(female) \wedge Salary(very\ low)$
$Sex(female) \wedge Salary(very\ low) \wedge District(Benesov)$
$Sex(female) \wedge Salary(very\ low) \wedge District(Benesov, Beroun)$
$Sex(female) \wedge Salary(very\ low) \wedge District(Benesov, Beroun, Blansko)$
$Sex(female) \wedge Salary(very\ low) \wedge District(Benesov, Beroun, Breclav)$
$Sex(female) \wedge Salary(very\ low) \wedge District(Benesov, Beroun, Brno)$
$\vdots$
$Sex(female) \wedge Salary(very\ low) \wedge District(Zdar, Zlin, Znojmo)$
$Sex(female) \wedge Salary(very\ low) \wedge District(Zlin)$
$Sex(female) \wedge Salary(very\ low) \wedge District(Zlin, Znojmo)$
$Sex(female) \wedge Salary(very\ low) \wedge District(Znojmo)$
$Sex(female) \wedge Salary(low)$
$\vdots$
$Sex(female) \wedge Salary(very\ high) \wedge District(Znojmo)$
$Sex(male)$
$\vdots$
$Sex(male) \wedge Salary(very\ high) \wedge District(Znojmo)$

Fig. 12. Antecedents order example

Let us remark that bit string representation of analysed data is used also in granular computing approach [9]. Informally speaking, a binary representation of a granule [9] corresponds to a card of category. The algorithm used in [9] is however different from the algorithm used in the 4ft-Miner procedure.

5 Time and Space Complexity

The core of machine operations used in running 4ft-Miner are the bit string Boolean operations $\dot{\wedge}, \dot{\vee}, \dot{\neg}$ and the bit string function $Count(\xi)$ see Sect. 4. Their operation time is linearly dependent on the length of particular bit strings i.e. cards of categories, steps in traces of literals and steps in traces of antecedent, succedent and condition. It implies that the operation time of the 4ft-Miner procedure must be approximately linearly dependent on the number of rows of the analysed data matrix.

This conclusion is confirmed by an experiment with the task described in Sect. 3.2. The task is run with similar parameters as the task desccribed in Fig. 2 for two further data matrices – *LoanDetails*10* and *LoanDetails*20*.

Data matrix *LoanDetails*10* is the data matrix *LoanDetails* magnified by factor 10. This means that *LoanDetails*10* contains 10 copies of each row of data matrix *LoanDetails*. In the same way, data matrix *LoanDetails*20* is the matrix *LoanDetails* magnified by factor 20.

The two tasks are run with the same definition of the set of relevant questions except the parameter *Base* of the 4ft-quantifier $\Leftrightarrow_{p,Base}$. We used the 4ft-quantifier $\Leftrightarrow_{0.95,250}$ for the data matrix *LoanDetails*10* instead of the quantifier $\Leftrightarrow_{0.95,25}$ for the data matrix *LoanDetails*. This was to achieve the same number of actually verified association rules. The optimisation of the solved task is based on skipping all subsets of potentially true association rules for which it is clear that their a-frequency cannot be 250 or more. These subsets of potentially interesting rules can be skipped on the frequency analysis of current steps in the traces of condition and of antecedent (see Sect. 4). In the same way we used the 4ft-quantifier $\Leftrightarrow_{0.95,500}$ for the data matrix *LoanDetails*20* instead of the quantifier $\Leftrightarrow_{0.95,25}$ for the data matrix *LoanDetails*.

We used PC with Intel Pentium4 on 3 GHz and 512 MB RAM. The results of experiments are given in Table 4.

Table 4. 4ft-Miner operation time

Data Matrix	Rows	4ft Quantifier	No. of Rules Verified	No. of Rules Found	exec. time	Ratio to *LoanDetails*
LoanDetails	6 181	$\Leftrightarrow_{0.95,25}$	5 380 263	290	8 min 6 sec	1.0
*LoanDetails*10*	61 810	$\Leftrightarrow_{0.95,250}$	5 380 263	290	39 min 22 sec	4.9
*LoanDetails*20*	123 620	$\Leftrightarrow_{0.95,500}$	5 380 263	290	87 min 34 sec	10.8

We can conclude that the dependency of the operation time of the 4ft-Miner procedure on the number of rows of the analysed data matrix is in our experiment a little worse than linear.

Also the representation of analysed data by necessary strings of bits is not a problem. Table 5 shows the memory required for data matrices *LoanDetails*, *LoanDetails*10* and *LoanDetails*20*.

We would like to point out that a lot of practically important tasks for mining association rules concerns data of a similar size as our example. This means that these tasks can be solved without problems at common PC's. Moreover in many cases we receive the solution in several minutes or even in several seconds. Therefore the 4ft-Miner procedure is also suitable for teaching purposes.

Table 5. Bit strings representation memory requirements

Attribute	Categories	Memory Required for Data Matrix		
		LoanDetails	*LoanDetails*10*	*LoanDetails*20*
Sex	2	1.5 kb	15 kb	30 kb
Age	5	3.8 kb	38 kb	76 kb
Salary	5	3.8 kb	38 kb	76 kb
District	77	58.1 kb	581 kb	1.13 MB
Crimes95	6	4.5 kb	45 kb	90 kb
Crimes96	6	4.5 kb	45 kb	90 kb
Unemployment95	8	6.0 kb	60 kb	120 kb
Unemployment96	10	7.5 kb	75 kb	150 kb
Quality	2	1.5 kb	15 kb	30 kb
Amount	6	4.5 kb	45 kb	90 kb
Repayment	10	7.5 kb	75 kb	150 kb
Months	5	3.8 kb	38 kb	76 kb
total	142	107 kb	1.04 MB	2.09 MB
the number of rows		6 181	61 810	123 620

6 LISp-Miner System

The 4ft-Miner procedure is an important part of the academic LISp-Miner system for KDD research and teaching `http://lispminer.vse.cz`. The LISp-Miner system is developed by a group of teachers and students. There are several research activities related to the LISp-Miner system.

The first activity concerns applications of the bit string approach used in the implementation of the 4ft-Miner procedure. A system of software modules were developed for dealing with corresponding data structures and algorithms [8]. These modules were further modified to be an efficient tool for the implementation of further data mining procedures. Examples of such procedures are the procedures KL-Miner [17] and SDS-Miner [8]. These modules were also used in implementation of a new version of the machine learning procedure KEX [2] that is also part of the LISp-Miner system [8]. The bit string approach can also be used in relational data mining [14].

There is also research into the logical foundations of KDD see [5, 13, 15, 16]. This research concerns the observational calculi introduced in [3]. The other research direction concerns the presentation of the results of data mining in a natural language [19].

Acknowledgement

The work described here has been supported by the project COST ACTION 274 – TARSKI, by the project 201/05/0325 of Czech Science Foundation, and by the project IGA 17/04 of University of Economics, Prague.

References

1. Aggraval R. et al. (1996) Fast Discovery of Association Rules. In: Fayyad UM et al. (eds) Advances in Knowledge Discovery and Data Mining. AAAI Press, Menlo Park (CA)
2. Berka P., Ivánek J. (1994) Automated knowledge acquisition for PROSPECTOR-like expert systems. In: Bergadano F., de Raedt L. (eds) Proceedings of ECML'94. Springer, Berlin Heidelberg New York
3. Hájek P., Havránek T. (1978) Mechanising Hypothesis Formation – Mathematical Foundations for a General Theory. Springer, Berlin Heidelberg New York
4. Hájek P. (guest editor) (1978) International Journal of Man-Machine Studies, special issue on GUHA, 10
5. Hájek P. (guest editor) (1981) International Journal of Man-Machine Studies, second special issue on GUHA, 15
6. Hájek P., Havránek T., Chytil M. (1983) GUHA Method. Academia, Prague (in Czech)
7. Hájek P., Sochorová A., Zvárová J. (1995) GUHA for personal computers. Computational Statistics & Data Analysis 19: 149–153
8. Karban T., Rauch J., Šimůnek M. (2004) SDS-Rules and Association Rules. In: Haddad H.M., Omicini A., Wainwright R.L., Liebrock L.M. (eds) Proceedings of the 2004 ACM SAC. ACM Press, New York
9. Louie E., Lin T.Y. (2000) Finding Association Rules using Fast Bit Computation: Machine-Oriented Modeling. In: Ras Z, Ohsuga S (eds) Foundations of Intelligent Systems. Springer, Berlin Heidelberg New York
10. Rauch J. (1971) Application of the three valued logic in the GUHA method. Diploma Thesis, Faculty of Mathematics and Physics, Charles University, Prague (in Czech)
11. Rauch J. (1978) Some Remarks on Computer Realisations of GUHA Procedures. International Journal of Man-Machine Studies 10: 23–28
12. Rauch J. (1997) Logical Calculi for Knowledge Discovery in Databases. In: Zytkow J., Komorowski J. (eds) Principles of Data Mining and Knowledge Discovery. Springer, Berlin Heidelberg New York
13. Rauch J. (1998) Four-Fold Table Calculi and Missing Information, In: Wang P. (ed) Proc. Joint Conference on Information Sciences '98 Vol. II. Duke University, Durham, North Carolina
14. Rauch J. (2002) Interesting Association Rules and Multi-relational Association Rules. In: Lee H.C., Lai F. (eds) Communications of Institute of Information and Computing Machinery. IICM, Taiwan
15. Rauch J. (2005) Definability of Association Rules in Predicate Calculus. In: Lin T.Y., Ohsuga S., Liau C.J., Hu X. (eds) Foundations and Novel Approaches in Data Mining. Springer, Berlin Heidelberg New York (*to appear*)
16. Rauch J. (2005): Logic of Association Rules Applied Intelligence 22 (1): 9–28 2005
17. Rauch J., Šimůnek M., Lín V. (2005) Mining for Patterns Based on Contingency Tables by KL-Miner – First Experience. In: Lin T.Y., Ohsuga S., Liau C.J., Hu X. (eds) Foundations and Novel Approaches in Data Mining. Springer, Berlin Heidelberg New York (*to appear*)

18. Šimůnek M. (2003) Academic KDD Project LISp-Miner. In Abraham A. et al. (eds) Advances in Soft Computing – Intelligent Systems Design and Applications, Springer, Berlin Heidelberg New York
19. Strossa P., Černý Z., Rauch J. (2004) Reporting Data Mining Results in a Natural Language. In: This book

Direct Mining of Rules from Data with Missing Values

Vladimir Gorodetsky, Oleg Karsaev and Vladimir Samoilov

St. Petersburg Institute for Informatics and Automation, 39, 14-th Liniya, St.Petersburg, 199178, Russia
{gor,ok,samovl}@mail.iias.spb.su

Summary. The paper presents an approach to and technique for direct mining of binary data with missing values aiming at extraction of classification rules, whose premises are represented in a conjunctive form. This approach does not assume an imputation of missing values. The idea is (1) to generate two sets of rules serving as the upper and low bounds for any other sets of rules corresponding to all arbitrary assignments of missing values, and then, (2) based on these upper and low bounds of the rules' sets, on testing procedure and on a classification criterion to select a subset of rules to be used for classification. The approach is primarily oriented to the application domains where an imputation is either cannot be theoretically justified or is impossible at all. Examples of such applications are given by domains where information used for classification is composed of asynchronous data streams of various frequencies and thus possessing different "life time", or such information is missing due to peculiarities of information collection system. Instead of missing value imputation, the proposed approach uses training dataset to cut down the potential rules set via forming its low and upper bounds with the subsequent testing the rules of the upper bound against the new dataset with missing values and selection of the most appropriate rules. The approach was applied to learning of intrusions detection in computer network based on asynchronous data streams incoming from multiple data sources. Experimental results confirm that the proposed approach to direct mining of data with missing values can yield good results.

1 Introduction

In the last three decades a large number of diverse powerful approaches, methods and techniques for data mining and Knowledge Discovery in Data bases (KDD) was developed. They are of wide use practically in any area of information technology, scientific research and industry that demonstrate ever increasing interest to extending practical application of data mining and KDD in new applications.

Unfortunately, as a rule, most of the popular and powerful data mining and KDD techniques cannot deal *directly* with real-life data due to the fact that

V. Gorodetsky et al.: *Direct Mining of Rules from Data with Missing Values*, Studies in Computational Intelligence (SCI) **6**, 233–264 (2005)
www.springerlink.com

most data bases are incomplete, contain wrong items and noise. The available data mining and KDD software tools usually contain special means aiming at *data cleaning* and *outlier detection*. The above means help to cope with certain kind of distortions in real-life data bases, thus, making the "traditional" data mining and KDD techniques applicable. However, incompleteness of data bases, particularly, presence of missing values, remains in challenge. The reason is that, opposite to the distortion like noise and presence of outliers, the problem of data mining with missing values is of fundamental difference. An ambiguity of missing values forces KDD miner either to predict these values, thus making the task more definite or to select somehow a solution from all possible ones. In any case, the latter leads to the computationally intensive search.

In the last two decades the main effort of a researcher dealing with data mining with missing values concerned the methods of the first type, i.e. the methods oriented to determining a reasonable assignment ("imputation") of the missing values. Significantly fewer investigations dealt with "*direct*" mining of data with missing values not presupposing the use of missing values imputation. A straightforward approach ignoring the examples with missing values is an exception. But as a rule such an approach significantly impoverish training and testing datasets, thus, making impossible a creation of powerful classification mechanisms. For instance, if percentage of missing values is 5% for each attribute and the dataset comprises 40 attributes then at average only 13% of the original sample can be used for training and testing [14]. However, in practice the percentage of missing values can be much higher, say, up to (20–30)% and over. In these cases the total number of examples without missing values can be about zero. Other existing approaches of the similar kind, i.e. approaches mainly not focused on missing values imputation, are outlined in Sect. 2 surveying the related works.

This paper proposes an approach to and technique for lattice-based direct mining of binary datasets with missing values aiming at extraction of classification rules with premises represented in a conjunctive form. The idea is to extract the sets of rules serving as the upper and low bounds for any other sets of rules corresponding to arbitrary assignments of missing values, and then, based on these upper and low bounds of rules' sets, on the results of testing procedure and on a classification quality criterion, to select a set of rules to be used in the classification mechanism. Let us emphasize that this method does not assume the imputation of missing values.

The rest of the paper is organized as follows. Section 2 briefly surveys related works indicating the basic ideas of dealing with missing values proposed in the previous research. Section 3 briefly describes a number of newly arisen applications of high practical importance where data mining and KDD with missing values is a central subtask. A peculiarity of these applications is a limited possibility to impute missing values due to restricted size of training and testing dataset and also due to distributed and asynchronous nature of input data streams used in both learning and decision making procedures.

Section 4 presents data mining with missing values problem statement and outlines the respective algorithm of rules extraction for the case if training and testing datasets are binary. It also presents the GK2 algorithm used in this work for extraction of rules from binary data, the basic idea and main theorems constituting a basis for building upper and low bounds of the rules' sets corresponding to any arbitrary assignments of missing values. This section also demonstrates the basic algorithms by example. Section 5 describes the experiments and experimental results of computation of upper and low bounds of the rules' sets and peculiarities of this procedure. As a case study the learning of intrusion detection in computer network as applied to anomaly detection task is used. Testing procedure and use of its results to create a classification mechanism are also described in Sect. 5. In conclusion the paper results are evaluated, and a tentative plan of future works is outlined.

2 Related Works

One of the earliest works on data mining and KDD with missing values applied to decision tree mining is [20]. In it, while considering the use of C4.5 technique to induce decision trees based on data with missing values, the author emphasized three main problems that are 1) How to deal with missing values in selection of a test to partition the training dataset if different attributes contain various percentage of missing values; (2) How to treat the cases, in which values of X are unknown for the selected test based on attribute X; and (3) How to proceed with a test on an attribute whose value is unknown if decision tree is induced. Although these questions were set within context of the particular method, C4.5, with certain variations they remain important up to now for majority of approaches to data mining and KDD dealing with missing values. In general case the above questions can be formulated as follows:

(1) How to deal with missing values in training procedure?
(2) How to evaluate the training quality? For instance, how to assign a truth value to a rule tested on a case when certain attributes of the rule premise cannot be assigned a value?
(3) How to use the constructed classification mechanism (decision tree-based, rule-based, etc.) for classification of unseen data with missing values?

In most papers all these questions are reformulated into "*how to impute missing values or cut the dataset*?" at every of three aforementioned steps (1)–(3) of data mining procedure. Indeed, if this problem were solved then the data mining problem would be a "standard" one, thus implying an application of a broad spectrum of existing techniques.

In practice, a selection of any technique for missing values imputation should depend on "the missing mechanism" [14] and on the relationships between the data attributes. From statistical point of view, three classes of missing data mechanisms can exist. In the simplest case (in the first class) the data are missing completely at random: *probability that an attribute X value is missing* does not depend on Xvalue location in the respective attribute domain. In the second class of the missing mechanism, the fact that the value of an attribute$X$ is missing depends on the value of some other attribute Y. This fact provides a data miner with additional information, which can be used in a certain way in order to more trustworthy impute the missing values. In the third class, a probability that value of an attribute $X$is missing depends on its value. Information concerning this probability can be used for missing value imputation. For example, in some cases data are missing because the value to be measured is either too large or too small, and this is why it is simply immeasurable. This fact can provide useful information for correct dealing with missing values. Additional source of a priori information that can be used for justification of missing data imputation is provided by knowledge about relationships existing between data attributes within particular classes.

The researchers propose many different approaches to and techniques for missing values imputation. In the simplest approach the records with missing values are erased from a training dataset, but this can lead to the fact that the rest of training sample could contain too small number of examples. An example was given in Sect. 1 [14].

Most of existing approaches is of statistical nature and use both knowledge about missing mechanism and correlations between attributes of each example computed based on available dataset. Imputation by regression (*linear regression* and *logistic regression*) [24] including *multiple imputation* [24] is one of such approaches to dealing with unknown data values. It can be applied to predict the value of one or several variables. These regression-based procedures use non-missing attributes. The disadvantage of this approach is that in many cases estimating by regression is as complex as learning of classification [30].

Hot Deck imputation [26] is a local approach, where the dataset is preliminary clustered on the basis of a similarity measure and then, within each cluster, the missing values are assigned the cluster most frequent values. Although this class of procedures has no rigorous mathematical justification, practitioners reported that this approach works well in many applications.

An approach called Expectation Maximization ([5, 14]), EM, is an algorithm based on estimations of the parameters of the probability distribution of an incomplete sample evaluated by maximum likelihood procedure.

The aforementioned approaches are based on the use of attributes with non-missing values. Several approaches use direct manipulation with missing data. As a rule, they are used to build decision trees ([13, 21]. The authors of the work [12] proposed to assign missing values the most probable ones estimated on the basis of training dataset. The paper [20] proposes to treat the value "unknown" as a new possible value of each attribute and to deal with it in the same way as with other attributes. However, the last approach works satisfactory if the unknown value is located close to the upper or low bounds of the value domain.

The paper [30] is focused on binary classification of data with missing values. Its idea is to somehow generate for both classes the equal number of rules with premises in conjunctive form, to assess the rules on the basis of testing procedure and then to use these rules, for example, in a weighted voting mode, while only accounting for those rules, which take definite values over the unseen case to be classified.

Specific problems have to be solved when we deal with missing values in association rule mining task. Ignorance of the missing values leads either to the reduced rates of support and confidence or/and generation of inconsistent rules that cannot be distinguished from the useful ones. This problem and probable approaches to overcoming it are discussed in [22] and [23]. The idea of these two papers consists in the use of some kind of dataset preprocessing, which, given item set, allows to discover from original dataset a so-called valid dataset, which corresponds to maximal data subset not containing missing values in the given item set. One more approach to the same task is proposed in [16]. It intends to generate so-called approximate rules, ~AR-rules. This algorithm is built on the well known Apriory algorithm and uses two main steps to deal with missing and noisy data. At the first step missing values are imputed via replacing with a probability distribution over possible values represented in terms of their frequencies in the dataset. At the second step this imputed dataset is used to mine association rules.

3 Examples of Applications

In the previous section it was indicated that one of the important aspects of data mining and KDD from databases with missing values is the nature of missingness. Indeed, the values of data can be missing due to imperfection of sensors and other components of measurement mechanism (errors, failures, e.g., transient failure, communication failure, etc.). In this case a reasonable data imputation procedure is pertinent. It is quite different if information cannot be collected due to certain reasons, like conditions of natural environment (e.g., if airborne observation equipment is used then information can be unavailable due to meteorological factors or due to masking). In this case imputation might be not relevant at all. Much more specific and complicate case, where missing values imputation is not applicable, can arise as a result of temporal and/or spatial nature of data, asynchronous and distributed mode of data collection, and discrete and distributed mode of decision making. An abstract example of such kind is as follows.

Let us consider what is often called a *situation* [6]. Generally, situation is understood as a state of a complex system constituted, for example, by a number of semi-autonomous objects ("*situation objects*", for short) having their particular goals (intents) and operating in a coordinated mode to achieve a common goal. Let us note that "situation object" can be a "physical" object (say, group of aircrafts) or an "abstract" one (e.g., components of software in which traces of attacks against computer network are manifested). Situation can be characterized by its "state" taking value from a finite set of labels. States of situation objects and respectively state of situation on the whole are of dynamic nature, i.e. they are varying in time.

There exists a distributed sensor system observing either particular objects or definite regions, where the situation objects can be located. Observations are distributed and asynchronous, and can be performed at different discrete time instants. The observations constitute the input of a decision making system aiming at classification of the state of the situations in discrete, as a rule, irregular time instants. This task is also well known as "*situation assessment*" [6][1]. As a rule, an observation system is capable to provide the situation assessment system only with a series of "snapshots" of the overall "picture", where a part of data can be missing. Here it is important to note that situation is specified not only in terms of *states* of particular situation objects but also in terms of a number of relationships given over them (spatial, temporal, and other, most of which is application-specific). Therefore, loss of information about an object unavoidably leads to missing of the values of attributes specifying relationships between the object in question and other ones.

According to the modern view of the architecture of the situation assessment system [7], the decision making procedure in it is organized at least at two levels. At the first level, local decisions corresponding to assessments of the particular object states are produced. At the second level these decisions are combined. Combining of decision is a conventional decision making task, whose input is formed by local decisions assessing the states of particular situation objects and relationships between them. As a rule, this input contains missing values and the size of sample presenting former experience and available for situation assessment learning is limited and does not provide statistically reliable information needed for missing values imputation.

Let us explain the above abstract consideration by example that is a detection of intrusions in computer network. This task is a typical application from the situation assessment scope containing missing values in both training and testing samples.

Currently the coordinated distributed attacks performed by a team of malefactors from distributed hosts constitute main threats for computer networks and information. "Traces" of an attack can appear in input traffic or in different data generated by a computer network assurance system. For example (see Fig. 1, and also [9]), these traces are displayed in *IP* packets of traffic, in *Tcpdump* generated via traffic preprocessing, in audit data trail, in sequences of system calls of operating system, in data resulting from

[1] Situation assessment task is a central subtask of the well known and very important modern task called "situational awareness" [6].

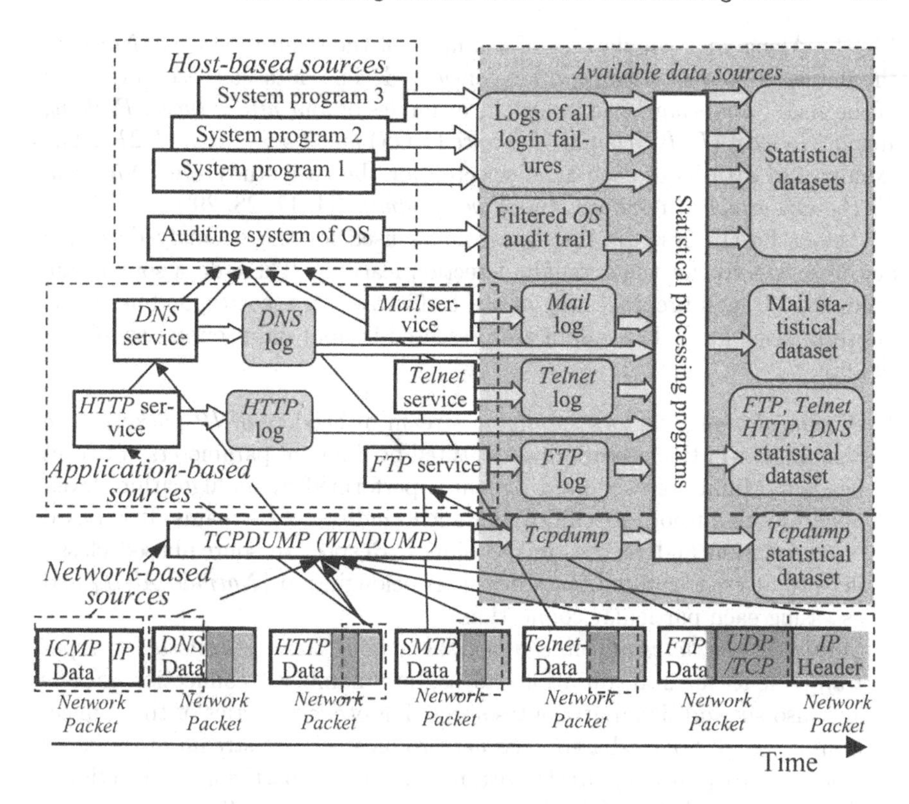

Fig. 1. Multiplicity of data sources (given in *grey* color) within a host which can alert intrusions (The content of the figure is taken from [9])

monitoring of application servers, queries to databases and directories, in data specifying users' profiles, etc. Such data are generated in different hosts of computer network. The timely detection of an illegitimate user's activity to assess a security status of computer network is potentially feasible only in case of information from different sources fusion. Formally, intrusion detection is a classification task, that has to detect intrusions based on *combining of particular alerts* produced by analyzers (local classifiers) dealing with particular data sources. Assessment of the security–related status of a computer network intending to detect malicious activity presents an interesting example of situation assessment tasks.

Let us describe a case study from intrusion detection scope, which is further used to validate an approach proposed for dealing with missing values in data mining and KDD procedures. The purpose of the description given below is also to explain why temporal nature of data from different sources leads to missing values in input data of an intrusion detection system (IDS).

In the case study the anomaly detection task is considered. In it classification of user's activity is produced on the basis of network-based data (see

Fig. 1). Along with the dataset of examples of the security status "*Normal*" the dataset of class "*Abnormal*" is considered. The dataset presenting cases of the class "*Abnormal*" comprises the examples of four *attack types*: *Probing, Remote to local* (*R2L*); *Denial of service* (*DOS*) and *User to root* (*U2R*). The examples of attacks of each type selected for the case study are *SYN-scan, FTP-crack attack, SYN flood*, and *PipeUpAdmin* ([4, 17, 28, 29]).

Let us briefly describe the data sources used for the anomaly detection and, respectively, for the anomaly detection learning. These data sources are produced through preprocessing of the traffic raw data, corresponding to the network-based level. These data are presented by four data sources as follows ([8, 9]):

1. *Stream of binary vectors specifying* stream of headers of *IP* packets. The components of this vector are constituted by different parameters of packet headers. Mining of such data stream is performed by an algorithm developed by the authors specifically for such kind of data streams. It is based on correlation and regression techniques. Respective source-based classifier produces stream of binary decisions belonging to {*Normal, Alert*} and assessing each particular connection.
2. *Statistical attributes of connections (sessions of users) manifested in input traffic.* As features the duration, status, total number of connection packets and also six additional attributes specifying other statistics of the connection are used. Source-based classifier produces binary stream of decisions belonging to {*Normal, Alert*} with regard to each particular connection.
3. *Statistical attributes of traffic (users' activity) during the short time (5 sec) intervals.* This data source is presented by four features specifying integral characteristics of input traffic that are total numbers of connections and services of different types during 5 sec. Source-based classifier operating with these data also produces stream of binary decisions belonging to {*Normal, Alert*} with regard to connections occurred within particular sliding windows of 5 sec time interval.
4. *Statistical attributes of traffic (users' activity) for long time intervals.* In this data source, the same statistics as for short time interval are used as the features. The length of time interval is varied and corresponds to 100 connections.

The training and testing samples for each of above four types were produced on the basis of processing of *Tcpdump/ Windump* data (see Fig. 1)[2]. *TCPtrace* utility was used for this purpose as well as several other programs.

Thus, the strategy of anomaly detection is organized in two steps. First, source-based classifiers asynchronously produce binary decisions taking values from the set {*Normal, Alert*}. These decisions constitute a binary vector of

[2] This preprocessing was carried out by Ph.D. student of SPIIRAS M. Stepashkin under supervision of Prof. I. Kotenko.

features processed at the upper level, whose purpose is to combine these decisions and produce the final one. This strategy corresponds to a well known meta-classification approach [19].

Meta-classification procedure possesses certain peculiarities caused by the fact that IDS is a real-time system, and source-based classifiers produce their decisions asynchronously, because each source-based classifier produces its decision at the time when it receives all the data needed for making decisions, which is irregular process. Figure 2 demonstrates this property graphically.

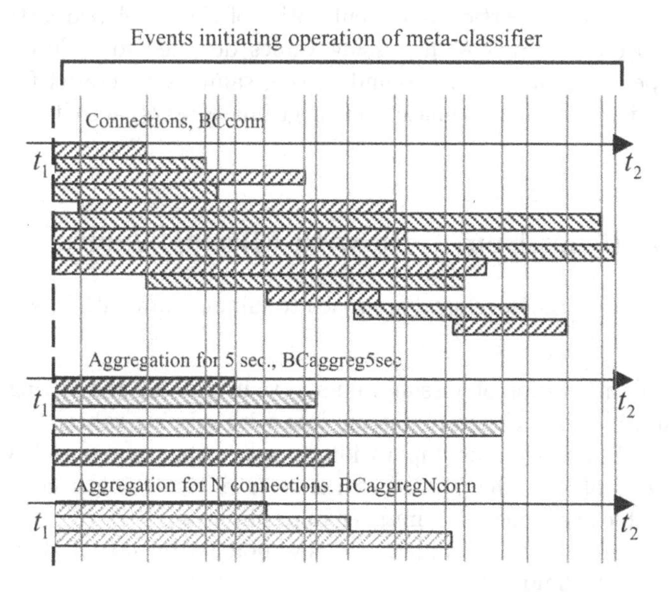

Fig. 2. Explanation of the model of meta-data used for training of meta-classifier and decision making within interval $[t_1, t_2]$. Vertical lines correspond to the times of arrival of new decisions of source-based classifiers that correspond to the times of producing decisions by meta-classifier

Let us call a decision newly produced by a source-based classifier an input *event* for meta-classifier and represent it as *<Decision of source-based classifier X, Time>*. At a certain time some base classifiers have already produced decisions though other ones have not. To combine such decisions, meta–classifier can act in two modes: (1) to wait when all base classifiers produce the decisions and then combine them, and (2) update the combined decision at the time of a new input event arrival. Let us omit here a thorough analysis of both approaches, although this aspect is very interesting, and only explain and justify shortly the second variant selected in this work.

For this variant, there exists a risk that some of the decisions previously produced by the source-based classifiers are already not relevant to the situation ("too archaic"). To take this fact into account, for each base classifier (data source) we introduce a so called *life time* of decisions produced by it. "Life times" are various for different sources, and their particular values are elicited from experts. If time interval elapsed from the moment of a decision producing exceeds its life time then in the meta-data the value corresponding to this decision is reckoned as missing. Therefore, in this case meta-classification is performed on the basis of data with missing values.

Thus, temporal and asynchronous nature of data sources leads to the data mining and decision making formal model with missing values.

The intrusion detection-based case study described above is used below for exploration of the properties and peculiarities of the developed lattice-based approach to data mining with missing values destined for rule extraction. Detailed specification of training and testing samples generated from input traffic containing both legitimate and illegitimate users' activity is done in Sect. 5.

4 Mining Rules from Data with Missing Values

The main ideas of the developed approach to mining data with missing values are as follows:

(1) To avoid imputation of missing values both in training and testing samples and also in classification of unseen data with missing values;
(2) Instead of missing value imputation, to assess the upper and low bounds of the sets of rules that can be extracted from the training dataset under all possible assignments of missing values in it; and
(3) To compute the above upper and low bounds by use of rigorous rule extraction techniques.

To better explain the ideas of the developed approach to extraction of rules from data with missing values, let us first outline general ideas of most existing mining techniques dealing with complete datasets. Although in this work we use GK2 algorithm developed by the authors [10, 11], for our purposes it is important to briefly remind about well known AQ basic algorithm, because GK2 differs from it only in algorithmic implementation of AQ basic step numbered below as step 3. It should be noted that AQ is the earliest proposed method of such kind [15], and up to now it remains a basic one for many other learning algorithms and systems (rough set-based [18], [27], RIPPER [3], CN2 [2], etc.). The last version of this method, AQ20, was implemented recently.

4.1 AQ Method

AQ method ([15]) aims at induction of rules in the form "*if <condition> then <class>*", where *<condition>* is a conjunction of attributes ("attribute test"). Several rules can also be represented as disjunction of rules. A conjunct is said to *cover* an example if it takes value "*true*" for this example.

The basic AQ induces rules for each class step by step. Simple description of this algorithm is as follows [15]:

1. Divide all examples into subsets ***PE*** of *positive* and ***NE*** of *negative* examples;
2. Choose randomly or by design one example from ***PE*** and call it the *seed*;
3. Find a set of *maximally general rules* (*MGR*) characterizing the *seed*. The generalization limit is defined by the set ***NE*** : a resulted description of the *seed* is not allowed to cover any example from ***NE***. The obtained set of rules is called a *star*;
4. According to a criterion select the best rules in the *star*;
5. If these rules jointly with the previously found rules cover all examples from ***PE*** then stop. Otherwise, find another *seed* among the uncovered examples in ***PE*** and go to 2.

Step 3 is performed by a special procedure called *star generation procedure* that is the central one in AQ and many other algorithms. The rule preference criterion is chosen according to the task at hand, and in many cases it combines several simple criteria, however, the rule coverage factor (the ratio of covered examples of the set ***PE***) usually is the first among them.

4.2 Extraction Rules from Propositional Data with Missing Values: The Main Theorem

Let us assume that training dataset is presented in binary scale and assigned one of three values: "0", "1" and "*" interpreted as "*false*", "*true*" and "*unknown*" (missing) respectively. Also assume that the column corresponding to the class label does not contain missing values.

Briefly the idea of dealing with missing values without their imputation is as follows. Assume that the selected *seed* does *not contain missing values,* and the set of counterexamples, ***NE***, does not contain cases equal to the selected *seed.* If we completely assigned missing values of training dataset in arbitrary way, we would be able to extract *maximally general rules* via a conventional technique, for example, through the AQ algorithm. Different variants of such assignments would lead to different sets of rules extracted. Let us denote the set of extracted rules for an arbitrary assignment of missing values as R_*.

It was found out that for each seed there exist two sets of *maximally general rules* serving as exact *low* and *upper* bounds for all possible sets of rules corresponding to any arbitrary assignments of missing values. They correspond to two special assignments of missing values in training dataset. *Given seed*, these bounds meet the following deducibility relations:

$$R_{low} \subseteq R_* \subseteq R_{upper} \qquad (1)$$

where R_{low}–the exact *low* bound of all the sets of *MGR*; R_{upper} – the exact *upper* bound of all the sets of *MGR,* and R_*–the set of *MGR* corresponding to an arbitrary assignment of the missing values.

Informally, it could be said that R_{upper} bound corresponds to the "*optimistic*" assignment of the missing values, whereas R_{low} corresponds to the "*pessimistic*" one. Let us explain conceptually how "*optimistic*" and "*pessimistic*" assignments can be built and then justify the above assignments design formally.

Let us denote an arbitrary example of the training dataset as $t(i)$, where i is the index of this example in dataset table. Let k be the index of the *seed* (index of chosen positive example), I_k^+ be the set of indexes assigned to attributes of *seed*. While searching for *MGR* corresponding to the chosen seed $t(k)$, we ignore all the columns of training dataset, whose indexes do not belong to the set I_k^+. After such reduction of the training dataset other positive examples and all negative examples can contain missing values for attributes with indexes from the set I_k^+. Let us denote the index set of missing values in particular negative example $t(l)$ by $I_{l,k}^-$, the index set of undefined attributes in particular positive example $t(r), r \neq k$, by $I_{r,k}^+$. Let us further consider *two variants of assignment of missing values* of attributes with indexes $i \in I_{l,k}^-$ (for negative examples) and with indexes $i \in I_{r,k}^+$ [3] (for positive examples):

$$t_i^l = \neg t_i^k, \quad i \in I_{l,k}^-, \quad l \in \boldsymbol{NE}\ , \tag{2}$$

$$t_i^r = t_i^k, \quad i \in I_{r,k}^+, \quad r \in \boldsymbol{PE}\ , \tag{3}$$

and

$$t_i^l = t_i^k, \quad i \in I_{l,k}^-\ , \quad l \in \boldsymbol{NE}\ , \tag{4}$$

$$t_i^r = \neg t_i^k, \quad i \in I_{r,k}^+, \quad r \in \boldsymbol{PE}\ . \tag{5}$$

The first assignment, (2)–(3), is such that it maximally "increases" the distinctions between the *seed* and negative examples, and maximally "increases" the similarities between the *seed* and other positive examples. On the contrary, the second one, (4)–(5), maximally "increases" the similarities between the *seed* and negative examples, and maximally "increases" the distinctions between the *seed* and other positive examples. Intuitively, the first assignment is called *optimistic* because it cannot decrease both the generalization level of the *MGR* generated from complete dataset and the values of the coverage factors of such *MGR*. In the second assignment that is called "*pessimistic*" both the generality of *MGR* generated from the complete dataset and values of their coverage factors cannot increase. The *Theorem 1* given below formulates formally the above facts introduced conceptually and indicates how to find the upper and low bounds of *MGR*.

[3] Informally, $I_{l,k}^-$ comprises the subset of indexes of missing values in the negative example $t(l)$, which are assigned in *seed* $t(k)$. Analogously, $I_{r,k}^+$ comprises the subset of indexes of missing values in the positive example $t(r)$, which are assigned in *seed* $t(k)$.

Theorem 1. *Let us assume that* seed $t(k)$ *does not contain missing values. Let* R_* *be the set of all* maximally general rules *for an arbitrary assignments of missing values in the negative and positive examples* $t(i)$, *whose indexes* $i \in I^-_{l,k}$, $l \in \boldsymbol{NE}$, *and* $i \in I^+_{r,k}$, $r \in \boldsymbol{PE}$ *respectively;* R_{upper} *be the set of all* maximally general rules *corresponding to the assignments of missing values in the positive and negative examples (2)–(3), and* R_{low} *be the set of all* maximally general rules *corresponding to the assignments of missing values (4)–(5).*

Then $R_{low} \subseteq R_* \subseteq R_{upper}$[4].

This *Theorem* provides general framework for mining data with missing values. In other words, it restricts the search through explicit indication of the set of rules within which the *maximally general rules* should be found. Unfortunately it says nothing about how, based on the above upper and low bounds, to select the set of rules in a particular application. In general case, the rule set under search results from analysis of particular rule properties, appropriate organization of the testing procedure and use of requirements to the classification mechanism properties. The main of these aspects will be analyzed below as applied to the case study introduced in Sect. 3.

To demonstrate the *Theorem 1* numerically let us first explain briefly an algorithm of rule extraction that is used for this purpose below.

4.3 GK2 Algorithm for Extraction Rules

Rule induction algorithm GK2 [10, 11] operates with relational Boolean data. For demonstration of this algorithm, the data sample presented in Table 1 is used.

Table 1. Boolean training data

$t(i)$	t^i_1	t^i_2	t^i_3	t^i_4	t^i_5	t^i_6	Class Q
t(1)	0	0	1	1	1	0	1
t(2)	0	1	1	0	1	0	0
t(3)	0	1	0	1	0	1	0
t(4)	1	1	0	1	0	0	0
t(5)	0	1	1	1	1	0	1
t(6)	0	1	1	0	1	1	0
t(7)	1	1	0	1	0	1	0
t(8)	1	0	1	0	1	0	1
t(9)	0	1	0	1	1	0	0
t(10)	1	1	0	0	0	0	1
t(11)	1	1	1	1	0	1	1
t(12)	0	1	0	0	0	1	0
t(13)	1	1	0	0	1	1	0
t(14)	1	1	1	1	1	0	0

Table 2. Matrix $S(1)$ for seed $t(1)$

	x^i_1	x^i_2	x^i_3	x^i_4	x^i_5	x^i_6	Class Q
$S_{2,1}$	1	0	1	0	1	1	0
$S_{3,1}$	1	0	0	1	0	0	0
$S_{4,1}$	0	0	0	1	0	1	0
$S_{6,1}$	1	0	1	0	1	0	0
$S_{7,1}$	0	0	0	1	0	0	0
$S_{9,1}$	1	0	0	1	1	1	0
$S_{12,1}$	1	0	0	0	0	0	0
$S_{13,1}$	0	0	0	0	1	0	0
$S_{14,1}$	0	0	1	1	1	1	0

[4] This Theorem is proved rigorously but here this proof is omitted for brevity.

The following notations are used hereinafter: $I_m=\{1,\ldots,m\}$ - index set of attributes (see Table 1); $I_k \subset I_m$ - a subset of attribute indexes; ***PE***, ***NE*** – the sets of indexes of positive and negative examples respectively; $t(k)$ -*k-th* example (row of the relational data table) of training dataset; $X = \{x_1,\ldots,x_m\}$ – the set of propositions corresponding to the data attributes. Each of them can be used with negation ($\bar{x}_i$) or without it (x_i); in both cases it is called "*literal*". A literal is denoted as $\tilde{x}_i$ and it can take value x_i or $\bar{x}_i$; $\Phi_X \{x_1,\ldots,x_m\}$ – the entire set of formulae that can be built over the proposition set $\{x_1,\ldots,x_m\}$ by use of connectives & (conjunction), $\vee$ (disjunction) and negation.

A premise <*condition*> of a rule, which GK2 induces in the form "*if* <*condition*> *then* <*predicted class*>", is a conjunction of the literals constituted by a subset of attributes. Formally, any rule R_j is specified as follows:

$$R_j = F_j \supset Q, F_j = \&_{i\in I_j}\tilde{x}_i \ , \tag{6}$$

where R_j stands for rule indexed by j, $F_j = \&_{i\in I_j}\tilde{x}_i$ is a rule premise, I_j is a subset of attribute indexes, Q is a class label, and $\supset$ denotes logical implication connective. Further it is default assumed that each rule to be extracted has to be *consistent* that is a rule premise F_j can take "*true*" value only over the positive training examples ***PE***. This also means that there is no pair of equal examples belonging to both ***PE*** and ***NE*** sets of the training sample.

If formula $F_j \supset Q$ (we speak about the formula as the whole but not about its premise) takes value "*true*" over all positive and negative examples of training dataset then it is called "*consistent*" with training dataset. However, if the formula is *consistent* with training dataset but its premise F_j covers no positive examples of class Q then it cannot be "interesting".

The purpose of GK2 is, given the *seed*, to search for *MGR*. In particular applications certain additional constraints (constraint on maximal length of rules, minimal value of coverage factor, etc.) can additionally be imposed.

Additional notations used below are as follows: Φ – the set of formulae F_j present in (6) (potential premises of rules $R_j = F_j \supset Q$), $\Phi \subset \Phi_X$; $U \subseteq \Phi$ – a subset of $F_j \in \Phi$ with positive coverage factor over examples of class Q[5]; $\boldsymbol{Z} \subseteq \boldsymbol{U}$ – the set of all premises of consistent formulae $F_j \supset Q$ having positive values of the coverage factor; $\boldsymbol{Z}^+ \subseteq \boldsymbol{Z}$ – the set of premises of the set of the *MGR*. corresponding to the given *seed*. Let us note that formulae sets $\boldsymbol{Z}^+ \subseteq \boldsymbol{Z} \subseteq \boldsymbol{U} \subseteq \Phi$ introduced are the *sets of premises* of rules if the seed belongs to the class Q. In the same way the set of premises $\boldsymbol{Z}^-$ of the *MGR* can be introduced for the seeds of the class $\bar{Q}$.

Let us introduce a definition of the partial order relation between premises F_j of rules $R_j \subset \boldsymbol{Z}$.

Definition 1. *Let* $F_j \in \mathbf{Z}$ *and* $F_r \in \mathbf{Z}$. *We say that* $F_j \precsim F_r$ *("formula* F_j *is less than or equal to* F_r*") if and only if* $I_j \subseteq I_r$, *and all common literals of*

[5] Let us note that formulae of the set $\boldsymbol{U}$ can be inconsistent.

these formulae have the same "sign" (i.e. each pair of common propositions both are either negative or positive).

In other words, formulae F_j and F_r are such, that the former comprises a subset of literals of the latter and their common literals are of the same "sign". Thus, formula F_j is "*shorter*" and this is why "*more general*" than formula F_r. It can be proved that this order imposes a lattice structure over all formulae $F_j \in \boldsymbol{Z}$, hence, $\boldsymbol{Z} = \langle \boldsymbol{Z}, \precsim \rangle$ is a lattice ([1]). According to the lattice properties, the lattice $\boldsymbol{Z}$ contains the subset of so-called "*minimal*" (shortest) formulae $\boldsymbol{Z}^+$, and this subset is such that Z^+ is deductively equivalent[6] to the complete formulae set $\boldsymbol{Z}$. It should be reminded that the formulae set $\boldsymbol{Z}^+$ comprises all the shortest ("*minimal*") premises of rules under search. Thus, the task of search for *MGR* is equal to the search for rules belonging to the subset $\boldsymbol{Z}^+$. GK2 is destined to search for the subset $\boldsymbol{Z}^+$.

Conceptually, the main course of GK2 can be as follows: (1) to search for the set $\boldsymbol{U}$, (2) to search for its subset $\boldsymbol{Z}$, and, finally, (3) to isolate the subset $\boldsymbol{Z}^+$ from the set $\boldsymbol{Z}$. The subset $\boldsymbol{Z}^+$ corresponds to the complete set of *MGR*.

Let us consider the theoretical basis of GK2 algorithm implementing step 3.2 of the AQ algorithm presented in Subsect. 4.2.

Let the seed $t(k) \in \boldsymbol{PE}$ correspond to the *k-th* line of the training dataset $\boldsymbol{T}$. Its formal specification in propositional logic looks as follows:

$$F_k = \&_{i \in I_k} \tilde{x}_i = \tilde{x}_1^k \tilde{x}_2^k \ldots \tilde{x}_m^k \text{ [7]}, \tag{7}$$

where

$$\tilde{x}_i = \{x_i, \quad if \quad t(k,i) = 1, \quad or \quad \bar{x}_i, \quad if \quad t(k,i) = 0\} \tag{8}$$

Formula $F_k \in \boldsymbol{U}$, because its coverage factor is positive, actually its coverage factor is no less than *1/*$|\boldsymbol{PE}|$, where $|\boldsymbol{PE}|$ is equal to the cardinality of the set $\boldsymbol{PE}$. It is also consistent according to the assumption that there is no identical example in $\boldsymbol{NE}$.

Each formula F_k defines a formula set U_k as follows:

$$U_k = \{F_{ki} \in \boldsymbol{U} : F_{ki} \precsim F_k\} . \tag{9}$$

where F_{ki} comprises a subset of literals constituting F_k, and the set $\boldsymbol{U}$, the entire set of formulae having positive values of coverage factor, is as follows:

$$\boldsymbol{U} = \cup_{k \in PE} U_k . \tag{10}$$

Let us recall that the set $\boldsymbol{U}$ can contain inconsistent premises, and our tentative purpose is to extract all consistent ones, i.e. to find the set $\boldsymbol{Z}$ of all *consistent* premises of rules having positive coverage factor.

---

[6] It is said that a formulae set $\boldsymbol{A}$ is deductively equivalent to the formulae set $\boldsymbol{B}$ if and only if each formula of B is inferable from the formulae set $\boldsymbol{A}$.

[7] In (7) and hereafter the conjunction symbol is omitted for brevity.

The premise (7) corresponds to one of the most specialized rules for the class Q. Generalization corresponds to deletion of certain literals from it, and this deletion either preserves the premise coverage factor value or increases it. Such a deletion has also to preserve consistency of the resulting premise. The last requirement restricts the possibility of rule generalization. Like AQ, the idea of GK2 is to search for consistent premises of minimal length.

GK2 algorithm aims to search for the complete set of *MGR* $\boldsymbol{Z}^+$ without search for the entire set of consistent rules with positive coverage factor $\boldsymbol{Z}$, that leads to considerable decrease of computations. Below this procedure is described in strictly formal way, but all its steps are explained informally by examples.

Let us specify an arbitrary *k-th* positive example $t(k)$ in terms of propositional formula like (7) indexing its literals by superscript k. The "*similarity*" *k-th* and *l-th* positive examples of the training sample corresponds to the coincidence of the literals with the same subscripts of the formulae F_k and F_l (both of them must be either positive or negative):

$$S(x_i^k, x_i^l) = \{1, \quad if \quad \tilde{x}_i^k = \tilde{x}_i^l, \quad and \quad 0, \quad otherwise.\} \tag{11}$$

Component-wise negation of the similarity vector sometimes is called a *distinction vector*. For each pair of examples, one can compute the binary *vector of similarities*.

Let us select an *example* $t(k) \in \boldsymbol{PE}$, and call it (like in AQ algorithm, [6]) a *seed*. Representation of this *seed* in terms of propositional formula looks like it is given in formula (7). Let us introduce the *similarity matrix* $\boldsymbol{S}(k)$ reflecting similarities between given *seed* $t(k)$ specified by F_k and all negative examples represented in the same form (7). The dimensionality of this matrix is equal to the number of attributes in the training data table $\boldsymbol{T}$ and its size is equal to the cardinality of the set $\boldsymbol{NE}$. Thus, in the Table 2 matrix $\boldsymbol{S}(k)$ indicates similarities between *seed* number k and negative examples in its positions assigned "1", and it indicates their distinctions in its positions assigned "0". An example of $\boldsymbol{S}$ matrix for *seed* $t(1)$ and negative examples, presented in Table 1, is given in Table 2.

The next step of GK2 is to introduce a lattice-like structure over the lines of matrix $\boldsymbol{S}(k)$,whose use decreases the amount of further computations.

Definition 2. *Let $S_{r,k}$ and $S_{l,k}$ be two lines of the matrix* $\mathbf{S}(k)$. *We say that $S_{r,k} \geq S_{l,k}$ if and only if this inequality is held for $S_{r,k}$ and $S_{l,k}$ component-wise and values "0" and "1" are considered here as integers, i.e.* $1 > 0$.

An example of such a partial ordering of lines of $\boldsymbol{S}$ matrix given in Table 2 in terms of Hasse diagram is depicted in Fig. 3.

The maximal elements of the set of lines of $\boldsymbol{S}$ matrix play an important role in the GK2 algorithm: they and only they have to be accounted for in the further search for *MGR* because they accumulate the *maximal similarities* (indicated as "1" in $\boldsymbol{S}$–matrix) between *seed* and total batch of the negative examples. In Fig. 3 the maximal elements are $S_{2,1}$, $S_{9,1}$, and $S_{14,1}$.

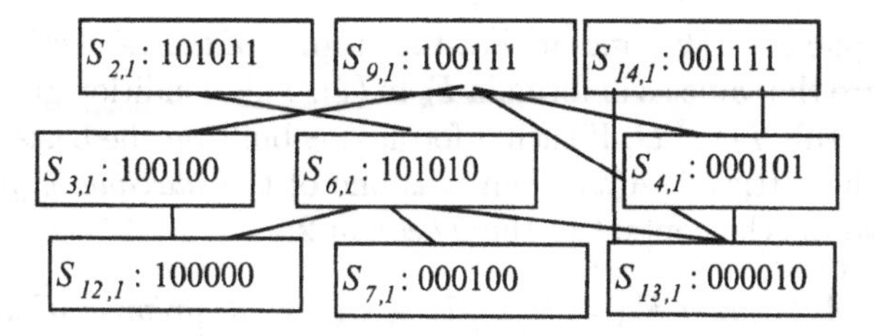

Fig. 3. Hasse diagram given over the lines of similarity matrix $S(1)$

Let $I_{\max}(k)$ stand for the set of maximal lines built for the *seed* k and $l \in I_{\max}(k)$ – index of a maximal line. Let $I_D(k,l)$ also stand for the set of indexes of attributes of the maximal line l assigned "0". In other words, each $I_D(k,l)$ comprises the list of positions of the maximal lines $S_{l,k}$, where it differs from the *seed k:*

$$I_D(k,l) = \{i : i \in I_m,\ S_{k,l}(i) = 0\}\ . \tag{12}$$

For example, the sets $I_D(k,l)$ corresponding to the maximal negative lines in Fig. 3 are as follows:

$$I_D(2,1) = \{2,4\}, I_D(9,1) = \{2,3\}, I_D(14,1) = \{1,2\}\ . \tag{13}$$

The sets $I_D(k,l)$ indicate minimal distinctions between the *seed* and negative examples and therefore indicate the literals - candidates for keeping in the premise F_k (7) to preserve consistency.

Thus, given seed, a generalization algorithm intended to search for all *MGR* can be described as follows:

1. Given *seed* k, build a lattice of auxiliary family set $MA(k) = \langle\{A_s(k)\}, \subseteq\rangle$ where all $A_s(k)$ meet two conditions:
   a. $A_s(k) \in I_m$, i.e. it is a subset of the attribute indexes;
   b. $A_s(k)$ has nonempty intersection with each $I_D(k,l)$:
   $$[\forall l \in I_{\max}(k)]\{A_s(k) \cap I_D(k,l) \neq \emptyset\}\ .$$
2. Find the minimal sets of the lattice *MA(k)*.
3. Map each minimal set of the lattice *MA(k)* the *maximally generalized rule* like it is prescribed in the *Theorem 2* given below.

Let us give an informal explanation of the algorithm described above. The major property of the sets $A_s(k) \subseteq I_m$ constituting the lattice *MA(k)* is that each of them must contain *at least one attribute from every set of distinctions* $I_D(k,l), I \in I_{\max}(K)$. Therefore, minimal elements of *MA(k)* correspond to the sets of minimal distinctions of the seed k for all lines of the similarity matrix $\boldsymbol{S}(k)$. Thus, the formula $F_{ki} = \&_{j \in I_k}\tilde{x}_j \in U_k$ is a shortest premise of a consistent rule with positive value of the coverage factor if the set of indexes I_k is one of $A_s(k)$, and at that the "signs" of the respective literals in premise F_{ki} are the same as in formula F_k in (7). This condition guarantees the consistency of rule $F_{ki} \supset Q$. If such a formula is built on the basis of minimal elements of the lattice $MA(k)$, then it is one of the *maximally general* one. This conclusion is the content of the *Theorem* 2:

Theorem 2. *A formula* $F_{ki} = \&_{j\in I_k}\tilde{x}_j \in U_k$ *is the premise of a consistent rule with positive value of coverage factor if and only if* $I_k \in \{A_s(k)\}$.

Corollary 1. *A formula* $F_{ki} = \&_{j\in I_k}\tilde{x}_j \in U_k$ *is the premise of a* maximally general rule *if and only if* $I_k \in \{A_s(k)\}$ *and* $A_s(k)$ *is a minimal element of the lattice* $MA(k)$.

It is worth noting that *there is no need to compute the complete family set* $\{A_s(k)\}$. It is actually possible to compute directly its minimal elements on the basis of an algorithm of *minimal binary cover* (see, for example, [16]). One of such algorithms was developed by the authors of this paper and used in the GK2 software implementation.

Thus, a brief description of the GK2 algorithm for extraction rules from binary data is as follows:

1. (Same as in AQ) Divide training data sample into subsets ***PE*** of *positive* examples and ***NE*** of *negative* examples;
2. (Same as in AQ) Select a *seed*, $t(k) \in \boldsymbol{PE}$.
3.1. Compute $\boldsymbol{S}(k)$ matrix according to (11).
3.2. Compute the set of indexes $I_{\max}(k) = \{S_{l,k}\}$ of maximal negative lines of the $\boldsymbol{S}(k)$ matrix.
3.3. Compute the index sets $I_D(k,l)$ using (12).
3.4. Find minimal elements of the lattice *MA(k)*.
3.5. Generate the premises of *MGR* as indicated in *Theorem 2*.

The further steps of GK2 are the same as for AQ algorithm.

Numerical Example

Let us continue the example of search for *MGR* based on sample given in Table 1. The results corresponding to the step 3.3 are given in (13). The next step, 3.4, has to generate the family set $\{A_s(K)\}$ in order to build the lattice *MA(k)* and to find its minimal elements. Although in the implemented software we use for this purpose an original algorithm of *minimal binary covers,* in example at hand $\{A_s(K)\}$ can just be built manually:

$$\{A_s(K)\} = \{\{2\},\{1,2\},\{2,4\},\{2,3\},\{1,3,4\},\{2,3,4\},\{1,2,3,4\}\} .$$

Hasse diagram of the lattice *MA(1)* structured according to the theoretic-set inclusion is given in Fig. 4. The *minimal* elements of the lattice *MA(1)* are {*2*} and {*1, 3, 4*}.

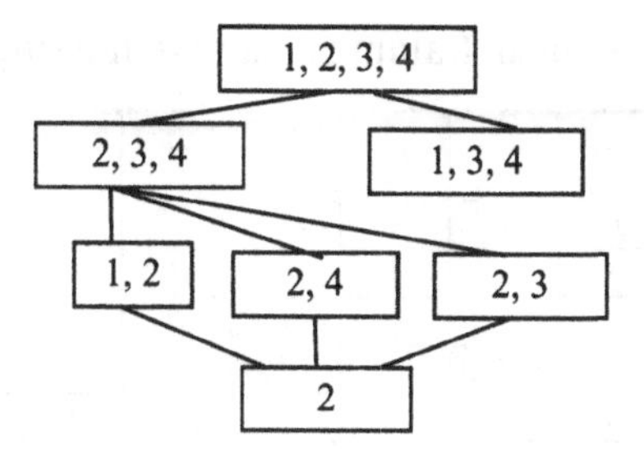

Fig. 4. Hasse diagram of the lattice *MA(1)* = <{{*2*},{*1,2*},{*2,4*},{*2,3*}, {*1,3,4*}, {*2,3,4*}, {*1,2,3,4*}}, ⊆>

The next step 3.5 is to generate *maximally general rules* on the basis of minimal elements of the lattice *MA(1)*) that results in the following:

$$R_1 = \bar{x}_2 \supset Q, R_2 = \bar{x}_1 x_3 x_4 \supset Q \ .$$

Let us notice that propositions x_1 and x_2 appear in rules R_2 and R_1 with negation because both of them have such "sign" in the formula F_1 built for the *seed* $t(1)$ (see Table 1).

The values of the coverage factors for these rules are as follows:

$$S(x_2, Q) = 0.4, S(\bar{x}_1 x_3 x_4, Q) = 0.4 \ .$$

4.4 Mining Rules from Data with Missing Values: An Example

Let the training sample with missing values be as given in Table 3. This dataset corresponds to the same data as in Table 1 but some attributes in the former are missing. The percentage of the missing data is about 25%. To perform a comparison of *MGR* that can be extracted from such a sample with the *MGR* extracted from completely defined sample, let us solve the task of extraction rules from data with missing values for the same *seed*, i.e. for the *seed* $t(1)$.

The *Theorem 2* assumes that seed does not contain missing values, therefore the first what is necessary to do is to delete from the training data the columns corresponding to the attributes, whose values are missing in the chosen *seed*, i.e. to delete columns corresponding to the attributes x_2 and x_5, therefore I_1^+={*1, 3, 4, 6*}. The columns to be deleted are given in Table 3 in grey color. Thus, the training dataset must be such as presented in Table 3 without columns indexed by the subscripts second and fifth.

According to the *Theorem 2*, to search for the set of *MGR* rules R_{upper}, it is necessary to assign the missing values to the positive and negative examples as made in Table 4. Respectively, to search for the MGR rules of the set R_{low}, it is necessary to assign the missing values as shown in Table 5. Below both tasks (search for R_{upper} and search for R_{low}) are being solved in parallel.

Table 3. Boolean training data with missing values

$t(i)$	t_1^i	t_2^i	t_3^i	t_4^i	t_5^i	t_6^i	Class Q
t(1)	0	*	1	1	*	0	1
t(2)	*	1	1	0	1	*	0
t(3)	0	1	0	1	0	*	0
t(4)	1	1	0	*	0	0	0
t(5)	0	*	*	1	1	0	1
t(6)	*	*	1	0	1	1	0
t(7)	*	1	0	1	0	*	0
t(8)	1	0	1	*	1	*	1
t(9)	0	1	0	*	1	0	0
t(10)	1	1	0	0	*	0	1
t(11)	1	*	1	1	0	1	1
t(12)	*	1	*	0	0	1	0
t(13)	*	1	0	0	1	1	0
t(14)	1	*	1	1	1	0	0

Table 4. Training data used for extraction of the set R_{upper}

$t(i)$	t_1^i	t_3^i	t_4^i	t_6^i	Class Q
t(1)	0	1	1	0	1
t(2)	1	1	0	1	0
t(3)	0	0	1	1	0
t(4)	1	0	0	0	0
t(5)	0	1	1	0	1
t(6)	1	1	0	1	0
t(7)	1	0	1	1	0
t(8)	1	1	1	0	1
t(9)	0	0	0	0	0
t(10)	1	0	0	0	1
t(11)	1	1	1	1	1
t(12)	1	0	0	1	0
t(13)	1	0	0	1	0
t(14)	1	1	1	0	0

Table 5. Training data used for extraction of the set R_{low}

$t(i)$	t_1^i	t_3^i	t_4^i	t_6^i	Class Q
t(1)	0	1	1	0	1
t(2)	0	1	0	0	0
t(3)	0	0	1	0	0
t(4)	1	0	1	0	0
t(5)	0	0	1	0	1
t(6)	0	1	0	1	0
t(7)	0	0	1	0	0
t(8)	1	1	0	1	1
t(9)	0	0	1	0	0
t(10)	1	0	0	0	1
t(11)	1	1	1	1	1
t(12)	0	1	0	1	0
t(13)	0	0	0	1	0
t(14)	1	1	1	0	0

The following computations correspond to the steps 3.1–3.5 of the GK2 algorithm:

Step 3.1 This step consists in computations of similarity matrices $\boldsymbol{S}(1)$ for data given in Table 4 and in Table 5. The results are given in Table 6 and Table 7 respectively. Let us denote these matrices as $S_{upper}(1)$ and $S_{low}(1)$ respectively.

Table 6. *S(1)* matrix used to search for the set MA_{upper}

t(i)	t_1^i	t_3^i	t_4^i	t_6^i	Class Q
t(1)	1	1	1	1	1
t(2)	0	1	0	0	0
t(3)	1	0	1	0	0
t(4)	0	0	0	1	0
t(5)	1	1	1	1	1
t(6)	0	1	0	0	0
t(7)	0	0	1	0	0
t(8)	0	1	1	1	1
t(9)	1	0	0	1	0
t(10)	0	0	0	1	1
t(11)	0	1	1	0	1
t(12)	0	0	0	0	0
t(13)	0	0	0	0	0
t(14)	0	1	1	1	0

Table 7. *S(1)* matrix used to search for the set MA_{low}

t(i)	t_1^i	t_3^i	t_4^i	t_6^i	Class Q
t(1)	1	1	1	1	1
t(2)	1	1	0	1	0
t(3)	1	0	1	1	0
t(4)	0	0	1	1	0
t(5)	1	0	1	1	1
t(6)	1	1	0	0	0
t(7)	1	0	1	1	0
t(8)	0	1	0	0	1
t(9)	1	0	1	1	0
t(10)	0	0	0	1	1
t(11)	0	1	1	0	1
t(12)	1	1	0	0	0
t(13)	1	0	0	0	0
t(14)	0	1	1	1	0

Steps 3.2. According to this step, it is necessary to order the negative lines of the matrices $S_{upper}(1)$ and $S_{low}(1)$ (see Fig. 5 and Fig. 6 respectively) and to find the maximal elements I_{max}^{upper} and $I_{\max}^{low}$ for both of them. The results are as follows:

$$I_{\max}^{upper} = \{3, 9, 14\} = \{1101, 1001, 0111\}\ ,$$
$$I_{\max}^{low} = \{2, 3, 14\} = \{1101, 1011, 0111\}.$$

Step 3.3. Computation of the index sets of distinctions for both cases. This step results in the following:

$$I_D^{upper}(1,3) = \{3,6\},\ I_D^{upper}(1,9) = \{3,4\}, I_D^{upper}(1,14) = \{1\}\ ;$$
$$I_D^{low}(1,2) = \{4\},\ I_D^{low}(1,3) = \{3\}, I_D^{low}(1,14) = \{1\}\ .$$

Step 3.4. Building the lattices $MA^{upper}(1)$ and $MA^{low}(1)$ and determining its minimal elements. For brevity, we present the final result of this step, i.e. the sets $MA_{\min}^{upper}(1)$ and $MA_{\min}^{low}(1)$ of minimal elements of the above partially ordered sets:

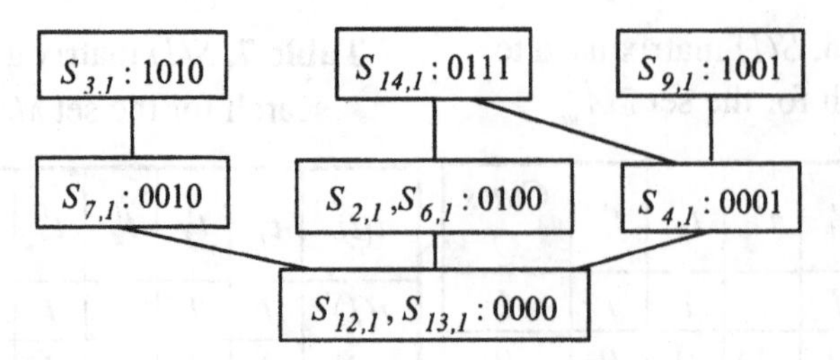

Fig. 5. Hasse diagram of the partially ordered set of the negative lines of similarity matrix $S_{upper}(1)$

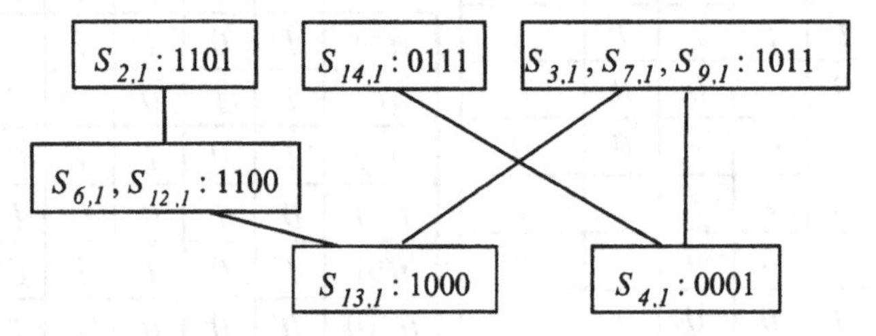

Fig. 6. Hasse diagram of the partially ordered set of the negative lines of similarity matrix $S_{low}(1)$

$$MA_{\min}^{upper}(1) = \{\{1,3,4\},\{1,3,6\}\}; MA_{\min}^{low}(1) = \{\{1,3,4\}\}$$

Step 3.5. The resulting sets of *maximally general rules* (upper and low bounds) are as follows:

$$R_{upper} = \{\bar{x}_1 x_3 x_4 \supset Q, \bar{x}_1 x_4 \bar{x}_6 \supset Q\} ;$$
$$R_{low} = \{\bar{x}_1 x_3 x_4 \supset Q\} .$$

Let us discuss the results of the example. It is obvious that $R_{low} \subset R_{upper}$. The coverage factors of the rules from the set R_{upper} are

$$S(\bar{x}_1 x_3 x_4, Q) = 0.4, S(\bar{x}_1 x_4 \bar{x}_6, Q) = 0.6 ,$$

and the same for the rule from the set R_{low} is

$$S(\bar{x}_1 x_3 x_4, Q) = 0.2 .$$

Let us compare the sets of *maximally general rules* computed for complete data, $R_* = \bar{x}_2 \supset Q, \bar{x}_1 x_3 x_4 \supset Q$ with R_{upper} and R_{low} sets of rules. The rule $\bar{x}_1 x_3 x_4 \supset Q$ is included in all sets, but the rule $\bar{x}_2 \supset Q \in Z^+$ does not appear in both R_{upper} and R_{low} . This fact does not contradict the *Theorem 2*, because in the last example the deducibility relation presented in formula (1) is held.

While comparing the values of the coverage factors, one can see that $S(\bar{x}_1 x_3 x_4, v) = 0.4$ preserves the same value for rules from R_* and R_{upper}, but both of them are greater than the coverage factor of the same rule belonging to the set R_{low}. All these conclusions are compatible with the *Theorem 2*.

5 Classification Mechanism Design and Experimental Results

This section demonstrates how to practically exploit the result of the *Theorem 1* for direct extraction of rules from binary data with missing values and also presents some experimental results obtained based on the case study described in Sect. 2, i.e. intrusion detection case study.

5.1 Training and Testing Datasets

The training and testing datasets were computed through preprocessing the real-life *Tcpdump/Windump* data containing *Normal* and *Abnormal* users' activity. For preprocessing purpose *TCPtrace* utility as well as several other programs were used. Through preprocessing of data traffic level four datasets were computed. General description of the structure of the training and testing datasets was given in Sect. 3 although in the experiments whose results are presented below these datasets were simplified.

It was assumed that detection of intrusions is performed according to two-level classification structure. At the first level four source-based classifiers produce decisions based on "local" data. The *output* of each local classifier (they are also called *base-classifiers*) is represented as a binary time-stamped stream of decisions taking values from the set (*Normal*, *Alert*). These four *asynchronous* streams of decisions extended by three additional attributes indicating the name of data sources form the *input* of the classifier of the second level, *meta-classifier*, whose task is to on-line combine asynchronous streams of decisions produced by the base classifiers. Meta–classifier produces its decisions at the time when it receives a new decision ("*event*") from any base classifier. Due to asynchronous nature of the base classifiers decision making and finite "life time" of decisions produced, some previous inputs of meta–classifier to the time when the latter has to update classification can become out of date. If so then at that time the respective inputs of meta–classifier are considered as missing. The nature of missingness was explained in Sect. 3 (see also Fig. 2).

We omit here descriptions of the base classifiers training and testing algorithms, the results of rule mining and mechanisms of base-classifiers' decision making, since these subjects are beyond the paper scope. For our purpose it is sufficient to describe the inputs of meta–classifier, *meta–data with missing values*, specified in terms of streams of binary attributes. Since the primary goal of the experiments is to explore the proposed idea of direct mining of data with missing values, and not to design an effective intrusion detection system, we simplified the mechanism that accounts for the data ageing resulting in missingness of values. The idea behind this is to compare the sets of rules and classification qualities for two statements of the same task. The first task is to mine a complete dataset (if to ignore the finite values of "life time"), generate classification rules and classification mechanism and to evaluate the quality of the resulting classification. The second task is similar but in it training and testing datasets were generated from the former datasets with the use of random missing data mechanism resulting in 20% of missing values. In the resulting datasets the number of missing values within particular examples varies from one to three and there exist about 20% of examples without missing values.

The resulting training datasets (both complete and with missing values) contain 128 examples; 14 of them correspond to the class "*Normal*" users' activity and 114 correspond to the class "*Abnormal*". The testing meta–data contain entirely different examples and consist of 11 examples of the class "Normal" and 233 examples of the class "Abnormal". Both tasks, with and without missing values are solved in parallel what allows to assess the developed direct mining algorithm. Hereinafter for brevity, the following notations of meta–classifier input data attributes are used:

BK_ConnPacket ≡ X1; BK_ConnAggreg ≡ X2; BK_Aggreg5sec ≡ X3; BK_Aggreg100con ≡ X4; InitConn ≡ X5; Init5sec ≡ X6; Init100conn ≡ X7.

5.2 Algorithm for Direct Mining of Rules from Data with Missing Values

The procedure of direct mining of rules from data with missing values developed based on the *Theorem 1* consists in the following steps:

1. Assign the missing values of the training dataset "pessimistically" and mine the rules of the set R_{low} using an algorithm (we used for this purpose the GK2 algorithm) for classes Q and $\bar{Q}$. Denote the resulting rule sets as $R_{low}(Q)$ and $R_{low}(\bar{Q})$ respectively.
2. Assign the missing values of the training dataset "optimistically" and mine the rules of the set R_{upper} for classes Q and $\bar{Q}$. Denote the resulting rules' sets as $R_{upper}(Q)$ and $R_{upper}(\bar{Q})$ respectively.
   Note. *In the experiments the length of rule premises was restricted by 3 literals.*
3. Assess the quality of the extracted rules of the sets $R_{upper}(Q)$, $R_{upper}(\bar{Q})$ using certain criteria through testing procedure.
   Notes. *(1) We do write about separate testing of the rules from the sets* $R_{low}(Q)$ *and* $R_{low}(\bar{Q})$ *due to deducibility relations* $R_{low}(Q) \subseteq R_{upper}(Q)$ *and* $R_{low}(\bar{Q}) \subseteq R_{upper}(\bar{Q})$ *following from the Theorem 1. (2) As the criteria of each rule quality we used a value of the coverage factor (in percent-*

age with regard to the number of the positive examples in training dataset) and percentage of false positives. (3) In most cases the rules from the sets $R_{low}(Q)$ *and* $R_{low}(\bar{Q})$ *possess more or less good quality.*

4. Based on evaluation criteria, select the best rules from the sets for a use in classification mechanism that jointly provide a reasonable tradeoff between the total number of the selected rules and quality.
   Note. *The selection procedure is carried out by experts, and the selection is based on application dependent requirements to the probabilities of correct classification, false positives (false alarms) and false negatives (missing of attack).*
5. Design classification mechanism and assess its performance quality.
6. If necessary, repeat some steps of the above algorithm to add new rules from the previously generated sets and/or generate additional rules.

5.3 Experimental Results

Let us describe experimental results in step by step fashion according to the above algorithm with comments at some of the steps. This description also aims to give more detailed explanations concerning the proposed algorithm for direct mining of data with missing values and to dwell upon some technical details of the algorithm implementation.

Steps 1 and 2 generate rules of the sets $R_{low}(Q)$, $R_{upper}(Q)$ $R_{low}(\bar{Q})$, and $R_{upper}(\bar{Q})$. Rule extraction procedure was restricted by search for rules, whose premises contain no more than three literals. Under this condition the total number of extracted rules for "*optimistic*" assignment of missing values is equal to 58 at that 35 of them belong to the rule set $R_{upper}(Q)$ (two of them also belong to $R_{low}(Q)$), and 23 – to $R_{upper}(\bar{Q}) \cup R_{low}(\bar{Q})$ at that $R_{low}(\bar{Q})$ contains eight rules not belonging to $R_{upper}(\bar{Q})$ but deducible from it.

Figure 7 demonstrates the sets of rules $R_{low}(\bar{Q})$ and $R_{upper}(\bar{Q})$, and the rules of the set $R_C(\bar{Q})$ extracted from the same training data though having no missing values. In this figure, the rules of the sets $R_{low}(\bar{Q})$, $R_C(\bar{Q})$ and $R_{upper}(\bar{Q})$ are marked with symbols "*L*" ("*Low*"), "*C*" ("*Complete*") and "U" ("*Upper*") respectively. It could be seen that the rules RS_{10}, RS_{21}, RS_{22}, RS_{23} are deducible from any of the rules RS_7, RS_8 or RS_{15}; the rules RS_4, RS_5 and RS_6 are deducible from the rule RS_{17}, and the rule RS_{24} is deducible from any of the rules RS_1, RS_2, or RS_{17}. Let also note that the rules of the set $R_C(\bar{Q})$ either belong to or are deducible from the set of rules $R_{upper}(\bar{Q})$. Therefore deducibility relation $R_{low}(\bar{Q}) \subseteq R_C(\bar{Q}) \subseteq R_{upper}(\bar{Q})$ is held.

The finally selected rules' sets $R(Q)$ and $R(\bar{Q})$, mapped by their relative *coverage factors* and *false positives* are presented in Table 8. Among the selected rules fourteen belong to the set $R(Q) \subseteq R_{upper}(Q)$ and eight – to $R(\bar{Q}) \subseteq R_{upper}(\bar{Q})$. The rules selected from the set $R_{upper}(\bar{Q})$ are highlighted grey in Fig. 8.

These rules were used to construct a classification mechanism, and it was constructed as a weighted voting algorithm where each rule of the set $R(Q)$

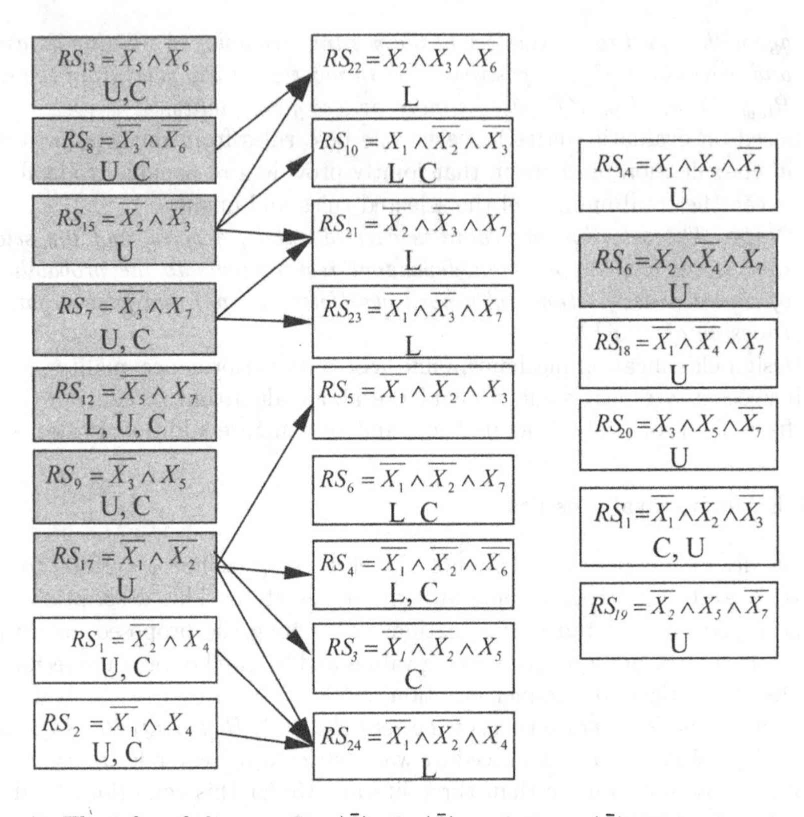

Fig. 7. The rules of the sets $R_{low}(\bar{Q})$, $R_C(\bar{Q})$, and $R_{upper}(\bar{Q})$. The rules selected for use in classification mechanism are given in *grey*

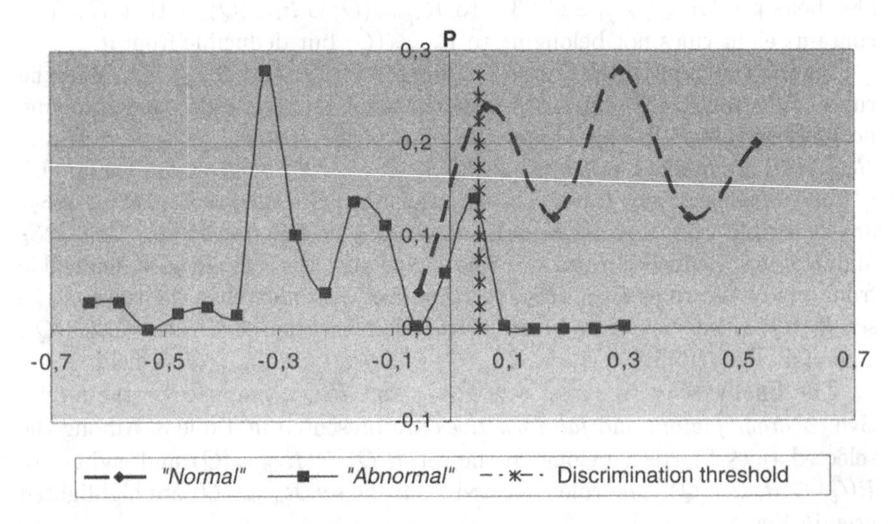

Fig. 8. Explanation of the choice of a threshold d' in classification mechanism (16)

Table 8. Qualities of the rules selected for a use in decision making mechanism

Rules for class Q (*Normal*)	R3	R4	R5	R7	R 10	R 12	R 15	R 19	R 20	R 22	R 24	R 25	R 29	R 38
Coverage %	18	9	18	36	18	18	18	18	27	27	27	18	9	18
Refusals %	27	64	55	45	27	45	54	36	36	54	54	54	45	36
False positives %	55	27	27	18	54	36,	27	45	36	18	18	27	45	45
Coverage %	67	47	49	50	61	49	48	57	67	52	46	48	52	47
Refusals %	33	53	51	50	37	49	51	41	31	46	53	51	47	52
False negatives %	0	0	0	0	0,4	0,4	0	1,3	0,9	0,9	0	0	0	0

Rules of class $\overline{Q}$ (Ab*normal*)	RS 7	RS 8	RS 9	RS 12	RS 13	RS 15	RS 16	RS 17
Coverage %	63	63	54	63	72	45	45	54
Refusals %	36	36	45	36	27	54	36	45
False positives %	0	0	0	0	0	0	18	0
Coverage %	9,4	19	21	27	27	30	22	33
Refusals %	39	41	39	36	36	41	48	33
False negatives %							28	33

was weighted proportionally to the difference between relative coverage and relative false positives and then normalized to 1. The same weighting procedure was applied to the rules of the set $R(\bar{Q})$.

5.4 Classification Mechanisms

Three classification mechanisms were tested. The first of them was built based upon the weighted rules set $R(Q)$:

$$\tilde{Q} = \begin{cases} Q, & if \quad W(X) \geq 0 , \\ \bar{Q} & otherwise , \end{cases} \tag{14}$$

where $W = \sum_{i=1}^{|R(Q)|} a_i v[R_i(Q)]$, a_i–weights of the rule $R_i(Q) \in R(Q)$, $|R(Q)|$– cardinality of the rule set $R(Q)$, and $v[R_i(Q)]$– truth value of the premise of the rule $R_i(Q)$.

The second mechanism is analogous to the first one but built on the basis of the weighted rules set $R_(\bar{Q})$:

$$\tilde{Q} = \begin{cases} \bar{Q}, & if \quad W(X) \geq 0 , \\ Q & otherwise , \end{cases} \tag{15}$$

where $W = \sum_{i=1}^{|R(\bar{Q})|} b_i v[R_i(\bar{Q})]$, b_i–weight of the rule $R_i(\bar{Q}) \in R(\bar{Q})$, $|R(\bar{Q})|$–cardinality of the rule set $R(\bar{Q})$, and $v[R_i(\bar{Q})]$–truth value of the premise of the rule $R_i(\bar{Q})$.

The third classification mechanism exploits both $R(Q)$ and $R(\bar{Q})$ sets of rules. Since the cardinalities of these sets are different, it is necessary to previously normalize them to provide the equal values of the sums of the weights of rules a_i and b_i:

$$\sum_{i=1}^{|R(Q)|} a_i' = \sum_{i=1}^{|R(\bar{Q})|} b_i' = 1 ,$$

that can be easily achieved by respective normalization of the previously computed weights a_i and b_i.

Thus, the third classification mechanism is as follows:

$$\tilde{Q} = \begin{cases} Q, & if \quad D(X) \geq \bar{d} , \\ Q & otherwise \end{cases} \tag{16}$$

where $D = \sum_{i=1}^{|R(Q)|} a_i' v[R_i(Q)] - \sum_{i=1}^{|R(\bar{Q})|} b_i' v[R_i(\bar{Q})]$, $\bar{d}$ – experimentally chosen threshold. Figure 8 demonstrates how the value of $\bar{d}$ was chosen. (in our case the chosen value is equal to 0,05).

The results of the testing of three above mentioned mechanisms are depicted in Fig. 8 and Fig. 9. In Fig. 8 two curves are depicted; the right one presents the estimation of the probability density for the value of d built for the cases when an input example of the classification mechanism (16) belongs to the class Q ("*Normal*"), whereas the left one presents the estimation of the probability density for the value of d built for the examples belonging to the class $\bar{Q}$ ("*Abnormal*"). The choice of the threshold value of $\bar{d}$ depends on the application-oriented requirements to the classification mechanism quality, which as a rule is selected as a tradeoff between probabilities of correct classification, false positive and false negative. Let us note that the square under right curve given $d \leq \bar{d}$ corresponds to the *false negatives* (in our case – false alarms) probability whereas the square under left curve given $d > \bar{d}$ corresponds to the *false positives* (in our case–missing of signals) probability.

Figure 9 proves that within the case study in question classification mechanism (16) reveals rather good performance quality. In particular, it provides the estimated probability of the correct classification close to 0,997 if tested on joined testing and training samples with missing values (the right most diagram in Fig. 9). Testing of this mechanism on joined testing and training samples without missing values results in the estimated probability of the correct classification close to 0,99 (the second diagram at the left most). This figure also demonstrates the results of testing classification mechanisms (14) and (15). It can be seen that their performance qualities are worse than performance quality of the mechanism (16).

In general, the presented experimental results allow to optimistically evaluate the developed approach to direct mining of rules from the dataset with missing values.

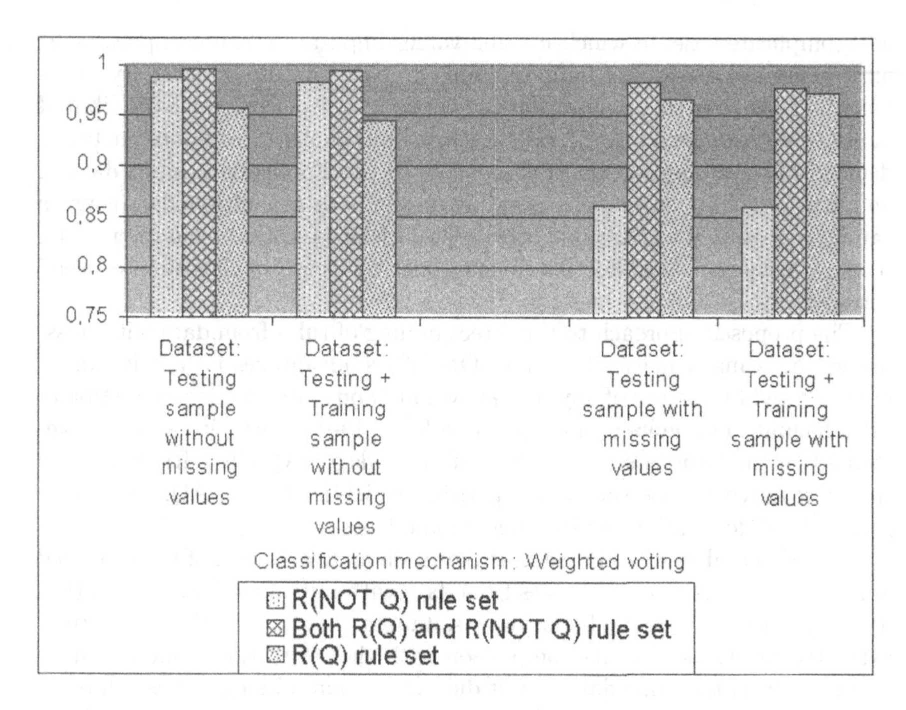

Fig. 9. The testing results demonstrating performance quality of three classification mechanisms. Vertical axis corresponds to the estimations of the probabilities of correct classification for four different testing samples

6 Conclusion

The paper develops an approach to and technique for the lattice based direct mining of binary data with missing values aimed at extraction of classification rules with premises represented in a conjunctive form. The proposed approach does not assume an imputation of missing values. The idea is to generate two sets of rules, which serve as the upper and low bounds for any other sets of rules corresponding to all possible assignments of missing values, and then, based on these upper and low bounds of the rules' sets, on testing procedure and on a classification criterion, to select a subset of rules to be used for classification. Instead of missing values imputation, the proposed approach uses training dataset to cut down the upper bound rule set via testing its rules on other dataset with missing values with the subsequent selection of the most appropriate rules.

The motivation of imputation avoidance is that there exist a number of important applications where information cannot be collected due to certain reasons; for example, due to conditions of natural environment (for example, if airborne observation equipment is used, the information can be unavailable due to meteorological factors or due to masking). Much more specific and complicate case, in which missing values imputation is not applicable at all, can arise as a result of temporal and/or spatial nature of data, asynchronous and distributed mode of data collection, and discrete and distributed mode of decision making. The detection of intrusions in computer network demonstrates an application of such kind. In this application, input data of an intrusion detection system operating on the basis of multiple data sources can be represented as temporal stream of decisions received from distributed solvers (agents) responsible for forming alerts in case of illegitimate users' activity.

The proposed approach to the direct mining of rules from data with missing values is mathematically sound. One of its advantages is that it can be implemented by the use of any well known methods intended for extraction of rules having the premises in conjunctive form. Particularly, it can be implemented on the basis of such well known methods as AQ, CN2, RIPPER, etc. In our research we use the GK2 algorithm which is close to AQ. The paper gives a brief description of GK2 algorithmic basis.

Experimental results aiming at exploration and validation of the proposed approach to direct mining of rules from data with missing values confirm that it can yield strong and reliable results. Indeed, analysis of the set of rules extracted by its use and its comparison with the set of rules that would be extracted from the same dataset if it did not contain missing values exhibited their considerable overlapping. Example of such a comparison given in Fig. 7 demonstrates this fact, particularly, here only one rule, RS_{16}, belonging to the upper bound rules' set does not belong to the set of rules that could be extracted from the same data having no missing values.

The approach will be further tested on different applications from various areas with the focus on applications from a situational awareness scope. An interesting future task in regard to direct mining of data with missing values is the use of dependencies between attributes to evaluate the predictive power of the rules extracted by direct mining along with their evaluations on the basis of testing procedures.

Acknowledgement

This research is supported by European Office of Aerospace Research and Development (AFRL/IF), Project #1993P, and Russian Foundation of Basic Research (grant #04-01-00494a).

References

1. Birkhoff, G.: Lattice Theory. Providence, Rhode Island (1963)
2. Clark, P. Niblett, T.: The CN2 Induction Algorithm. Machine Learning Journal, Vol. 3 (1989) 261–283

3. Cohen, W.W.: Fast Efficient Rule Induction. Machine Learning: The 12th Intern. Conference, CA, Morgan Kaufmann (1995)
4. Cole, E.: Hackers Beware. New Riders Publishing (2002)
5. D.A.P., L.N.M., R.D.B.: Maximum Likelihood from Incomplete Data via the EM Algorithm. Journal of the Royal Statistical Society. Series B, Vol.39 (1977)
6. Endsley, M.R., Garland, D.G. (Eds.): Situation Awareness Analysis and Measurement. Mahwah, NJ: Lawrence Erlbaum (2000)
7. Goodman, I., Mahler R., Nguen, H.: Mathematics of Data Fusion. Kluwer Academic Publishers (1997)
8. Gorodetsky, V., Karsaev, O., Kotenko, I., Samoilov, V.: Multi-Agent Information Fusion: Methodology, Architecture and Software Tool for Learning of Object and Situation Assessment. In Proceedings of the International Conference "Information Fusion-04", Stockholm, Sweden, June 28 to July 1 (2004) 346–353
9. Gorodetsky, V., Karsaev, O., Samoilov, V.: Multi-agent Technologies for Computer Network Security: Attack Simulation, Intrusion Detection and Intrusion Detection Learning. International Journal of Computer Systems Science and Engineering. Vol.18, No.4, (2003) 191–200
10. Gorodetsky, V., Karsaev, O.: Mining of Data with Missing Values: A Lattice-based Approach. International Workshop on the Foundation of Data Mining and Discovery in the 2002 IEEE International Conference on Data Mining, Japan, (2002) 151–156
11. Gorodetsky, V., Karsaev, O.: Algorithm of Rule Extraction from Learning Data. Proceedings of the 8th International Conference "Expert Systems & Artificial Intelligence" (EXPERSYS-96), (1996) 133–138
12. Kononenko, I., Bratko, I., and Roskar: Experiments in Automatic Learning of Medical Diagnostic Rules. Technical Report. Josef Stefan Institute, Ljubjana, Yugoslavia (1984)
13. Liu, W.Z., White, A.P., Thompson, S.G. Bramer, M.A.: Techniques for Dealing with Missing Values in Classification. Advances in Intelligent Data Analysis (Liu, X., Cohan, P., Bertold, M., Eds). Proceedings of the Second International Symposium on Intelligent Data Analysis. Springer Verlag, (1997) 527-536
14. M.Magnani. Techniques for Dealing with Missing Values in Knowledge Discovery Tasks. (2003) http://magnanim.web.cs.unibo.it/data/pdf/missingdata.pdf
15. Michalski, R.S.: A Theory and Methodology of Inductive Learning. Machine Learning, vol.1, Eds. Carbone,l J.G., Michalski, R.S. Mitchel, T.M., Tigoda, Palo Alto, (1983) 83–134
16. Nayak, R., Cook, D.J.: Approximate Association Rule Mining. Proceedings of the Florida Artificial Intelligence Research Symposium (2001)
17. Northcutt, S., McLachlan, D., Novak, J.: Network Intrusion Detection: An Analyst's Handbook. New Riders Publishing (2000)
18. Pawlak, Z.: Rough Sets. International Journal of Computer and Information Sciences, #11, (1982) 341–356
19. Prodromidis, A., Chan, P., Stolfo, S.: Meta-Learning in Distributed Data Mining Systems: Issues and Approaches, Advances in Distributed Data Mining, AAAI Press, Kargupta and Chan (eds.), (1999) http://www.cs.columbia.edu/~sal/hpapers/ DDMBOOK.ps.gz
20. Quinlan, J.R.: Unknown Attribute Values in Induction. Proceedings of the Sixth International Workshop on Machine Learning (B.Spatz, ed.), Morgan Kaufmann, (1989) 164–168

21. Quinlan, J.R.: Induction of Decision Trees. Machine Learning. Vol. 1, (1986) 81–106
22. Ragel, A., Cremilleux, B.: Mvc–a Preprocessing Method to Deal with Missing Values. Knowledge-Based Systems, (1999) 285–291
23. Ragel, A., Cremilleux, B.: Treatment of Missing Values for Association Rules. Proceedings of the 2nd Pacific–Asia Conference on Research and Development in Knowledge Discovery and Data Mining (PAKDD-98), X. Wu, R. Kotagiri, and K. B. Korb, eds., LNCS, Vol. 1394, Berlin, Springer Verlag, (1998) 258–270
24. Rubin, D.B.: A Overview of Multiple Imputation. Survey Research Section, American Statistical Association (1988) 79–84
25. Rubin, D.B.: Multiple Imputation for Nonresponse in Surveys. John Willey and Sons (1987)
26. Sandle, I.G.: Hot-deck Imputation Procedures. Incomplete Data in Sample Survey, Vol.3 (1983)
27. Skowron, A.: Rough Sets in KDD. In Proceedings of 16th World Computer Congress, Vol. "Intelligent Information Processing", Beijing (2000) 1–17
28. Scambray, J., McClure, S.: Hacking Exposed Windows 2000: Network Security Secrets. McGraw-Hill (2001)
29. Scambray, J., McClure, S., Kurtz, G.: Hacking Exposed. McGraw-Hill (2000)
30. Weiss, S., Indurkhya, N.: Decision Rule Solutions for Data Mining with Missing Values. IBM Research Report RC-21783 (2000)

Cluster Identification Using Maximum Configuration Entropy

C.H. Li[1]

Department of Computer Science, Hong Kong Baptist University, Hong Kong
chli@comp.hkbu.edu.hk

Summary. Clustering is an important task in data mining and machine learning. In this paper, a normalized graph sampling algorithm for clustering that improves the solution of clustering via the incorporation of a priori constraint in a stochastic graph sampling procedure is adopted. The important question of how many clusters exists in the dataset and when to terminate the clustering algorithm is solved via computing the ensemble average change in entropy. Experimental results show the feasibility of the suggested approach.

1 Introduction

Clustering has been an important area of research in pattern recognition and computer vision. Unsupervised clustering can be broadly classified into whether the clustering algorithm is hierarchical or non-hierarchical. Hierarchical methods often model the data to be clustered in the form of a tree, or a dendrogram [1]. The lowest level of the tree is usually each datum as a cluster. A dissimilarity measure is defined for merging clusters at a lower level to form a new cluster at a higher level in the tree. The hierarchical methods are often computationally intensive for large number of samples and is difficult to analyze if there is no logical hierarchical structure in the data.

Non-hierarchical methods divide the samples into a fixed number of groups using some measure of optimality. The most widely used measure is the minimization of the sum of squared distances from each sample to its cluster center. The k-means algorithm, also known as Forgy's method [2] or MacQueen [3] algorithm is a classical algorithm for non-hierarchical unsupervised clustering. However, the k-means algorithm tends to cluster data into even populations and rare abnormal samples in medical problems cannot be properly extracted as individual clusters.

Graph based clustering has received a lot of attention recently. A graph theoretic approach to solving clustering problem with the minimum cut criteria has been proposed [4]. A factorization approach has been proposed for

C.H. Li: *Cluster Identification Using Maximum Configuration Entropy*, Studies in Computational Intelligence (SCI) **6**, 265–276 (2005)
www.springerlink.com

clustering [5], the normalized cuts have been proposed as a generalized method for clustering [6] and [7]. Graph-based clustering is also shown to be closely related to similarity-based clustering [8].

While previous approach to minimum cuts problem find solutions with high efficiency for generic data [9]. In a recent paper, we introduced a normalized sampling algorithm that incorporates information on cluster properties which improves the solutions for clustering problem [10]. Entropy constraints are incorporated into the minimum cuts framework for controlling the clustering process in finding the most relevant clusters. In this paper, we will investigate the use of entropy for determining the number of clusters and the termination criteria for the clustering process.

2 Graph Based Clustering

In this section, we will briefly review the graph based clustering process. The cluster data is modeled using an undirected weighted graph. The data to be clustered are represented by vertices in the graph. The weights of the edge of vertices represents the similarity between the object indexed by the vertices. Let x_i be the feature vector corresponding to the i-th object to be clustered and there are N data to be clustered. The similarity matrix $S \in R^{N^2}$ where s_{ij} represents the similarity between the vector x_i and x_j. A popular choice of the similarity matrix is of the form

$$s_{ij} = \exp\left(-\frac{|x_i - x_j|}{\sigma^2}\right) \quad (1)$$

where $|x|$ is an euclidean distance of x and σ is a scale constant. The exponential similarity function have been used for clustering in [5], [6] and [7]. The graph G is simply formed by using the objects in the data as nodes and using the similiarity s_{ij} as the values (weights) of the edges between nodes i and j. The full similarity graph consists of N pairs of similarity values which maybe quite large when the number of data is large. A reduction of computation complexity can be obtained if we can use the following graphs for representing the similarity relationship. The thresholded similarity graph G is obtained by removing similarities values which are close to zero, thus the set of edges consists of all pairs of vertices with similarities larger than a threshold θ,

$$E = \{(i,j) : s_{ij} > \theta\} \ . \quad (2)$$

The threshold θ can be selected as three or four times the values of σ which can guarantee the similarities values of those removed edges are close to zero. However, the disadvantage of the thresholded graph is that the number of removed nodes cannot be guaranteed and have high dependence on the distribution of the data. Another similarity graph that can drastically reduce the number of edges is that k-nearest neighbor (kNN) similarity graph. The kNN graph consists of the edges that are the k-th nearest neighbours of each other, thus the set of edges are given by

$$E = \{(i, j) : j \in N_i\} . \tag{3}$$

The kNN similarity graph has the advantage that the number of edges is bounded by kN which has linear dependence on N.

For a partition of the vertices V into tow disjoint sets A and B, the cut is the total similarity between vertices in A and B,

$$f(A, B) = \sum_{i \in A} \sum_{j \in B} s_{ij} . \tag{4}$$

The minimum cut criterion is the finding of the partitions A and B such that the cut is minimized. As $f(A, B)$ measures the similarity between the disjoint sets A and B, minimizing this criterion has the intuitive advantage of minimizing the similarity between two groups of vertices. Two partitions that have small similarity between each other can be considered as good partitions.

However, the minimum cuts partitioning often finds solutions that consists of sporadic outlying data point or clusters which are often not too useful for categorization [6]. Thus the minimum cuts criterion has to be augmented with other constraint for effective clustering.

3 Maximum Entropy Algorithm

While the minimum cuts provides good measure of similarity between two clusters, another measure is needed for measuring the usefulness of the partitioning. The entropy is often used as information criteria in compression and communications [11]. In recent years, the maximum entropy principle has been increasingly popular as an aid for inference in various pattern recognition and machine learning applications. The maximum entropy principle was originally proposed by Jaynes [12] to the inference of unknown probability distribution. With the use of the concentration theorem and the study of multiplicities, it has been shown that distributions of higher entropy have higher multiplicities and are thus more likely to be observed [13]. Applications of maximum entropy includes image reconstruction [14] and image segmentation [15].

The maximum entropy algorithm has also found to be suitable for Bayesian inference methodology. In the Bayesian statistics approach to pattern recognition, the specification of the a priori information is often required to establish the a priori probability of the data. The maximum entropy principle allows us to select a distribution with maximum entropy that satisfies the given constraint of the problem.

4 Graph Contraction Algorithm

The contraction algorithm for minimum cuts [9] consists of an iterative procedure that contract two connected nodes i and j:

- Randomly select an edge (i, j) from G with probability proportional to S_{ij}
- Contract the edge (i, j) into a meta node ij
- Connect all edges incident on i and j to the meta node ij while removing the edge (i, j)

This contraction is repeated until all nodes are contracted into a specified number of clusters. The probability of drawing an edge connecting $\{i, j\}$ is given by,

$$P(\{i,j\}) = \frac{s_{ij}}{\sum_i \sum_j s_{ij}} . \tag{5}$$

Thus two vertices that have high similarity values will be likely to be merged together to form a cluster. The contraction algorithm iteratively select vertices with high similarity values for merging until the desired termination criterion is reached. This contraction algorithm will be referred to as unit sampling method as the contraction algorithm in [9] considers metanode as one unit irrespective of the content or the number of vertices inside the metanode.

Figure 1 shows a schematic representation of the contraction process. The figure on the left shows the graph before contraction. The two nodes in the center marked with crosses are selected to be contracted. The figure on the right shows the graph after contraction. In the center of the right figure, a single metanode is created from the contraction of two nodes as shown in the left figure. Notice the edges connecting to the nodes before contraction are all retained after contraction.

Define the configuration entropy of the system by

$$H = -\sum_{i=1}^{N} \frac{n_i}{N} \log\left(\frac{n_i}{N}\right) \tag{6}$$

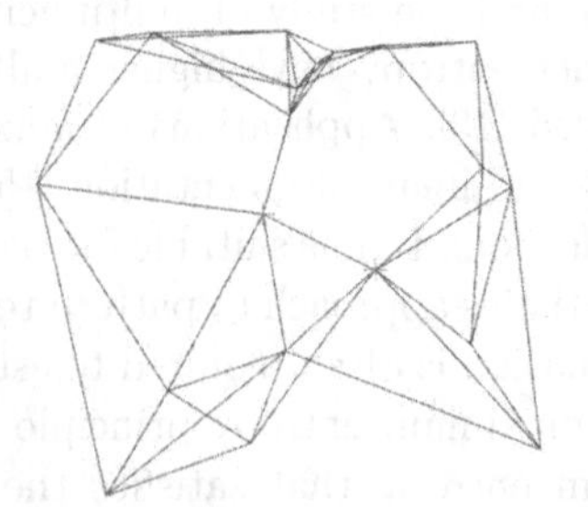

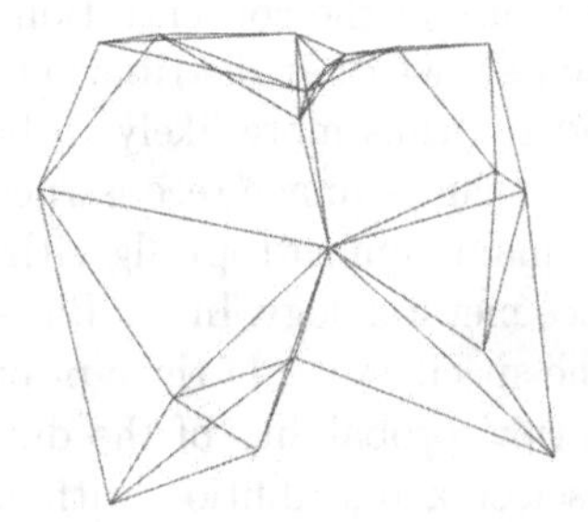

Fig. 1. Graph Contraction. *Left figure*: two nodes selected for contraction; *Right figure*: Graph after contraction

where n_i is the number of vertices in the metanode i. There are a number of observations configuration entropy. In the starting phase of the contraction, each node contain only one data and there is no metanode. The entropy in the starting phase of contraction is given by

$$H = -\sum_{i=1}^{N} \frac{1}{N} \log\left(\frac{1}{N}\right) \tag{7}$$

and equals to $\log(N)$. This is the maximum entropy for distributing N objects. The graph contraction process leads to a monotonic decreases in entropy, which is shown as follows. Consider the entropy change dH in a contraction process of merging the metanodes a and b

$$dH = -\left(\frac{n_a}{N}\log\left(\frac{n_a}{N}\right) + \frac{n_b}{N}\log\left(\frac{n_b}{N}\right) - \frac{n_a+n_b}{N}\log\left(\frac{n_a+n_b}{N}\right)\right) \tag{8}$$

where n_a and n_b is the number of data in the metanodes a and b respectively. This expression can be simplified to

$$dH = -\left(\frac{n_a}{N}\log\left(1+\frac{n_b}{n_a}\right) + \frac{n_b}{N}\log\left(1+\frac{n_a}{n_b}\right)\right) \tag{9}$$

which is negative number for all positive real values of n_a and n_b. The entropy at the end of the contraction, when all nodes are merged into one meta node, is zero. There are considerable advantages in using the entropy as measurement for the states of the clustering. The computation complexity for calculation the entropy for each contraction is incremental and can be implemented in constant time.

There are various implications from this study of the change in entropy. The first implication is that this change of entropy allows us to select suitable criteria for determining the number of clusters. This will be discussed in the following section. The second implications of this change in entropy provides another motivation for improving the sampling method. As the clustering algorithm aims at finding high entropy cluster with small cuts values. The sampling procedure for selecting nodes for merging can be improved if the higher entropy values configurations are preferentially selected as will be discussed in the next section.

5 Normalized Contraction

A major problem for the unit sampling method for contraction is that as the contraction proceeds, the metanodes with a large number of nodes are likely to be selected as their similarities are accumulated over a number of nodes. Although this selection process is effective for minimum cuts problem, such selection may leads to a large central cluster with small outliers near the data boundary. As the clustering process aims to discover clusters with significant proportions and distinctly different similarity, a more accurate representation of the similarity between two metanodes is needed. The normalized similarity between two metanodes is given by s'_{ij} where

$$s'_{ij} = \frac{s_{ij}}{n_i n_j} . \tag{10}$$

where n_i and n_j is the number of vertices in the metanodes i and j. The normalized graph sampling algorithm selects an edge $\{i, j\}$ for contraction according to the following probability,

$$P(\{i, j\}) = \frac{s'_{ij}}{\sum_i \sum_j s'_{ij}} . \tag{11}$$

Thus two metanodes that have high similarity values and smaller number of vertices will be more likely to be merged together to form a cluster. Another major advantage of the normalized sampling approach is that for each contraction, the probability of selecting two nodes with smaller number of data inside the node is higher. From (9), smaller n_a and n_b will cause a smaller change in entropy. Thus this normalized sampling scheme has a higher probability of forming lower entropy configurations. The net effect in multiple contractions is that the entropy of the normalized sampling remains high until the last few iterations where there most of the metanodes contain a large number of datapoint and the merging of such metanodes causes a dramatic decreases in entropy.

Figure 2 shows the comparison of the entropy change between the two sampling schemes of a contraction process for contracting 500 data. The details of this data (dataset I) and the experiment is described in the experimental results section. The starting entropy is log(500) and final entropy when all data is merged into a single node is 0. At the beginning of the contraction, both schemes have similar entropy. Starting at one third of iterations, the normalized sampling scheme start to give higher entropy compared to the unit sampling scheme without normalization.

To take a closer look between the entropy change between successive iterations, the entropy change for the two sampling schemes are shown in Fig. 3 and Fig. 4 respectively. We can see that large entropy changes in the unit sampling scheme occurs at sporadic locations during the contraction process while the large entropy changes occurs mostly at the last few iterations. The confinement of large entropy changes towards the last few iterations allows the development of an efficient two stage algorithm.

6 Entropy Criteria for Clustering

The entropy concentration property of the normalized sampling is epse The graph-contraction based clustering process starts with n data which contracts

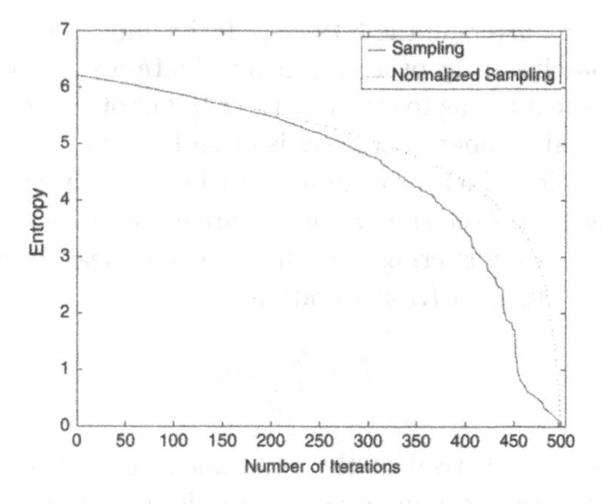

Fig. 2. Entropy of contraction process for sampling scheme and normalized sampling scheme

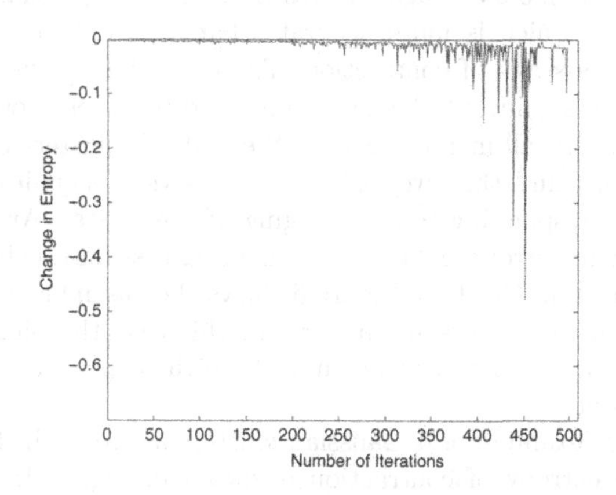

Fig. 3. Entropy Change for unit sampling scheme

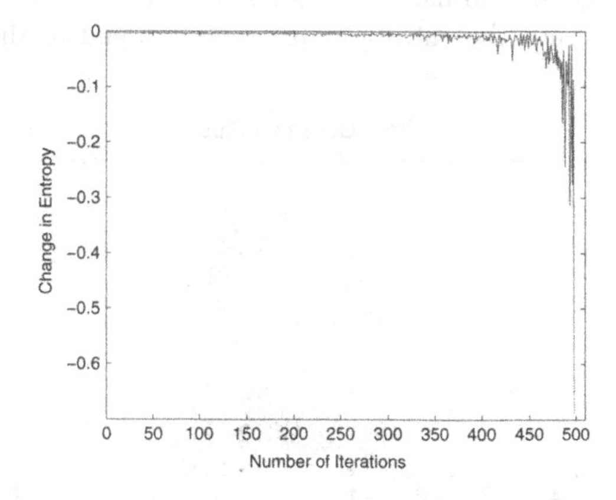

Fig. 4. Entropy Change for normalized sampling scheme

to a system of $n-k$ clusters in the k-th step. If the number of clusters is known a priori, then the clustering process can terminate accordingly. However, in some situations, the user has to estimate the number of clusters in the dataset. Although the actual number of clusters is often hard to be estimated, we can obtain some statistics which provide us with information about the possible cluster arrangement. Two clusters measures are found to be valid in this graph sampling approach to clustering. The first cluster measure is the ensemble average change in entropy $d\hat{H}_t$ at iteration t

$$d\hat{H}_t = \sum_{i \in E} dH_t \tag{12}$$

where i denotes the i-th trial of the experiment E, where each of the experiment is started with a unique pseudo-random seed. The iterations with maximum change in entropy will correspond to likely positions for clustering. The use of the ensemble average ensures that the average change in entropy can be measured which is more accurate than just measuring the entropy change of a single series of contraction. There will be significant changes in the entropy when significant clusters are merged together. However, as there are multiple clusters, all merges between the multiple clusters will cause large change in entropy and thus we expect to see several significant changes in entropy, each corresponding to one merging of the cluster. And the first significant changes in entropy indicates the merging first major cluster. Figure 5 shows three Gaussian Clusters. Figure 6 shows the ensemble average entropy of contraction on the Gaussian clusters. In this case the biggest change in entropy occurs in iterations 700 which leads to the conclusion that there is 3 clusters in the data.

Take another example of 4 Gaussian clusters as shown in Fig. 5, the ensemble average entropy of contraction is shown in Fig. 8. In this case, the maximum change in entropy corresponds to two clusters. This scenario represents the merging of two pairs of clusters into two clusters. To get a clear picture, we zoom into the entropy figure near the end of the iterations as

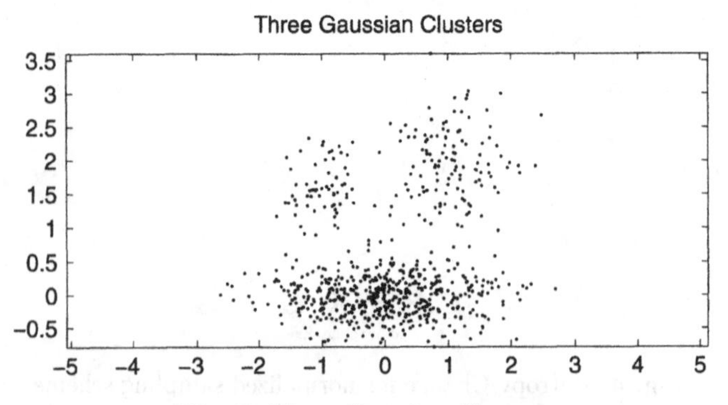

Fig. 5. Three Gaussian Clusters

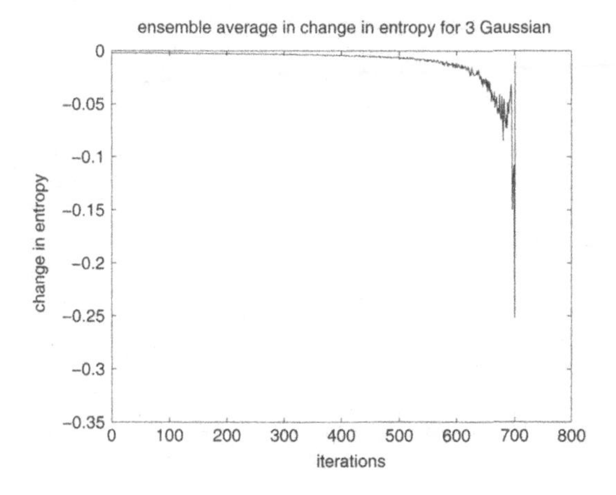

Fig. 6. Ensemble average entropy of contraction on the Gaussian clusters

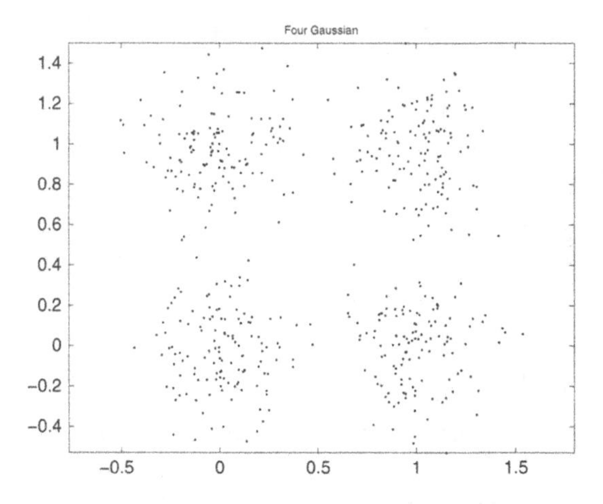

Fig. 7. Four Gaussian clusters

shown in Fig. 9. The first significant change occurs at iterations 507, which corresponds to estimating four clusters in the dataset. The largest change of entropy occurs when the each of the two pairs of clusters merged into two clusters.

7 Results on Real Datasets

In this section, the graph clustering algorithm is applied to UCI datasets. The similarity graph employed is the 8-nearest neighbour graphs. The scale constant σ is set to the mean of all 8-nearest neighbour distances of the data. The contractions are repeated 100 times to obtain the ensemble average value.

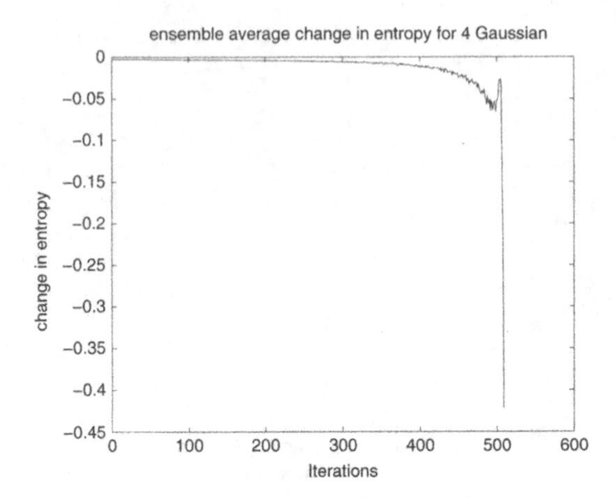

Fig. 8. Ensemble average entropy of contraction on the 4 Gaussian clusters

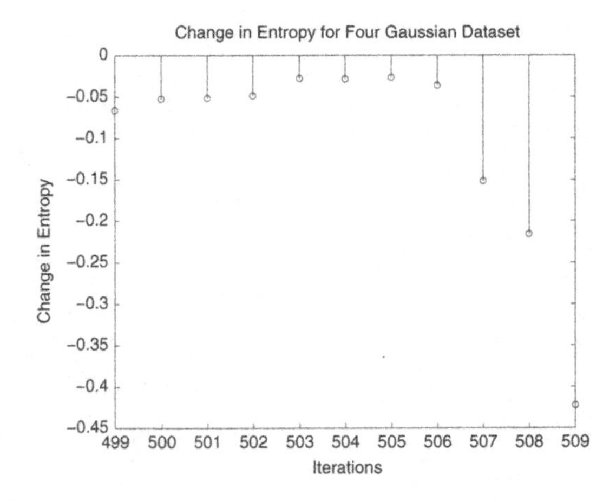

Fig. 9. Zoom-in view of Ensemble average entropy of contraction on the Gaussian clusters

In the vehicle dataset, there are 846 vehicles outlines from four different types of vehicles. Using the normalized contraction algorithm, we plot the average entropy of the contraction process. As the entropy drops concentrate at the end of the contraction, only the last ten iterations are plotted, corresponding to the formation of eleven clusters at the first data-point to two clusters at the last data-point. Figure 10 shows the average change in entropy in the vehicle dataset. The largest change in entropy occurs at the last 3 iterations which corresponds to predicting correctly that there are 4 clusters.

In the image segmentation dataset, there are 2310 data, each with 19 features. There are 7 different types of images in the dataset. Figure 11 shows the average change in entropy of the image dataset. The largest change in

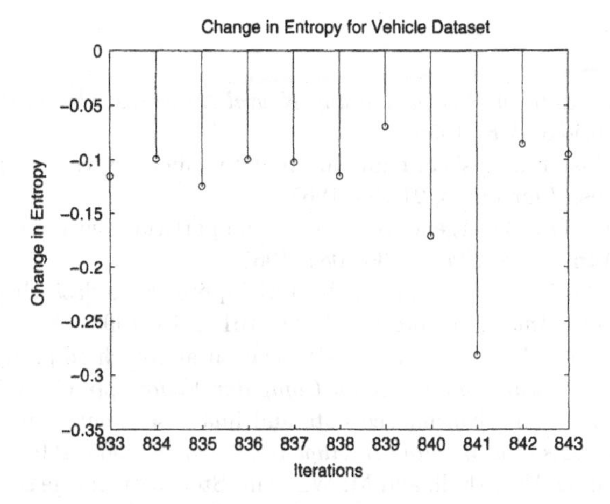

Fig. 10. Zoom-in view of Ensemble average entropy on the vehicle dataset

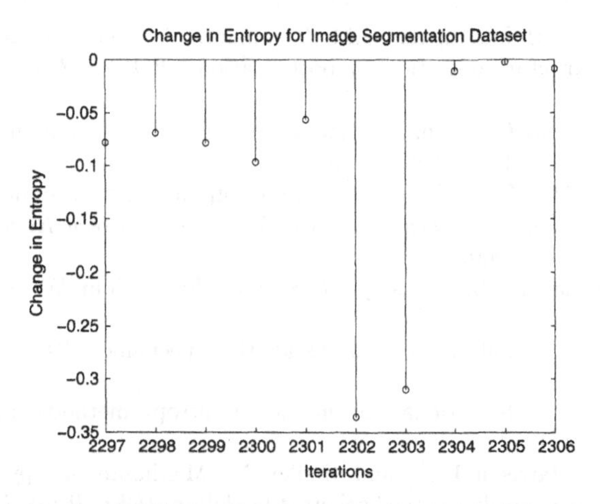

Fig. 11. Zoom-in view of Ensemble average entropy on the image segmentation dataset

entropy corresponds to predicting the number of clusters to be six, while the actual number of clusters is 7. However, this is a noisy dataset which is non-separable. The number of predicted cluster is very close to the actual value.

Although the exact number of clusters in an unknown dataset is often difficult to be established, however measures can be established to cluster datasets by similarity. Furthermore, the study in this paper suggest that the study of entropy helps to analyze the possible structure within the data. Results using synthetic datasets and the real datasets suggest that the average change in entropy is an effective indicator for the possible cluster structure in the datasets.

References

1. B.D. Ripley. *Pattern Recognition and Neural Networks.* Cambridge University Press, Cambridge, UK, 1996.
2. E. Forgy. Cluster analysis of multivariate data: efficiency vs. interpretablility of classifications. *Biometrics*, 21:768, 1965.
3. J. MacQueen. On convergence of k-means and partitions with minimum average variance. *Ann. Math. Statist.*, 36:1084, 1965.
4. Wu and Leahy. An optimal graph theoretic approach to data clustering. *IEEE Trans. Pattern Anal. Machine Intell.*, 11:1101–1113, 1993.
5. P. Perona and W.T. Freeman. A factorization approach to grouping. In *Proceedings of European Conference on Computer Vision*, pp. 655–670, 1998.
6. J. Shi and J. Malik. Normalized cuts and image segmentation. *IEEE Trans. Pattern Analysis and Machine Intelligence*, 21(8):888–905, 2000.
7. Y. Gdalyahu, D. Weinshall, and M. Werman. Stochastic image segmentation by typical cuts. In *Proceedings of the IEEE CVPR 1999*, volume 2, pp. 596–601, 1999.
8. J. Puzicha, T. Hofmann, and J.M. Buhmann. A theory of proximity based clustering: structure detection by optimization. *Pattern Recognition*, 33:617–634, 2000.
9. D.R. Karger and C. Stein. A new approach to the minimum cut problem. *Journal of the ACM*, 43(4):601–640, 1996.
10. C.H. Li and P.C. Yuen. Normalized graph sampling in medical image database. In *Proceedings of the International Conference on Pattern Recognition.* IEEE Computer Society, 2002.
11. Cover and Thomas. *Elements of Informaton Theory.* John Wiley & Sons, New York, 1991.
12. E.T. Jaynes. Information theory and statistical mechanics. *Phys. Rev.*, 106:620–630, 1957.
13. E.T. Jaynes. On the rationale of maximum entropy methods. *Proc. of IEEE*, 70(9):939–952, 1982.
14. Battle D.J., Harrison R.P., and Hedley M. Maximum-entropy image-reconstruction from sparsely sampled coherent field data. *IEEE Trans. Image Processing*, 6(8):1139–1147, 1997.
15. C.H. Li and C.K. Lee. Minimum cross entropy thresholding. *Pattern Recognition*, 26:617–625, 1993.

Mining Small Objects in Large Images Using Neural Networks

Mengjie Zhang

School of Mathematical and Computing Sciences, Victoria University of Wellington, P.O. Box 600, Wellington, New Zealand
mengjie@mcs.vuw.ac.nz

Summary. Since the late 1980s, neural networks have been widely applied to data mining. However, they are often criticised and regarded as a "black box" due to the lack of interpretation ability. This chapter describes a domain independent approach to the use of neural networks for mining multiple class, small objects in large images. In this approach, the networks are trained by the back propagation algorithm on examples which have been cut out from the large images. The trained networks are then applied, in a moving window fashion, over the large images to mine the objects of interest. During the mining process, both the classes and locations of the objects are determined. The network behaviour is interpreted by analysing the weights in learned networks. Visualisation of these weights not only gives an intuitive way of representing hidden patterns encoded in learned neural networks for object mining problems, but also shows that neural networks are not just a black box but an expression or a model of hidden patterns discovered in the data mining process.

1 Introduction

As more and more data is collected as electronic form, the need for data mining is increasing extraordinarily. Due to the high tolerance to noisy data and the ability to classify unseen data on which they have not been trained, neural networks have been widely applied to data mining [1, 2, 3].

However, neural networks have been criticised for their poor interpretability, since it is difficult for humans to interpret the symbolic meaning behind the learned network weights and the network internal behaviour. For this reason, a neural network is often regarded as a black box classifier or a prediction engine [2]. In this chapter, we argue that it is not quite true, particularly for data mining in image data. We use the "weight matrices" to interpret the "hidden patterns" in learned networks for multiple class object mining problems.

This chapter addresses the problem of mining a number of different kinds of small objects in a set of large images. For example, it may be necessary to find all tumors in a database of x-ray images, all cyclones in a database of

M. Zhang: *Mining Small Objects in Large Images Using Neural Networks*, Studies in Computational Intelligence (SCI) **6**, 277–303 (2005)
www.springerlink.com

satellite images or a particular face in a database of photographs. The common characteristic of such problems can be phrased as "Given $subimage_1$, $subimage_2, \ldots, subimage_n$ which are examples of the object of interest, find all images which contain this object and its location(s)". Figure 2 shows examples of problems of this kind. In the problem illustrated by Fig. 2(b), we want to find centers of all of the 5 cent and 20 cent coins and determine whether the head or the tail side is up. Examples of other problems of this kind include target detection problems [4, 5, 6] where the task is to find, say, all tanks, trucks or helicopters in an image. Unlike most of the current work in the object mining/detection area, where the task is to find only objects of one class [4, 7, 8], our objective is to mine objects from a number of classes.

Neural networks have been applied to object classification and mining problems [7, 9, 10, 11]. In these approaches, various features/attributes such as brightness, colour, size and perimeter are extracted from the sub-images of the objects and used as inputs to the networks. These features are usually quite different and specific for different problem domains. Extracting and selecting good features is often very time consuming and programs for feature extraction and selection often need to be hand-crafted. The approach described in this chapter directly uses raw pixels as inputs to the networks.

1.1 Goals

The overall goal of this chapter is to develop a domain independent approach to the use of neural networks for mining multiple class objects in large images and to investigate a way of interpreting weights in learned networks. Instead of using specific image features, this approach directly uses raw image pixels as inputs to neural networks. This approach will be examined on a sequence of object mining problems of increasing difficulty. Specifically, we investigate:

- Will this approach work for a sequence of multiclass object mining problems of increasing difficulty?
- Will the performance deteriorate as the degree of difficulty of the detection problems increases?
- Can the weights in learned networks be interpreted in some ways and "hidden patterns" be successfully discovered and represented?
- Will the number of training examples affect object mining performance?

1.2 Structure

In the remainder of this chapter, we first describe the background, the image databases and our neural network approach, then present a set of experimental results. After making a detailed analysis and interpretation of the learned network weights, we conclude this approach and give a number of future directions.

2 Background

2.1 Object Mining and Its Dimensions

The term *object mining* here refers to the detection of small objects in large images. This problem is also known as *object detection*, which includes both *object classification* and *object localisation*. *Object classification* refers to the task of discriminating between images of different kinds of objects, where each image contains only one of the objects of interest. *Object localisation* refers to the task of identifying the positions of all objects of interest in a large image.

We classify the existing object mining systems into three dimensions based on whether the approach is segmentation free or not, domain independent or specific, and the number of object classes of interest in an image.

According to the number of independent stages used in the detection procedure, we divide the object mining methods into two categories:

- **Segmentation based approach**, which uses multiple independent stages for object mining. Most research on object mining involves four stages: *preprocessing, segmentation, feature extraction* and *classification* [12, 13, 14], as shown in Fig. 1. The preprocessing stage aims to remove noise or enhance edges. In the segmentation stage, a number of coherent regions and "suspicious" regions which might contain objects are usually located and separated from the entire images. The feature extraction stage extracts domain specific features from the segmented regions. Finally, the classification stage uses these features to distinguish the classes of the objects of interest. The algorithms or methods for these stages are generally domain specific. Learning techniques such as decision trees, neural networks and genetic algorithms/programming are usually applied to the classification stage. In general, each independent stage needs a program to fulfill that specific task and accordingly multiple programs are needed for object mining problems. Success at each stage is critical to achieving good final detection performance. Detection of trucks and tanks in visible, multispectral infrared and synthetic aperture radar images [5] and recognition of tanks in cluttered images [9] are two examples.
- **Single pass approach**, which uses only a single pass to mine the objects of interest in large images. There is only a single program produced for the whole object mining procedure. The major property of this approach is that it is segmentation free. Mining tanks in infrared images [6] and

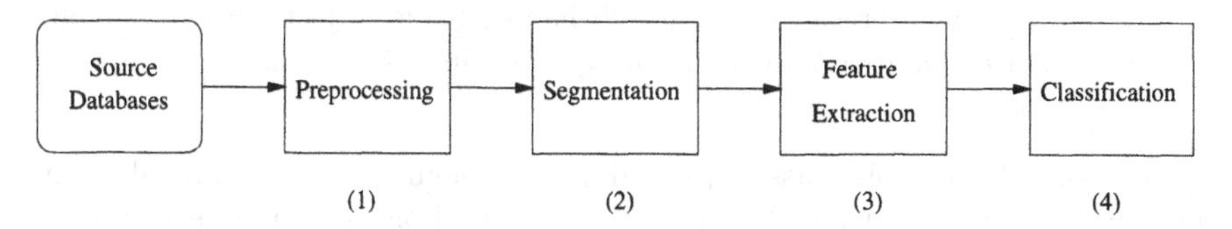

Fig. 1. A typical procedure for object detection

detecting small targets in cluttered images [11] based on a single neural network are examples of this approach.

While most recent work on object mining/detection problems concentrates on the segmentation based approach, this chapter focuses on the single pass approach.

In terms of the generalisation of the object mining systems, there are two major approaches:

- **Domain specific object mining**, which uses specific image features as inputs to the detector or classifier. These features, which are usually highly domain dependent, are extracted from entire images or segmented images. In a lentil grading and quality assessment system [15], for example, features such as brightness, colour, size and perimeter are extracted and used as inputs to a neural network classifier. This approach generally involves a time consuming investigation of good features for a specific problem and a hand-crafting of the corresponding feature extraction programs.
- **Domain independent object mining**, which usually uses the raw pixels (low level features) directly as inputs to the detector or classifier. In this case, feature selection, extraction and the hand-crafting of corresponding programs can be completely removed. This approach usually needs learning and adaptive techniques to learn features for the object mining task. Directly using raw image pixel data as input to neural networks for detecting vehicles (tanks, trucks, cars, etc.) in infrared images [4] is such an example. However, long learning/evolution times are usually required due to the large number of pixels. Furthermore, the approach generally requires a large number of training examples [16].

While most recent work focuses on domain specific approaches, this chapter uses a domain independent approach.

Regarding the number of object classes of interest in an image, there are two main types of object mining problems:

- **One class object mining problem**, where there are multiple objects in each image, however they belong to a single class. In nature, these problems contain a binary classification problem: *object* vs *non-object*, also called *object* vs *background*. Examples are detecting small targets in thermal infrared images [11] and detecting a particular face in photograph images [17].
- **Multiple class object mining problem**, where there are multiple object classes of interest each of which has multiple objects in each image. Detection of handwritten digits in zip code images [18] is an example of this kind.

In general, multiple class object mining problems are more difficult than one class mining problems. This chapter is focused on mining/detecting multiple objects from a number of classes in a set of images, which is particularly difficult. Most research in object mining/detection which has been done so far belongs to the one class object mining problem.

2.2 Performance Evaluation

Performance in object mining/detection is measured by detection rate and false alarm rate. The detection rate is the number of objects correctly reported as a percentage of the total number of real objects in the image(s). The false alarm rate, also called false alarms per object or *false alarms/object* [11], refers to the number of non-objects incorrectly reported as objects as a percentage of the total number of real objects. For example, a mining/detection system looking for grey squares in Fig. 2 (a) may report that there are 25. If 9 of these are correct the detection rate will be $(9/18) * 100 = 50\%$. The false alarm rate will be $(16/18) * 100 = 88.9\%$. Note that the detection rate is always between 0 and 100%, while the false alarm rate may be greater than 100% for difficult object detection problems.

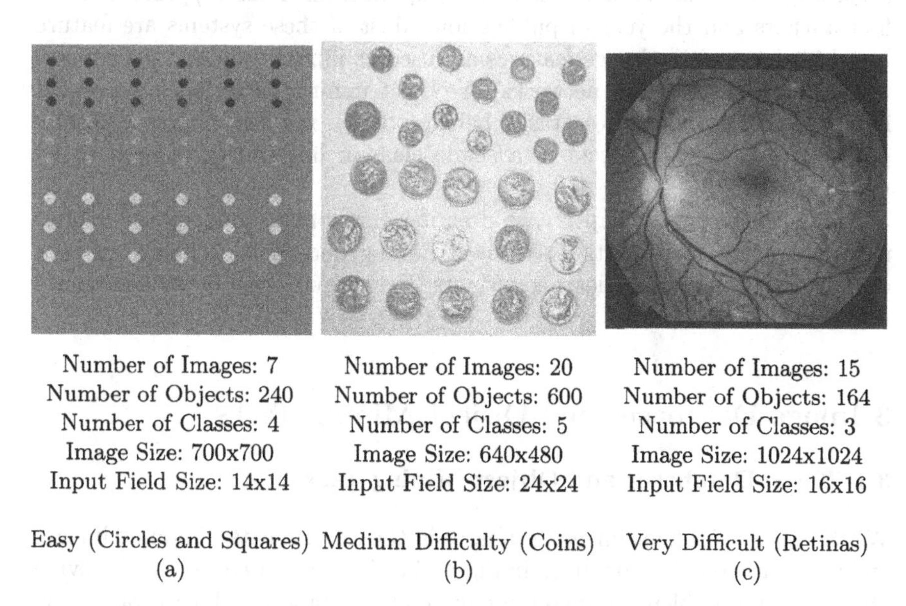

Number of Images: 7
Number of Objects: 240
Number of Classes: 4
Image Size: 700x700
Input Field Size: 14x14
Easy (Circles and Squares)
(a)
Number of Images: 20
Number of Objects: 600
Number of Classes: 5
Image Size: 640x480
Input Field Size: 24x24
Medium Difficulty (Coins)
(b)
Number of Images: 15
Number of Objects: 164
Number of Classes: 3
Image Size: 1024x1024
Input Field Size: 16x16
Very Difficult (Retinas)
(c)

Fig. 2. Object mining/detection problems of increasing difficulty

The main goal of object mining is to obtain a high detection rate and a low false alarm rate. There is, however, a tradeoff between them for a detection system. Trying to improve the detection rate often results in an increase in the false alarm rate, and vice versa. It is important to note that mining objects in images with very cluttered backgrounds is an extremely difficult problem and that false alarm rates of 200–2,000% are common [8, 11].

Also note that most research which has been done in this area so far either only presents the results of object classification where all the objects have been properly localised and segmented, or only gives the object localisation results where all objects of interest belong to a single class. The object mining results presented in this chapter are the performance for the whole object mining/detection problem (both object localisation and classification).

2.3 Neural Networks for Object Classification and Detection

Since the late 1980s, the use of neural networks in object classification and mining has been investigated in a variety of application domains. These domains include military applications, human face recognition, agricultural product classification, handwritten digit recognition and medical image analysis. The types of the neural networks used include multilayer feed forward networks [19], self organising maps [20], higher order networks [21, 22] and ART networks [23, 24].

A summary of neural networks for object classification is given in Table 1 according to the kinds of the networks, application domains/problems, the first authors and the year of publication. Most of these systems are feature based in that specific image features are used as inputs to neural networks.

A summary of neural networks for object mining/detection is presented in Table 2. Typically, these tasks belong to the *one-class object detection* problems, where the objects in a single class in large pictures need to be mined/detected.

In the rest of this chapter, we describe our domain independent neural network approach to multi-class object mining/detection problems, present our results, and interpret learned weights (hidden patterns) in trained neural networks.

3 Image Databases and Object Mining Tasks

3.1 Three Databases and Object Mining Tasks

We used three different image databases in the experiments. Example images and key characteristics are given in Fig. 2. The images were selected to provide object mining problems of increasing difficulty. Database 1 (Easy) was generated to give well defined objects against a uniform background. The pixels of the objects were generated using a Gaussian generator with different means and variances for each class. There are three classes of small objects of interest in this database: black circles (*class*1), grey squares (*class*2) and white circles (*class*3). The coin images (database 2) were intended to be somewhat harder and were taken with a CCD camera over a number of days with relatively similar illumination. In these images the background varies slightly in different areas of the image and between images and the objects to be detected are more complex, but still regular. There are four object classes of interest: the head side of 5 cent coins (class *head*005), the head side of 20 cent coins

Table 1. Neural networks for object classification

Kind of Network	Applications/Problems	Authors	Year	Source
Feed Forward Networks	Classification of mammograms	Verma	1998	[25]
	Target recognition with sonar	Barshan	2000	[26]
		Ayrulu	2001	[27]
	Classification of missiles, Planes and helicopters	Howard	1998	[28]
	Maneuver target recognition	Wong	1998	[29]
	Classification of tanks and Trucks on laser radar images	Troxel	1988	[30]
	Underwater target classification	Azimi-Sadjadi	2000	[31]
	Mine and mine-like Target detection	Miao	1998	[32]
	Handwritten character recognition	Verma	1998	[33]
	Agricultural product recognition	Winter	1996	[15, 34]
	Multispectral remote sensing Data classification	Lee	1997	[35]
	Natural object classification	Singh	2000	[36]
	Helicopter Classification	Stahl	2000	[37]
Shared Weight Neural Networks	Handwritten optical character recognition	Soulie	1993	[38]
	Zip code recognition	LeCun	1989	[18, 39]
	Digit recognition	de Redder	1996	[40, 41]
	Lung nodule detection			
	Microcalcification classification	Lo	1995	[42]
Auto-associative Memory Networks	Face recognition,	Abdi	1988	[43] [44]
		Valentin	1994	[45]
ART Networks	Vehicle recognition	Bernardon	1995	[46]
	Tank recognition	Fogler	1992	[47]
Serf-Organising Maps	Handwritten word recognition	Wessels	2000	[48]
		Dehghan	2000	[49]
Probability Neural Networks	Cloud classification	Tian	1998	[50]
	Radar target 'detection'	Kim	1992	[51, 52]
Gaussian Basis Function Networks	3D hand gesture recognition	Ahmad	1993	[53]
Neocognitron Networks	Bend point and end point recognition	Fukushima	1998	[54]
High Order Neural Networks	2D and 3D helicopter (F18) recognition	Spirkovska	1994	[55]
	Apple sorting (classification)	Hecht-Nielsen	1992	[56]
	Recognition of bars, triangles and squares	Cross	1995	[57]
Hybrid/Multiple Neural Networks	River identification	Liu	1998	[58]
	Character recognition	LeCun	2001	[59]
	Review	Egmont-Petersen	2001	[60]

Table 2. Object mining/detection based on neural networks

Kind of Network	Applications/Problems	Authors	Year	Source
Feed Forward Neural Networks	Target detection in thermal infrared images	Shirvaikar et al.	1995	[11]
Shared Weight Neural Networks	Vehicle detection	Khabou et al.	2000	[61]
		Won et al.	2000	[62]
Probabilistic Decision-Based Neural Networks	Face detection and recognition	Lin	2000	[63]
Other/ Hybrid/Multiple Neural Networks	Aircraft detection and recognition	Waxman et al.	1995	[5]
	Vehicle detection	Bosch et al.	1998	[64]
	Face detection	Feraud et al.	2001	[65]
		El-Bakry et al.	2000	[66]
	Triangle detection	Ahmad et al.	1990	[67]
	Detection/extraction of weak targets for Radars	Roth	1989	[68]
	Review of target detection	Rogers et al.	1995	[10]
	Review of target recognition	Roth	1990	[8]

(class *head*020), the tail side of 5 cent coins (class *tail*005) and the tail side of 20 cent coins (class *tail*020). All the objects in each class have a similar size. They are located at arbitrary positions and with different rotations. The retina images (database 3) were taken by a professional photographer with special apparatus at a clinic and contain very irregular objects on a highly cluttered background. The objective is to find two classes of retinal pathologies – haemorrhages (class *haem*) and micro aneurisms (class *micro*). To give a clear view of representative samples of the target objects in the retina images, one sample piece of these images is presented in Fig. 3. In this figure, haemorrhage and micro-aneurism examples are labeled using white surrounding squares. These objects are not only located in different places, but the sizes of the objects in each class are different as well, particularly for the haemorrhages. In addition, there are also other objects of different classes, such as veins (class *vein*) with different shapes and the retina "edges" (class *edge*). The backgrounds (class *other*) are varied, some parts are quite black, some parts are very bright, and some parts are highly cluttered. Note that in each of the databases the background (*non-object*) counts as a class (class *other*), but not a class of interest.

The object mining tasks here are to determine the locations and classes of the relatively small objects of interest in these large images by learning good neural networks. To implement these tasks, the next subsection briefly describes a number of different data sets used in this approach.

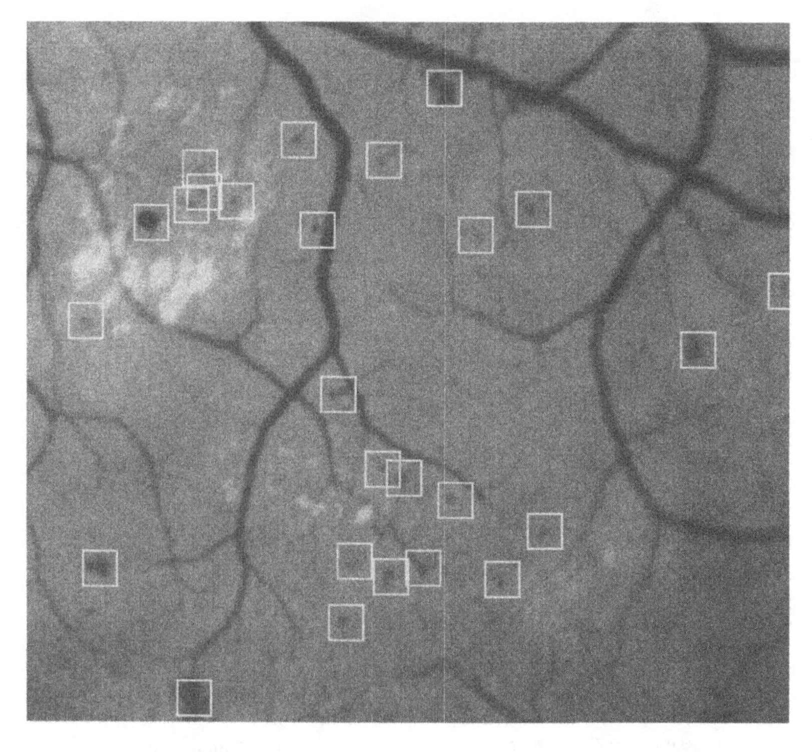

Fig. 3. An enlarged view of one piece of the retina images

3.2 Image Data Sets

To avoid confusion, we define a number of terms related to the image data.

A set of entire images in a database constitutes an *image data set* for a particular problem domain. In this chapter, it is randomly split into two parts: a *detection training set*, which is used to learn a detector, and a *detection test set*, which is used for measuring object mining performance. *Cutouts* refer to the object examples which are cut out from a detection training set. These cutouts form a *classification data set*, which is randomly split into two parts: a *classification training set* used for network training, and a *classification test set* for network testing. The classification test set plays two roles: one is to measure object classification performance, the other is used as a "tuning/validation set" for monitoring network training process in order to obtain good parameters of the learned network for object mining in large images. An *input field* refers to a square within a large image. It is used as a moving window for the network sweeping (detection) process. The size of the input field is the same as that of the cutouts for network training. The relationships between the various data sets are shown in Fig. 4.

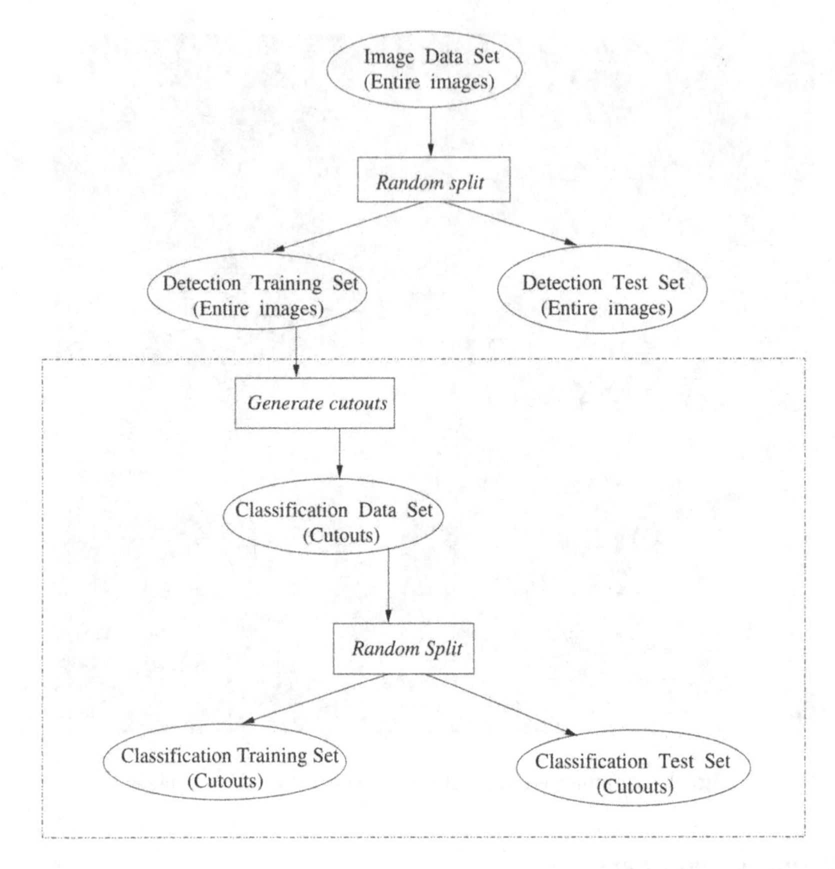

Fig. 4. Relationship between the classification and detection data sets

4 The Object Mining Method

4.1 Overview

An overview of the approach is shown in Fig. 5. A brief outline of the method is as follows.

1. Assemble a database of images in which the locations and classes of all the objects of interest are manually determined. Divide these full images into two sets: a *detection training set* and a *detection test set*.
2. Determine an appropriate size ($n \times n$) of a square which will cover all single objects of interest and form the input field of the networks. Generate a classification data set by cutting out squares of size $n \times n$ from the detection training set. Each of the squares, called *cutouts* or *sub-images*, only contains a single object and/or a part of the background. Randomly split these cutouts into a classification training set and a classification test set.

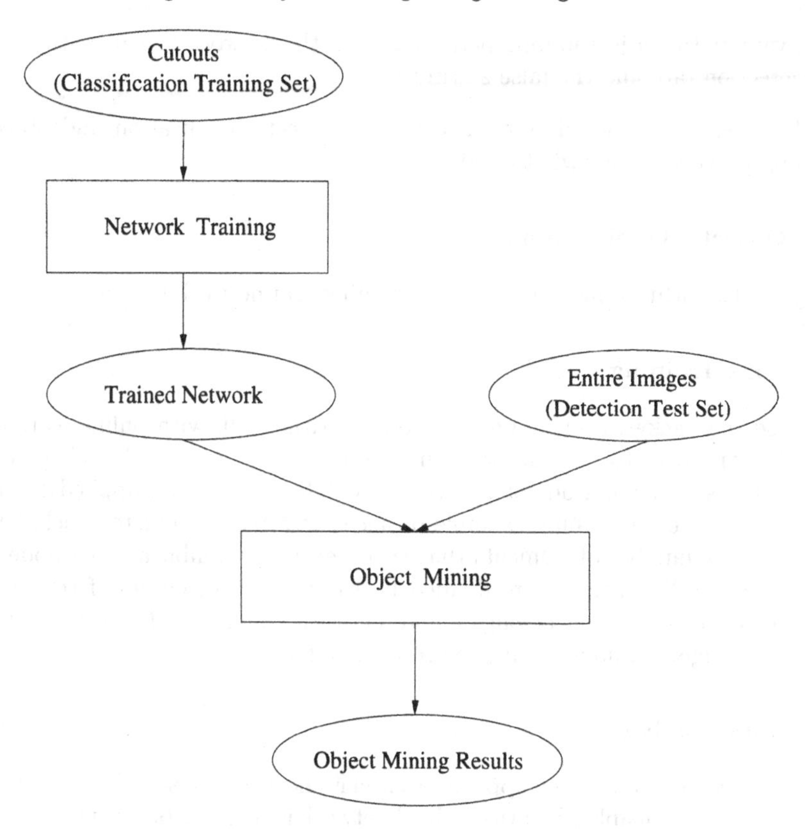

Fig. 5. Overview of the object mining approach

3. Determine the network architecture. A three layer feed forward neural network is used in this approach. The $n \times n$ pixel values form the inputs of the training data and the classification is the output. One hidden layer is used in this approach.[1] The number of hidden nodes is empirically determined.
4. Train the network by the back propagation algorithm [19] on the classification training data. The trained network is then tested on the classification test set to measure the object classification performance. The classification test data is also used to avoid overtraining the network. This step is designed to find the best trained network for object mining/detection.
5. Use the trained network as a moving window template to mine the objects of interest in the detection test set. If the output of the network for a class exceeds a given threshold then report an object of that class at the current location.

[1] The theoretical results provided by Irie and Miyake [69] and Funahashi [70] have proved that any continuous mapping can be approximated by a network with a single hidden layer.

6. Evaluate the object mining performance of the network by calculating the detection rate and the false alarm rate.

The remainder of this section describes object classification and object mining process in somewhat detail.

4.2 Object Classification

Object classification involves network training and network testing.

Network Training

We use the backward error propagation algorithm [19] with online learning and the fan-in factor [18, 40] to train the networks. In online learning (also called the stochastic gradient procedure) weight changes are applied to the network after each training example rather than after a complete epoch. The fan-in is the number of elements that either excite or inhibit a given node of the network. The weights are divided by the number of inputs of the node to which the connection belongs before network training and the size of the weight change of a node is updated in a similar way.

Network Testing

The trained network is then applied to the classification test set. If the test performance is reasonable, then the trained network is ready to be used for object mining. Otherwise, the network architecture and/or the learning parameters need to be changed and the network re-trained, either from the beginning or from a previously saved, partially trained network. During network training and testing, the classification is regarded as correct if the output node with the largest activation value corresponds to the desired class of an object.

4.3 Object Mining

After network training is successfully done, the trained network is used to determine the classes and locations of the objects of interest in the detection test set, which is not used in any way for network training. Classification and localisation are performed by the procedures: *network sweeping*, *finding object centres* and *object matching*.

Network Sweeping

During network sweeping, the successfully trained neural network is used as a template matcher, and is applied, in a moving window fashion, over the large images to mine the objects of interest. The template is swept across and down these large images, pixel by pixel in every possible location.

After the sweeping process is done, an *object sweeping map* for each detected object class will be produced. An object sweeping map corresponds to a grey level image. Sample object sweeping maps for *class1*, *class2* and *class3* together with the original image for the easy object mining problem are shown in Fig. 6. During the sweeping process, if there is no match between a square in an image and the template, then the neural network output is 0, which corresponds to black in the sweeping maps. A partial match corresponds to grey on the centre of the object, and a good match is close to white. The object sweeping maps can be used to get a qualitative indication of how accurate the object mining step is likely to be. In the sweeping maps, if a pixel value at a location is greater than or equal to a *threshold*, then report an object at that location. Figure 6 reveals that *class1* and *class3* objects will be detected very accurately, but there will probably be errors in class2.

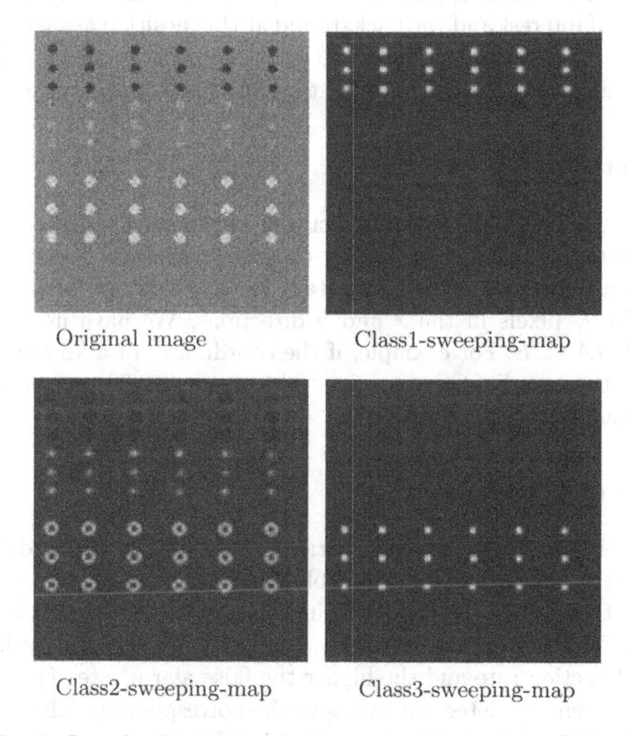

Fig. 6. Sample object sweeping maps in object mining/detection

Finding Object Centres

We developed a *centre-finding algorithm* to find the centres of all objects detected by the trained network. For each class of interest, this algorithm is used to find the centres of the objects based on the corresponding object sweeping map. The centre-finding algorithm is given as follows. For each object sweeping map,

Step 1 Set a *threshold* for the class (see Sect. 4.4).
Step 2 Set all of the values in the sweeping map to zero, if they are less than the threshold. For all the non-zero values, repeat Step 3 and Step 4:
Step 3 Search for the largest value, save the corresponding position (x, y), and label this position as an object centre.
Step 4 Set all values in the square input field of the labeled centre (x, y) to zero.

If two or more "objects" for different classes at a position are found, the decision will be made according to the network activations at that position. For example, if one object for *class2* and one for *class3* at position `(260, 340)` for the easy object mining problem are reported and the activations for the three classes of interest and the background at this position are `(0.27, 0.57, 0.83, 0.23)`, then the object for *class3* will be considered the mined/detected object at that position since the activation for *class3* is the biggest.

Object Matching

Object matching compares all the object centres reported by the centre finding algorithm with all the desired known object centres and reports the number of objects correctly mined/detected. Here, we allow location error of $TOLERANCE$ pixels in the x and y directions. We have used a value of 2 for $TOLERANCE$. For example, if the coordinates of a known object are $(19, 21)$ and the coordinates of a mined object are $(21, 22)$, we consider that the object has been correctly detected.

4.4 Choice of Thresholds

During the object mining process, various thresholds result in different detection results. The higher the threshold, the fewer the objects that can be detected by the trained network, which results in a lower detection rate but also a lower false alarm rate. Similarly, the lower the threshold selected, the higher the detection rate and the higher the false alarm rate. Thus there is a trade-off between the detection rate and the corresponding false alarm rate. Ideally there will be a threshold that gives 100% detection rate with 0% false alarms. If such a threshold cannot be found we use the one that gives 100% detection rate with fewest false alarms as the best object mining result. The threshold can be found by exhaustive search. We use the following heuristic procedure to speed up the search:

1. Initialise a threshold (T) to 0.7, apply the centre-finding algorithm and object matching process and calculate the detection rate (DR) and the corresponding false alarm rate (FAR) to the detection test set.

2. If the DR is less than 100%, decrease the T and calculate the new DR and FAR. Repeat this to obtain all the possible DRs and FARs until a new DR reaches 100% or a new T is less than or equal to 0.40.
3. From the point of step 1, if the FAR is not zero, increase T in order to obtain a new point (a DR with its corresponding FAR). Repeat this procedure until either the FAR is zero, or DR is zero, or the T is greater than or equal to 0.999.

The constants 0.7, 0.40 and 0.999 were empirically determined and worked well for all the image databases. To illustrate the relationship between the detection rate/false alarm rate and the threshold selection, the detection results of one trained network for *class2* in the easy images are presented in Table 3.

Table 3. Choice of thresholds: Object mining results for *class2* in the easy images

Easy Pictures (Basic Method)	Object Classes							
	Class2							
Threshold	0.54	0.56	0.57	0.595	0.625	0.650	0.673	0.700
Detection Rate(%)	100	96.67	93.33	90.00	86.67	83.33	80.00	76.67
False Alarm Rate(%)	90.5	35.1	13.8	11.9	11.2	8.1	5.6	3.7
Threshold	0.725	0.747	0.755	0.800	0.835	0.865	–	–
Detection Rate(%)	73.33	70.00	66.67	63.33	60.00	56.67	–	–
False Alarm Rate(%)	3.1	2.5	2.3	1.0	0.25	0	–	–

5 Results

5.1 Object Classification Results

To classify the object cutouts for a particular problem, the number of hidden nodes of the network needs to be empirically determined. We tried a series of numbers of hidden nodes for the three object mining problems: 1, 2, 3, 4, 5, 6, 7, 8, 9, 10, 13, 15, 20, 30, 50, 100, 150, 200, 300, and 500. The experiments indicated that the networks 196-4-4,[2] 576-3-5 and 256-5-3 with a set of learning parameters gave the best performance for the easy, the coin, and the retina problems, respectively.

Network training and testing results for object classification are presented in Table 4. In all cases, the network training and testing procedure was repeated 15 times and the average results are presented. Line 1 shows that the best network for the easy images is 196-4-4, the average number of epochs

[2] 196-4-4 refers to a feed forward network architecture with 196 input nodes, 4 hidden nodes and 4 output nodes. In this chapter, we use a similar way to express other network architectures.

Table 4. Object classification results for the three databases

Image Databases	Network Architecture	Training Epochs	Training Accuracy	Test Accuracy
Easy Images	196-4-4	199.40	100%	100%
Coin Images	576-3-5	234.6	100%	100%
Retina Images	256-5-3	475.8	81.62%	71.83%

used to train the network is 199.40, and the trained network can achieve 100% accuracy on the cutouts of both the classification training set and test set. For the coin images, we can also achieve the ideal performance for object classification. However, this is not the case for the retina images, where only 71.83% accuracy was obtained on the test set of the cutouts.

5.2 Object Mining Results

This sub section describes object mining performance of this approach for the three image databases. For each problem, the 15 trained networks obtained in object classification are used to mine objects in the detection test set and the average results are presented.

Easy Images

The best detection rates and the corresponding false alarm rates for the three classes in the easy images are presented in Table 5. The best detection rates achieved for all the three classes are 100%, showing that this approach can successfully mine all the objects of interest in this database. At this point, mining objects in *class1* (black circles) and *class 3* (white circles) did not produce any false positives (false alarm rates are zero), while mining objects in *class2* (grey squares) resulted in a 91.2% false alarm rate on average.

Table 5. Object mining results for the easy images

Easy Images	Object Classes		
	Class1	Class2	Class3
Detection Rate(%)	100	100	100
False Alarm Rate(%)	0	91.2	0

Coin Images

The object mining results for the coin images are described in Table 6. As in the easy images, the trained neural networks achieved 100% detection rates for all the classes, showing that this approach correctly mined all the objects

Table 6. Object mining/detection results for the coin images

Coin Images	Object Classes			
	head005	tail005	head020	tail020
Detection Rate (%)	100	100	100	100
False Alarm Rate (%)	0	0	182.0	37.5

of interest from different classes in the coin images. In each run, it was always possible to find a threshold for the network output for class *head005* and *tail005*, which resulted in mining all of the objects of these classes with no false alarms. However, mining classes *head020* and *tail020* was a relatively difficult problem. The average false alarm rates for the two classes at a 100% detection rate were 182% and 37.5% respectively.

Retina Images

Compared with the performance of the easy and coin images, the results of the two classes *haem* and *micro* in the very difficult retina images are disappointing. All the objects of class *micro* were correctly mined (a detection rate of 100%), however, with a very high false alarm rate. The best detection rate for class *haem* was only 73.91%. Even at a detection rate of 50%, the false alarm rate was still quite high.

ROC Curves

To give an intuitive view, the extended ROC curves (Receiver Operating Characteristic curves [71, 72]) for class *class2* in the easy images, class *head020* in the coin images, and classes *haem* and *micro* in the retina images are presented in Fig. 7 (a), (b), (c) and (d) respectively.

5.3 Discussion

As can be seen from the object mining results obtained here, it was always possible to detect all objects of interest in the easy and the coin images. This reflects the fact that the objects in the two databases are relatively simple or regular and the background is uniform or relatively uniform. While mining objects in the easy images only resulted in a few false alarms, mining objects in the coin images resulted in a relatively higher false alarm rate. This is mainly because the object mining problems in the coin images are more difficult than in the easy images.

Due to the high degree of difficulty of the object mining problems, the results for the retina images were not good. For class *micro*, while all objects were correctly detected, a very high number of false alarms were produced. This is mainly because these objects are irregular and complex and the background is highly cluttered. For class *haem*, it was not possible to detect all

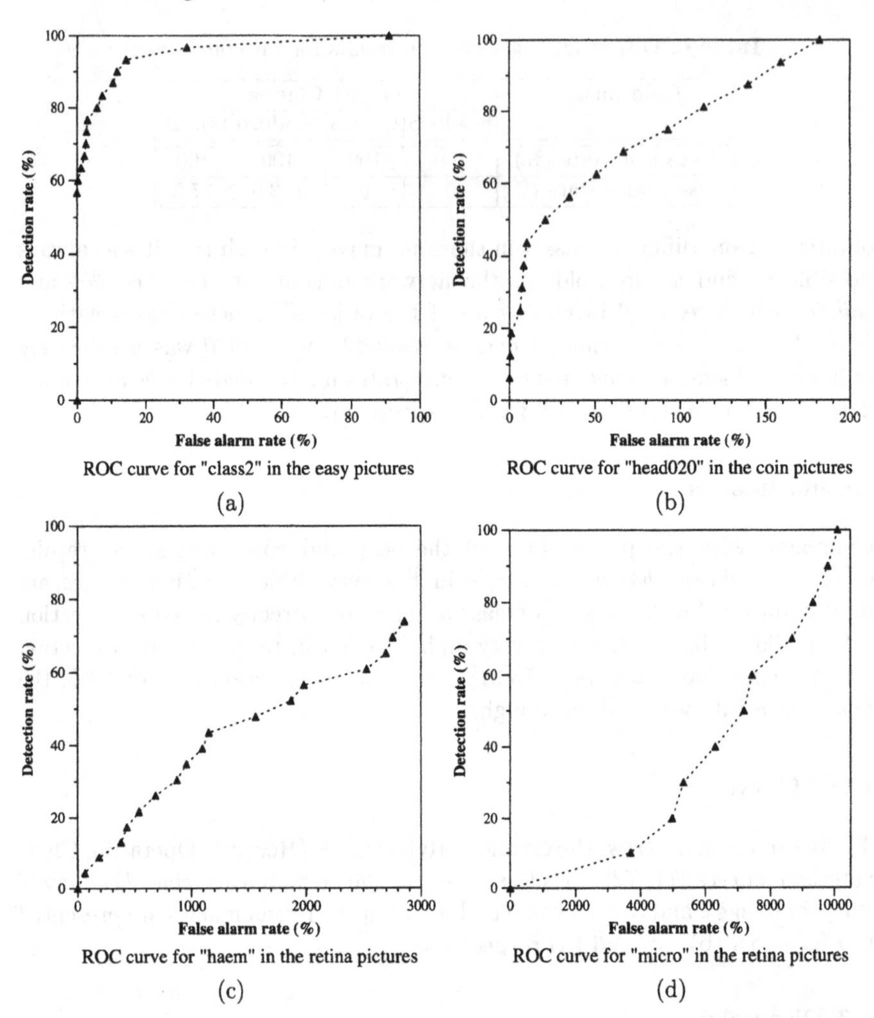

Fig. 7. Extended ROC curves for some "difficult" classes in the three databases

objects of interest (the best detection rate was 73.91%). This is mainly due to the size variance of these objects. The sizes of objects in class *haem* vary from 7×7 to 16×16 pixels – these objects are too different to be considered the same class. Another reason is that the training object examples in the retina images are not sufficient (only 164, see Fig. 2). Due to the insufficient number of training examples, we did not even achieve good object classification results (see Table 4).

Comparing the results obtained in object classification and object mining, it can be found that object classification results are "better" than the object mining results for all the three databases. In the coin images, for example, the trained networks achieved perfect object classification results, that is, 100%

accuracy on classifying all the objects in the four classes. While mining 5 cent coins achieved ideal performance, that is, all objects of the two classes (*head005, tail005*) were correctly mined with zero false alarm rate, mining the two 20 cent coin classes (*head020, tail020*) produced a number of false positive objects. The results for the other databases showed a similar pattern. This is due mainly to the fact that multiclass object mining task is generally much more difficult than only the classification task on the same problem domains since multiclass object mining includes both object classification and object localisation.

6 Visualisation and Analysis of Learned Network Weights

To analyse why this approach can be used for multiclass object mining, this section interprets the network internal behaviour through visual analysis of weights in trained networks. For presentation convenience, we use the trained networks for mining regular objects in the coin images. Most other networks contained similar patterns.

Figure 8 shows the network architecture (576-3-5) used in the coin images. The weight groups between the input nodes and the hidden nodes and between

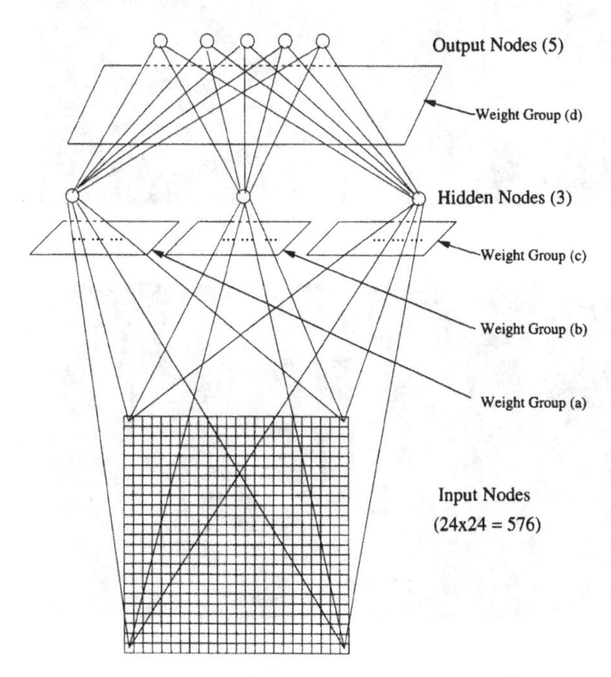

Fig. 8. Network architecture with four weight groups for object mining/detection in the coin images

the hidden nodes and the output nodes are also presented. The weight matrices (a), (b), (c) and (d) shown in Fig. 9 correspond to weight groups (a), (b), (c) and (d) in the network architecture in this figure.

Figure 9 shows the weights from a trained 576-3-5 network which has been successfully applied to the coin images. In this figure the full squares represent positive weights and the outline squares represent negative weights, while the size of the square is proportional to the magnitude of the weight. Matrices (a), (b) and (c) show the weights from the input nodes to the first, the second and the third hidden nodes. The weights are shown in a 24×24 matrix (corresponding to the 24×24 input field) to facilitate visualisation. Figure 9 (d) shows the weights from the hidden nodes to the output nodes and the biases of these output nodes. The five rows in this matrix correspond to the classes *other, tail020, head020, tail005* and *head005*. The first three of the four columns correspond to weights from the three hidden nodes (associated with weight matrices (a), (b) and (c)) to the five output nodes. The last column corresponds to the biases of the five output nodes.

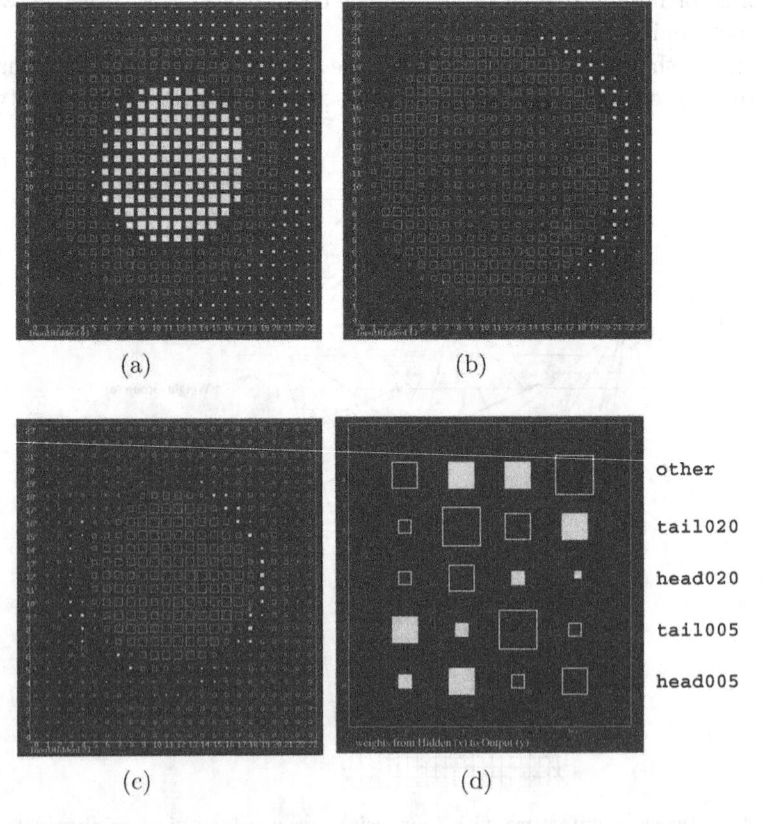

Fig. 9. Weights in a trained network for object mining/detection in the coin images

Inspection of the first column of Fig. 9 (d) reveals that weight matrix (a) has a positive influence on 5 cent coins and a negative influence on 20 cent coins. It has a strong influence on class *tail005* but a weak influence on class *head005*. The same matrix has a strong negative effect on class *other* (background). Inspection of the second column reveals that weight matrix (b) has a positive effect on the 5 cent coins and a negative effect on the 20 cent coins. Moreover, it has a strong influence on class *head005* but a week influence on class tail005. Also it has a very strong positive influence on class *other*. This indicates that the combination of matrices (a) and (b) not only can separate the 5 cent coins from the 20 cent coins and the background, but also can discriminate the 5 cent tails from the 5 cent heads. Inspection of the third column reveals that the weight matrix (c) has a strong negative influence on the tails of both 5 cent and 20 cent coins and a week influence on the heads of both 5 cent and 20 cent coins. It also strongly supports the background. The fourth column suggests that the biases also play a complementary role for mining objects, particularly for class *tail020*, class *other* and class *head005*. If we regard the nodes of the hidden layers as representing feature detectors learnt by the network, then Figs. 9(a)–(c) are a visual representation of these features. Visually these features "make sense" as there are regions corresponding to the 5 cent coins, the annulus remaining to the background when a 5 cent coin is "removed" from the centre of a 20 cent coin.

For object mining in large images described here, the weight matrices can be considered "hidden patterns" encoded in learned neural networks. Visualisation of the learned network weights reveals that patterns contained in learned networks can be intuitively represented, which strongly supports the idea that neural networks are not just a "black box", but a model or an expression of patterns discovered during learning.

7 Conclusions

This chapter presented a pixel based approach to mining multiple class objects in large images using multilayer feed forward neural networks. The back propagation algorithm was used to train the network on the sub-images which had been cut out from the large images. The trained network was then applied, in a moving window fashion, over the entire images to mine/detect the objects of interest. Object mining performance was measured on the large images in the detection test set. The experimental results showed that this approach performed very well for mining a number of simple and regular objects against a relatively uniform background. It did not perform well on the very difficult object mining problems in the retina images. As expected, the performance degrades when the approach is applied to object mining problems of increasing difficulty.

Visualisation of the weights in trained neural networks resulting from this approach revealed that trained networks contained feature detectors which "made sense" for the problem domain and could discriminate objects from different classes. This provides a way of revealing hidden patterns in learned neural networks for object mining problems. This approach also shows that trained neural networks are not just a black box, but a model or an expression of patterns discovered in the learning process.

The approach has the following advantages:

- Raw image pixel data are used as inputs to neural networks, and accordingly traditional specific feature extraction and selection can be avoided.
- It is a domain independent approach and can be directly applied to multiclass object mining problems in different areas.
- Multiple class objects can be mined (classified and localised) in large images with a single trained neural network.
- Patterns encoded in learned neural networks can be visually represented by using weight matrices.

The approach also has a number of disadvantages, which need to be addressed in the future:

- This approach can mine translation and limited rotation invariant objects but does not seem to successfully detect objects with a large size variance such as class *haem* in the retina images.
- In this approach, the network was trained based on the object cutouts but the trained network was directly applied to the entire images. This might be one of the main reasons which resulted in many false positives in some difficult object mining problems. We are investigating ways of refining the trained networks based on the entire images to improve object mining performance.
- The insufficient number of training examples led to poor object mining performance in the retina images. It would be very interesting to investigate whether this approach can achieve good results using sufficient training examples in the future – this will reveal whether this approach can be used to mine/detect complex and irregular objects against a highly cluttered background.

Acknowledgement

We would like to thank Dr. Peter Andreae and Dr. Victor Ciesielski for a number of useful discussions.

References

1. Fayyad, U., Piatetsky-Shapiro, G., Smyth, P. (1996) From data mining to knowledge discovery in databases. AI Magazine **17** 37–53

2. Groth, R. (2000) Data Mining: Building Competitive Advantage. Prentice Hall PTR
3. Jan, J., Kamber, M. (2001) Data Mining: Concepts and Techniques. Morgan Kaufmann Publishers
4. Gader, P.D., Miramonti, J.R., Won, Y., Coffield, P. (1995) Segmentation free shared weight neural networks for automatic vehicle detection. Neural Networks **8** 1457–1473
5. Waxman, A.M., Seibert, M.C., Gove, A., Fay, D.A., Bernandon, A.M., Lazott, C., Steele, W.R., Cunningham, R.K. (1995) Neural processing of targets in visible, multispectral IR and SAR imagery. Neural Networks **8** 1029–1051
6. Won, Y., Gader, P.D., Coffield, P.C. (1997) Morphological shared-weight networks with applications to automatic target recognition. IEEE Transactions on neural networks **8** 1195–1203 ISSN 1045–9227.
7. Roitblat, H.L., Au, W.W.L., Nachtigall, P.E., Shizumura, R., Moons, G. (1995) Sonar recognition of targets embedded in sediment. Neural Networks **8** 1263–1273
8. Roth, M.W. (1990) Survey of neural network technology for automatic target recognition. IEEE Transactions on neural networks **1** 28–43
9. Casasent, D.P., Neiberg, L.M. (1995) Classifier and shift-invariant automatic target recognition neural networks. Neural Networks **8** 1117–1129
10. Rogers, S.K., Colombi, J.M., Martin, C.E., Gainey, J.C., Fielding, K.H., Burns, T.J., Ruck, D.W., Kabrisky, M., Ocley, M. (1995) Neural networks for automatic target recognition. Neural Networks **8** 1153–1184
11. Shirvaikar, M.V., Trivedi, M.M. (1995) A network filter to detect small targets in high clutter backgrounds. IEEE Transactions on Neural Networks **6** 252–257
12. Caelli, T., Bischof, W.F. (1997) Machine Learning and Image Interpretation. Plenum Press, New York and London
13. Faugeras, O. (1993) Three-Dimensional Computer Vision – A Geometric Viewpoint. The MIT Press
14. Gose, E., Johnsonbaugh, R., Jost, S. (1996) Pattern Recognition and Image Analysis. Prentice Hall PTR, Upper Saddle River, NJ 07458
15. Winter, P., Sokhansanj, S., C. Wood, H., Crerar, W. (1996) Quality assessment and grading of lentils using machine vision. In: Agricultural Institute of Canada Annual Conference, Saskatoon, SK S7N 5A9, Canada, Canadian Society of Agricultural Engineering
16. Baum, E.B., Haussler, D. (1989) What size net gives valid generalisation? Neural Computation **1** 151–160
17. Lin, S.H., Kung, S.Y., Lin, L.J. (1997) Face recognition/detection by probabilistic decision-based neural network. IEEE Transactions on Neural Networks **8** 114–132
18. LeCun, Y., Boser, B., Denker, J.S., Henderson, D., Hubbard, R.E.H.W., Jackel, L.D. (1989) Backpropagation applied to handwritten zip code recognition. Neural Computation **1** 541–551
19. Rumelhart, D.E., Hinton, G.E., Williams, R.J. (1986) Learning internal representations by error propagation. In Rumelhart, D.E., McClelland, J.L., the PDP research group, eds.: Parallel distributed Processing, Explorations in the Microstructure of Cognition, Volume 1: Foundations. The MIT Press, Cambridge, Massachusetts, London, England
20. Kohonen, T. (1988) Self-organization and Associative Memory. 3rd edn. Springer, Berlin Heidelberg New York

21. Giles, G.L., Maxwell, T. (1987) Learning, invariances, and generalisation in high-order neural networks. Applied Optics **26** 4972–4978
22. Pitts, W., McCulloch, W.S. (1947) How we know universals: The perception of auditory and visual foems. Bulletin of Mathematical Biophysics **9** 127–147 University of Chicago Press, Chicago.
23. Carpenter, G.A., Grossberg, S. (1987) A massively parallel architecture for a self-organising neural pattern recognition machine. Computer Vision, Graphics and Image Processing **37** 54–115
24. Carpenter, G.A., Grossberg, S. (1987) Stable self-organisation of pattern recognition codes for analog input patterns. Applied Optics **26** 4919–4930
25. Verma, B. (1998) A neural network based technique to locate and classify microcalcifications in digital mammograms. In: 1998 IEEE World Congress on Computational Intelligence – IJCNN'98, Anchorage, Alaska. IEEE
26. Barshan, B., Ayrulu, B., Utete, S.W. (2000) Neural network-based target differentiation using sonar for robotics applications. IEEE Transactions on Robotics and Automation **16** 435–442
27. Ayrulu, B., Barshan, B. (2001) Neural networks for improved target differentiation and localization with sonar. Neural Networks **14** 355–373
28. Howard, A., Padgett, C., Liebe, C.C. (1998) A multi-stage neural network for automatic target detection. In: 1998 IEEE World Congress on Computational Intelligence – IJCNN'98, Anchorage, Alaska
29. Wong, Y.C., Sundareshan, M.K. (1998) Data fusion and tracking of complex target maneuvers with a simplex-trained neural network-based architecture. In: 1998 IEEE World Congress on Computational Intelligence – IJCNN'98, Anchorage, Alaska
30. Troxel, S.E., Rogers, S.K., Kabrisky, M. (1988) The use of neural networks in PSRI target recognition. In: IEEE International Conference on Neural Networks, Sheraton Harbor Island, San Diego, California Vol I, 593–600
31. Azimi-Sadjadi, M.R., Yao, D., Huang, Q., Dobeck, G.J. (2000) Underwater target classification using wavelet packets and neural networks. IEEE Transactions on Neural Networks **11** 784–794
32. Miao, X., Azimi-Sadjadi, M.R., Dubey, A.C., Witherspoon, N.H. (1998) Detection of mines and minelike targets using principal component and neural network methods. IEEE Transactions on Neural Networks **9** 454–463
33. Verma, B. (1998) A feature extraction technique in conjunction with neural network to classify cursive segmented handwritten characters. In: 1998 IEEE World Congress on Computational Intelligence – IJCNN'98, Anchorage, Alaska, IEEE
34. Winter, P., Yang, W., Sokhansanj, S., Wood, H. (1996) Discrimination of hard-to-pop popcorn kernels by machine vision and neural network. In: ASAE/CSAE meeting, Saskatoon, Canada
35. Lee, C., Landgrebe, D.A. (1997) Decision boundary feature extraction for neural networks. IEEE Transactions on Neural Networks **8** 75–83
36. Singh, S., Markou, M., Haddon, J. (2000) Natural object classification using artificial neural networks. In: Proceedings of the IEEE-INNS-ENNS International Joint Conference on Neural Networks (IJCNN'00), Volume III, Como, Italy
37. Stahl, C., Aerospace, D., Schoppmann, P. (2000) Advanced automatic target recognition for police helicopter missions. In Sadjadi, F.A., ed.: Proceedings of SPIE Volume 4050, Automatic Target Recognition X.

38. Soulie, F.F., Viennet, E., Lamy, B. (1993) Multi-modular neural network architectures: applications in optical character and human face recognition. International journal of pattern recognition and artificial intelligence **7** 721–755
39. LeCun, Y., Jackel, L.D., Boser, B., Denker, J.S., Graf, H.P., Guyon, I., Henderson, D., Howard, R.E., Hibbard, W. (1989) Handwritten digit recognition: application of neural network chips and automatic learning. IEEE Communications Magazine 41–46
40. de Ridder, D. (1996) Shared weights neural networks in image analysis. Master's thesis, Delft University of Technology, Lorentzweg 1, 2628 CJ Delft, The Netherlands
41. de Ridder, D., Hoekstra, A., Duin, R.P.W. (1996) Feature extraction in shared weights neural networks. In: Proceedings of the Second Annual Conference of the Advanced School for Computing and imaging, ASCI, Delft 289–294
42. Lo, S.C.B., Chan, H.P., Lin, J.S., Li, H., Freedman, M.T., Mun, S.K. (1995) Artificial convolution neural network for medical image pattern recognition. Neural Networks **8** 1201–1214
43. Abdi, H. (1988) A generalized approach for connectionist auto-associative memories: interpretation, implications and illustration for face processing. In Demongeot, J., Herve, T., Rialle, V., Roche, C., eds.: Artificial Intelligence and Cognitive Sciences. Manchester University Press 149–165
44. Valentin, D., Abdi, H., O'Toole (1994) Categorization and identification of human face images by neural networks: A review of linear auto-associator and principal component approaches. Journal of Biological Systems **2** 413–429
45. Valentin, D., Abdi, H., O'Toole, A.J., Cottrell, G.W. (1994) Connectionist models of face processing: A survey. Pattern Recognition **27** 1208–1230
46. Bernardon, A., Carrick, J.E. (1995) A neural system for automatic target learning and recognition applied to bare and camouflaged SAR targets. Neural Networks **8** 1103–1108
47. Rogler, R.J., Koch, M.W., Moya, M.M., Hostetler, L.D., Hush, D.R.: (1992) Feature discovery via neural networks for object recognition in SAR imagery. In: International Joint Conference on Neural Networks (IJCNN'92), Baltimore, Maryland, IEEE IV 408–413
48. Wessels, T., Omlin, C.W. (2000) A hybrid system for signature verification. In: Proceedings of the IEEE-INNS-ENNS International Joint Conference on Neural Networks (IJCNN'00), Volume V, Como, Italy
49. Dehghan, M., Faez, K., Ahmadi, M. (2000) A hybrid handwritten word recognition using self-organizing feature map, discrete hmm, and evolutionary programming. In: Proceedings of the IEEE-INNS-ENNS International Joint Conference on Neural Networks (IJCNN'00), Volume, Como, Italy
50. Tian, B., Azimi-Sadjadi, M.R., Haar, T.H.V., Reinke, D. (1998) A temporal adaptive probability neural network for cloud classification from satellite images. In: 1998 IEEE World Congress on Computational Intelligence – IJCNN'98, Anchorage, Alaska, IEEE
51. Kim, M.W., Arozullah, M. (1992) Generalised probabilistic neural network based classifier. In: International Joint Conference on Neural Networks (IJCNN'92), Baltimore, Maryland, IEEE III–648–653
52. Kim, M.W., Arozullah, M. (1992) Neural network based optimum radar target detection in non-Gaussian noise. In: International Joint Conference on Neural Networks (IJCNN'92), Baltimore, Maryland, IEEE III 654–659

53. Ahmad, S., Tresp, V. (1993) Some solutions to the missing feature problem in vision. In Hanson S.J., C.J., C.L., G., eds.: Advances in Neural Information Processing Systems. Morgan Kaufmann Publishers, San Mateo, CA.
54. Fukushima, K., Kimira, E., Shouno, H. (1998) Neocognitron with improved bend-extractors. In: 1998 IEEE World Congress on Computational Intelligence – IJCNN'98, Anchorage, Alaska
55. Spirkovska, L., Reid, M.B. (1994) Higher-order neural networks applied to 2D and 3D object recognition. Machine Learning **15** 169–199
56. Hecht-Nielsen, R. (1992) Neural networks and image analysis. In Carpenter, Grossbury, eds.: Neural networks for vision and image processing. MIT Press 449–460
57. Cross, N., Wilson, R. (1995) Neural networks for object recognition. Technical report, Department of Computer Science, University of Warwick, Coventry
58. Liu, X., Wang, D., Ramirez, J. R. (1998) Extracting hydrographic objects from satellite images using a two-layer neural network. In: 1998 IEEE World Congress on Computational Intelligence – IJCNN'98, Anchorage, Alaska, IEEE
59. LeCun, Y., Bottou, L., Bengio, Y., Haffner, P. (2001) Gradient-based learning applied to document recognition. In: Intelligent Signal Processing, IEEE Press. 306–351
60. Egmont-Petersen, M., de Ridder, D., Handels, H. (2001) Image processing with neural networks - a review. Pattern Recognition. Preprint, to appear in Pattern Recognition, 2001.
61. Khabou, M., Gader, P. D., Keller, J. M. (2000) Ladar target detection using morphological shared-weight neural networks. Machine Vision and Applications **11** 300–305
62. Won, Y., Nam, J., Lee, B. H. (2000) Image pattern recognition in natural environment using morphological feature extraction. In Ferri, F.J., Iñesta, J.M., Amin, A., Pudil, P., eds.: Advances in Pattern Recognition, Joint IAPR International Workshops SSPR 2000 and SPR 2000, Lecture Notes in Computer Science, Vol. 1876, Springer. 806–815
63. Lin, S.H. (2000) An introduction to face recognition technology. Special Issue on Multimedia Technologies and Informing Science, Part II, Information Science, the International Journal of an Emergin Discipline **3** 1–7
64. Bosch, H., Milanese, R., Labbi, A. (1998) Object segmentation by attention-included oscillations. In: 1998 IEEE World Congress on Computational Intelligence – IJCNN'98, Anchorage, Alaska. 1167–1171
65. Feraud, R., Bernier, O.J., Viallet, J.E., Collobert, M. (2001) A fast and accurate face detector based on neural networks. IEEE Transactions on Pattern Analysis and Machine Intelligence **23** 42–53
66. El-Bakry, H.M., Abo-Elsoud, M.A., Kamel, M.S. (2000) Fast modular neural nets for human face detection. In: Proceedings of the IEEE-INNS-ENNS International Joint Conference on Neural Networks (IJCNN'00), Volume III, Como, Italy
67. Ahmad, S., Omohundro, S. (1990) A network for extracting the locations of point clusters using selective attention. In: the 12th Annual Conference of the Cognitive Science Society, MIT Also in Technical Report #90–011
68. Roth, M.W. (1989) Neural networks for extraction of weak targets in high clutter environments. IEEE Transactions on System, Man, and Cybernetics **19** 1210–1217

69. Irie, B., Miyake, S. (1988) Capability of three-layered perceptrons. In: Proceedings of the IEEE 1988 International Conference on Neural Networks, New York, IEEE I 641–648
70. Funahashi, K.I. (1989) On the approximate realization of continuous mappings by neural networks. Neural Networks **2** 183–192
71. Metz, C.E. (1986) ROC methodology in radiologic imaging. Investigative Radiology **21** 720–732
72. Zhang, M. (2000) A Domain Independent Approach to 2D Object Detection Based on the Neural and Genetic Paradigms. PhD thesis, Department of Computer Science, RMIT University, Melbourne, Australia

Improved Knowledge Mining with the Multimethod Approach

Mitja Lenič, Peter Kokol, Milan Zorman, Petra Povalej, Bruno Stiglic, and Ryuichi Yamamoto

Laboratory for system design, Faculty of Electrical Engineering and Computer Science, University of Maribor, Maribor, Slovenia
Division of Medical Informatics, Osaka Medical College, Takatsuki City, Osaka

Automatic induction from examples has a long tradition and represents an important technique used in data mining. Trough induction a method builds a hypothesis to explain observed facts. Many knowledge extraction methods have been developed, unfortunately each has advantages and limitations and in general there is no such method that would outperform all others on all problems. One of the possible approaches to overcome this problem is to combine different methods in one hybrid method. Recent research is mainly focused on a specific combination of methods, contrary, multimethod approach combines different induction methods in an unique manner – it applies different methods on the same knowledge model in no predefined order where each method may contain inherent limitations with the expectation that the combined multiple methods may produce better results. In this paper we present the overview of an idea, concrete integration and possible improvements.

1 Introduction

The aggressive rate of growth of disk storage and thus the ability to store enormous quantities of data has far outpaced our ability to process and utilize it. This challenge has produced a phenomenon called data tombs – data is deposited to merely rest in peace, never to be accessed again. But the growing appreciation that data tombs represent missed opportunities in for example supporting scientific discovering, business exploitation or complex decision making has awaken the growing commercial interest in knowledge discovery and data mining techniques. That in order has stimulated new interest in the automatic knowledge extraction from examples stored in large databases – a very important class of techniques in data mining field.

Machine learning has long tradition in knowledge extraction, which is obviously needed in today's society that has relatively inexpensive and generally available means to collect and store the data from various environments. With

M. Lenič et al.: *Improved Knowledge Mining with the Multimethod Approach*, Studies in Computational Intelligence (SCI) **6**, 305–318 (2005)
www.springerlink.com

the variety of environments it is almost impossible to develop single induction method that would fit all possible requirements.

The question how the learning capacity can be increased and generalized has received much attention in particular in machine learning community [1]. The main problem is to learn how to learn and to be able to present new knowledge in the form easy understandable to the human.

Most of the work was concentrated on a single approach with single knowledge representation (see Sect. 2 on single approaches), but recently the explosion of hybrid methods can be observed. One of the possible reasons is that separate research communities like symbolic machine learning, computational learning theory, neural networks, statistics and pattern recognition, etc. have discovered each other. But also many hybrid approaches are somehow limited – thereafter we constructed a new so called multimethod approach trying out some original solutions. The first ideas were presented last year on the ICDM workshop [10], in the current paper we are presenting the improvements.

The brief overview of the idea of hybrid approaches is given in Sect. 3. Then brief overview of multimethod approach [10] is presented in Sect. 4. Concrete integration methods are presented in Sect. 5. In Sect. 6 experiments on some real-world problems from medical domain and results are presented.

Main contributions of our research presented in this paper are:

- an user friendly improved multimethod environment enabling us to "play" (combine, compare, evaluate, etc) with classical single or hybrid approaches and concepts by introducing the operational view on the methods.
- an innovative approach enabling not only to combine induction approaches but to integrate different approaches in a single model, i.e. construction of various parts of a single solution/knowledge with different methods, application of different discretization methods,
- improved exploration capabilities of search algorithms,
- introduction of dynamic problem adaptation based on induced population.

2 Single Method Approaches

Trough the time different approaches evolved, such as symbolic approaches, computational learning theory, neural networks, etc. Most methods are focused in finding a way to extract generalized knowledge from examples. They all use inductive inference, that is the process of moving from concrete examples to general model(s), where the goal is to learn how to extract knowledge from objects by analyzing a set of instances (already solved cases) whose classes are known. Instances are typically represented as attribute-value vectors. Learning input consists of a set of vectors/ instances, each belonging to a known class, and the output consists of a mapping from attribute values to classes. This mapping should accurately classify both the learning instances and also new unseen instances.

Comparing single approaches we cannot find a clear winner. Each method has its own advantages and some inherent limitations.

Decision trees [6] for example are easy understandable to the human and can be used even without a computer, but they have difficulties expressing complex nonlinear problems. On the other hand connectivistics approaches, that simulate cognitive abilities of the brain, can extract complex relations, but solutions are not understandable to humans, and therefore in such way not directly usable for data mining. Evolutionary approaches to knowledge extraction are also a good alternative, because they are not inherently limited to local solution [12] but are computationally expensive.

There are many other approaches, like representation of the knowledge with rules, rough-sets, case based reasoning, support vector machines, different fuzzy methodologies, ensemble methods [2] and all of them try to answer the question: How to find optimal solution i.e. learn how to learn.

3 Hybrid Approaches

Hybrid approaches rest on the assumption that only the synergetic combination of single models can unleash their full power. Each of the single methods has its advantages but also inherent limitations and disadvantages, which must be taken into account when using a particular method. For example: symbolic methods usually represent the knowledge in human readable form and the connectivistcs methods perform better in classification of unseen objects and are less affected by the noise in data as are symbolic methods. Therefore the logical step is to combine both worlds to overcome the disadvantages and limitations of a single one.

In general the hybrids can be divided according to the flow of knowledge into four categories [4]:

- sequential hybrid (chain-processing) – the output of one method is an input to another method. For example, the neural net is trained with the training set to reduce noise. Than the neural net and the training set are used to generate decision tree with the neural net's decision's.
- parallel hybrid (co-processing) – different methods are used to extract knowledge. In the next phase some arbitration mechanism should be used to generate appropriate results.
- external hybrid (meta processing) – one method uses another external one. For example meta decision trees [5], that use neural nets in decision nodes to improve the classification results.
- embedded hybrid (sub-processing) – one method is embedded in another. That is the most powerful hybrid, but the least modular one, because usually the methods are tightly coupled.

The hybrid systems are commonly static in the structure and cannot change the order of how the single methods are applied.

To be able to use embedded hybrids of different internal knowledge representation, it is commonly required to transform one method representation into another. Some transformations are trivial, especially when converting from symbolic approaches. The problem is when the knowledge is not so clearly presented like in a case of the neural network [3, 14]. The knowledge representation issue is very important in the multimethod approach and we solved it in original manner.

4 Multimethod Approach

Multimethod approach was introduced in [9, 10]. While studying presented approaches we were inspired by the idea of hybrid approaches and evolutionary algorithms. Both approaches are very promising in achieving the goal to improve the quality of knowledge extraction and are not inherently limited to sub-optimal solutions. We also noticed that almost all attempts to combine different methods use the loose coupling approach. Loose coupling is easier to implement, but methods work almost independent of each other and therefore a lot of luck is needed to make them work as a team.

Opposed to the conventional hybrids described in the previous section, our idea is to dynamically combine and apply different methods in no predefined order to the same problem or decomposition of the problem.

The main concern of the mutlimethod approach [9] is to find a way to enable dynamic combination of methods to the somehow quasi-unified knowledge representation. Multiple equally qualitative solutions like in evolutionary algorithms (EA), where each solution is obtained using application of different methods with different parameters. Therefore we introduce a population composed out of individuals/solutions that have the common goal to improve their classification abilities on a given environment/problem. We also enable coexistence of different types of knowledge representation in the same population. The most common knowledge representation models have to be standardized to support the applicability of different methods on individuals. In that manner the transformation support between each individual method does not need to be provided. The action is based on the assumption that it is highly improbable to find unified representation for all knowledge representations, therefore we decided to standardize the most popular representations like neural nets, decision trees, rules, etc. Standardization brings in general greater modularity and interchangeability, but it has the following disadvantages – already existing methods cannot be directly integrated and have to be adjusted to the standardized representation.

Initial population of extracted knowledge is generated using different methods. In each generation different operations appropriate for individual knowledge are applied to improve existing and create new intelligent systems. That enables incremental refinement of extracted knowledge, with different views on a given problem. For example, using different induction methods such as different purity measures can be simply combined in decision trees. That is also true for combining different learning techniques for neural networks. As long as knowledge representation is the same a combination of different methods is not a big obstacle.

The main problem is how to combine methods that use different knowledge representation (for example neural networks and decision trees). In such cases we have two alternatives: (1) to convert one knowledge representation into another using different already known methods or (2) to combine both knowledge representations in a single intelligent system. The first approach requires implementation of knowledge conversion (for example there are numerous methods that can convert neural networks into decision trees and vice versa). Such conversions are not perfect and some of the knowledge can be lost. But on the other hand it can give a different view on presented problem that can lead to better results.

The second approach, which is based on combining knowledge, requires some cut-points where knowledge representations can be merged. For example in decision tree such cut-points are internal nodes, where condition in an internal node can be replaced by another intelligent system (for example support vector machine – SVM). The same idea can be applied also in decision leafs (Fig. 1).

Using the idea of the multimethod approach we designed a framework that operates on a population of extracted knowledge representations – individuals. Since methods are usually composed out of operations that can be reused in other methods we introduced methods on the basis of operators. Therefore we introduced the operation on an individual as a function that transforms one or more individuals to a single individual. Operation can be a part of one or more methods, like pruning operator, boosting operator, etc. Operator based view provides the ability to simply add new operations to the framework (Fig. 2).

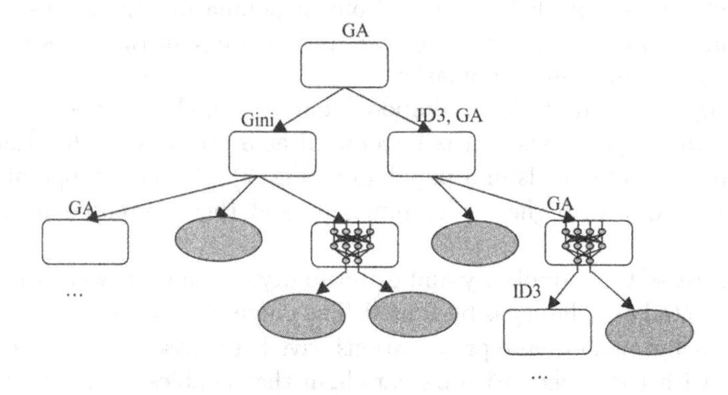

Fig. 1. An example of a decision tree induced using multimethod approach. Each node is induced with appropriate method (GA – genetic algorithm, ID3, Gini, Chisquare, J-measure, SVM, neural network, etc.)

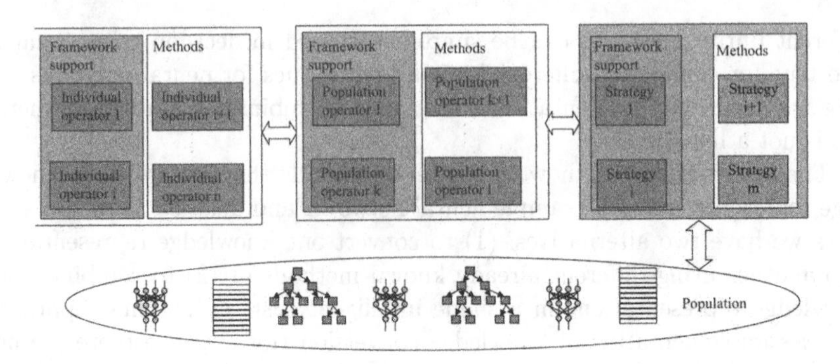

Fig. 2. Multimethod framework

4.1 Operators

Usually methods are composed out of operations that can be reused in other methods. We introduced the operation on an individual that is a function, which transforms one or more individuals into a single individual. Operation can be part of one or more methods, like pruning operator, boosting operator, etc.

The transformation to another knowledge representation is also introduced on the individual operator level. Therefore the transition from one knowledge representation to another is presented as a method. Operator based view provides us with the ability to simply add new operations to the framework (Fig. 2).

Representation with individual operations facilitates an effective and modular way to represent the result as a single individual, but in general the result of operation can be also a population of individuals (for example mutation operation in EA is defined on individual level and on the population level). The single method itself is composed out of population operations that use individual operations and is introduced as a strategy in the framework that improves individuals in a population.

A single method itself is composed out of population operations that use individual operations and is introduced as a strategy in the framework that improves individuals in a population (Fig. 2). Population operators can be generalized with higher order functions and thereby reused in different methods.

To increase the modularity and extensibility of the framework the idea of object oriented paradigm has been used. The polymorphism and inheritance in operations and individual representations have been introduced. We extended the idea with the aspect-oriented paradigm that enables clear separation of concerns and avoids tangled individual representation.

With this approach we achieved the modularity and extensibility of the framework without a lot of constrains to the implementation of methods.

The individual and population operations can be easily waved together with no additional effort.

The modular design of the framework enables application of some subparts of a method to another method. For example, independent of the induction method used, universal pruning and error reducing operations can be applied on a decision tree. The aspects of individual representation and methods/operations on the individual are strictly separated.

4.2 Meta Level Control

Important concern of the multimethod framework is how to provide the meta level services to manage the available resources and the application of the methods. We extended the quest for knowledge into another dimension that is the quest for the best application order of methods. The problem that arises is how to control the quality of resulting individuals and how to intervene in the case of bad results. Due to different knowledge representations, solutions cannot be trivially compared to each other and the assessment of which method is better is hard to imagine. Individuals in a population cannot be explicitly evaluated and the explicit fitness function cannot be easily calculated. Therefore the comparison of individuals and the assessment of the quality of the whole population cannot be given. Even if some criteria to calculate fitness function could be found, it would be probably very time consuming and computational intensive. Therefore the idea of classical evolutionary algorithms controlling the quality of population and guidance to the objective cannot be applied.

Almost all methods require some parameters that affect their operations and the way of knowledge extraction. By using different parameters, the generalization ability can dramatically increase. In our case the meta multimethod level is self adapting and does not require parameters from the user.

To achieve self-adaptive behaviour of the evolutionary algorithm the strategy parameters have to be coded directly into the chromosome [1]. But in our approach the meta level strategy does not know about the structure of the chromosome, and not all of the methods use the EA approach to produce solution. Therefore for meta level chromosomes the parameters of the method and its individuals are taken. When dealing with self-adapting population with no explicit evaluation/fitness function there is also an issue of the best or most promising individual [8]. But of course the question how to control the population size or increase selection pressure, must be answered effectively.

To overcome above problem we have introduced three basic types of operations on the population: operations for reducing the population size, operations for maintaining the population size and operations for increasing the population size. These operations can be introduced by the method (i.e. evolutionary induced knowledge) or are already integrated in the multimethod framework.

5 Implementation of Multimethod Approach: MultiVeDec

The multimethod framework MultiVeDec is implemented in Java. We've adapted existing methods to work inside the framework and unifying the knowledge representation inside a single methodology. For the individual representation the notion of polymorphism was introduced. The framework operations and method operations are designed in a way to accept different (sub)types of the individuals. That enables us to introduce generalized operators that can be executed on generalized and specialized knowledge representation.

We developed the meta level support for method combinations that enables easy weaving of concepts. For example, decision trees can be constructed using top-down induction algorithm using adaptive impurity measure that results in not only one, but multiple different induced decision trees that define the population of knowledge representation that can be then used for the optimization with evolutionary approach. In the evolutionary approach the methods are used as mutation operators on decision nodes, which define the subset of the problem and can be solved with different method. Each condition node is also a classifier that selects the child node based on given instance. In that manner we extended the approach presented in [5].

We've implemented symbolic knowledge extraction from SVMs. SVMs can be used to reduce the dimensions of the problem and can provide a decision boundary for discriminative subspace induction [11]. With that approach the amount of data can be reduced and thereby improve performance of other methods.

With evolutionary approaches we attend to find generalized knowledge specialized by the heuristic methods and thereby avoid local optimum. With incremental addition we also validated the modular design of the multimethod framework that is a good test bed for a new methods.

5.1 Concrete Combination Techniques

Basic idea of the multimethod approach is straightforward and is not hard to understand. Main issue represents different method integration and cooperation techniques. Basically, the multimethod approach searches for solutions in huge (infinite) search space and exploits the acquisition technique of each method integrated. Population of individual solutions represent different aspects of extracted knowledge. Conversion from one to another knowledge representation introduces new aspects. We can draw parallels between the multimethod approach and the scientific discovery. In the real life, based on the observed phenomena various hypotheses are constructed. Different scientific communities draw different conclusions (hypotheses) consistent with collected data. For example there are many theories about creation of the universe, but currently widely accepted theory is the theory of the big bang. During the following phase scientists discuss their theories, knowledge is exchanged, new aspects encountered and data collected is re-evaluated – in that manner existing hypothesis are improved and new better hypotheses are constructed.

In the next subsection we describe some interesting aspects and combination of methods in our implementation of multimethod approach.

Decision Trees

Decision trees are very appropriate to use as glue between different methods. In general condition nodes contain classifier (usually simple attribute comparison), what enables quick and easy integration of different methods. On other hand there exists many different methods for decision tree induction that all generate knowledge in a form of tree. Most popular method is method of greedy heuristic induction of decision tree that produces single decision tree with respect to purity measure (heuristic) of each split in decision tree. Altering purity measure may produce totally different results with different aspects (hypotheses) for a given problem. Clearly there is no single solution, but there are many equivalent solutions that satisfy different objective criteria. Therefore inducing a hypothesis (decision tree) can be viewed as search/optimisation problem. In that case induction of the hypothesis can be made with use of evolutionary algorithms (EA). When designing EA algorithm operators for mutation, crossover and selection have to be carefully chosen [12].

Crossover works on two selected individuals as an exchange of two randomly selected sub-trees. In this manner a randomly selected training object is selected first which is used to determine paths (by finding a decision through the tree) in both selected trees. Then an attribute node is randomly selected on a path in the first tree and an attribute is randomly selected on a path in the second tree. Finally, the sub-tree from a selected attribute node in the first tree is replaced with the sub-tree from a selected attribute node in the second tree and in this manner an offspring is created, which is put into a new population. Another aspect is separation of genotype and phenotype. When representing whole decision tree (conditions and leafs) as genotype, convergence to solution is quite poor. The reason is that decision nodes do not confirm to actual training objects. If decision leafs are removed from genotype and calculated into phenotype of solution, search space is reduced, which results in faster convergence of EA. In multimethod approach we went a step further: in addition we removed most of conditions from genotype. Genotype is represented with a small tree (from root), phenotype is calculated using different decision tree induction methods. In that way we've drastically reduced search space for EA and can therefore afford to use complex condition (even other classifiers) in condition nodes.

With the combination of presented crossover and other induction methods, which that works as a constructive operator towards local optimums, and mutations, which that works as a destructive operator in order to keep the needed genetic diversity, the searching for the solution tends to be directed toward the global optimal solution based on fitness criteria, which also represents another aspect of quest for good results. As commonly noticed, there is no single global criteria that can be used to compare induced individuals. Problem becomes a multi-objective optimization problem, which has to find a good hypothesis with high accuracy, sensitivity, specificity that is not very complex (small in size) and is easy understandable to human. For that reason multimethod approach can uses coevolution and parallel evolution of individuals with different optimization criteria.

There is another issue when combining two methods using other classifier for separation. For example on Fig. 3 there are two hypotheses h_1 and h_2 that could be perfectly separately induced using existing set of methods. None of the available methods in not able to acquire both hypotheses in a single hypothesis. Therefore we need a separation of problem using another hypothesis h3 that has no special meaning to induce successful composite hypothesis.

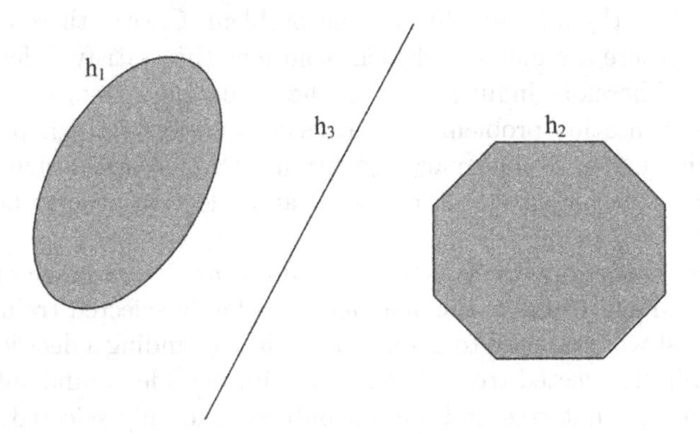

Fig. 3. Hypothesis separation

SVMs

Support Vector Machines (SVM) introduced by Vapnik [16] are statistical method that uses structural risk minimization principle that automatically deals with the accuracy/generalisation trade-off. The method gained the popularity in past few years especially on the field of document and text mining, but is not widely used in medical applications. The main reason for that may be the knowledge representation of SVM. SVM do not represent the knowledge in symbolic, in a way easy understandable to the human, but represents its knowledge in the form of support vectors and its weights used in classifications, but are therefore more explanatory than neural networks. Support vector represent important instances that represent boundary of decision class. Such instances can be then used by other methods to determine theirs decision boundary. With use of different kernels, support vectors can represent other important aspect of classification like centroids [17]. On other hand attribute relevance and separation hyper plane can be extracted and reused in other induction algorithms.

Neural Nets

In our research we focused on multi-layer feed-forward neural networks (MLP). For learning methods we used EA and backpropagation learning. Backpropagation networks are widely used in various fields, giving very good results.

The multi-layer perceptron architecture is the most popular one in practical applications. Each layer uses a linear combination function. The inputs are fully connected to the first hidden layer, each hidden layer is fully connected to the next, and the last hidden layer is fully connected to the outputs.

Finding the right topology for neural network with two hidden layers is a time consuming task and requires a lot of experiences, therefore we evolve topologies with EA. We also use different neuron activation functions with different parameters.

5.2 Problem Adaptation

In many domains we encounter very data with very unbalanced class distribution. That is especially true for applications in medical domain, where most of the instances are regular and only small percent of the instances are assigned to irregular class. Therefore for most of the classifier that want to achieve high accuracy and low complexity it is most rational to classify all new instances into majority class. But that feature is not desired, because we want to extract knowledge (especially when we want to explain decision making process) and want to determine reasons for separation of classes. To cope with presented problem we introduced instance reweighting method that works in similar manner like boosting but on different level. Instances that are rarely correctly classified gain on importance. Fitness criteria of individual take importance into account and forces competition between individual induction methods. Of course there is a danger of over adapting to the noise, but in that case overall classification ability would be decreased and other induced classifier can perform better classification (self adaptation).

6 Experimental Results

Our current data mining research is mainly performed in the medical domain, thereafter the knowledge representation which should be in a human understandable form is very important – so we are focused on the decision tree induction. To make objective assessment of our method a comparison of extracted knowledge used for classification was made with reference methods

C4.5, C5/See5 without boosting, C5/See5 with boosting, and genetic algorithm for decision tree construction [12]. The following quantitative measure were used:

$$accuracy = \frac{num\ of\ correctly\ classified\ objects}{num.\ of\ all\ objects}$$

$$accuracy_c = \frac{num\ of\ correctly\ classified\ objects\ in\ class\ c}{num.\ of\ all\ objects\ in\ class\ c}.$$

$$average\ classaccuracy = \frac{\sum_i accuracy_i}{num.of classes}$$

We decided to use average class accuracy instead of sensitivity and specificity that are usually used when dealing with medical databases. Experiments have been made with seven real-world databases from the field of medicine. For that reason we've only selected symbolic knowledge from whole population of resulting solutions.

Detailed description of databases can be found in [9]. Other databases have been downloaded from the online repository of machine learning datasets maintained at UCI.

We compared two variations of our multimethod approach (MultiVe- Dec) with four conventional approaches, namely C4.5 [7], C5/See5, Boosted C5 and genetic algorithm [12]. The results are presented in Table 1. Gray ma.

Table 1. A comparison of the multimethod approach to conventional approaches

	C4.5		C5		C5 Boost		Genetically induced decision trees		SVM(RBF)		MultiVeDec without problem adaptation		MultiVeDec	
	□	◊	□	◊	□	◊	□	◊	□	◊	□	◊	□	◊
av	76.7	44.5	81.4	48.6	83.7	50.0	90.7	71.4	83.7	55.7	93.0	78.7	90.0	71.4
breastcancer	96.4	94.4	95.3	95.0	96.6	96.7	96.6	95.5	96.5	97.1	96.6	95.1	96.9	97.8
heartdisease	74.3	74.4	79.2	80.7	82.2	81.1	78.2	80.2	39.6	42.7	78.2	78.3	83.1	82.8
Hepatitis	80.8	58.7	82.7	59.8	82.7	59.8	86.5	62.1	76.9	62.5	88.5	75.2	88.5	75.2
mracid	94.1	75.0	100.0	100.0	100.0	100.0	94.1	75.0	94.1	75.0	100.0	100.0	100.0	100.0
Lipids	60.0	58.6	66.2	67.6	71.2	64.9	76.3	67.6	25.0	58.9	75.0	73.3	75.0	73.3
Mvp	91.5	65.5	91.5	65.5	92.3	70.0	91.5	65.5	36.1	38.2	93.8	84.3	93.8	84.3

□ accuracy on test set ◊ average class accuracy on test set

Looking at the table we can see that in the case of the av database MultiVeDec without problem adaption achieved the best accuracy and average class accuracy, and that was also the case with the mracid, Hepatitis and Mvp databases. In the case of Lipids MultiVeDec has the best class accuracy on the testing set, and in the case of breastcancer and heartdisease the C5 Boost was the very near, but MultiVeDec better for few percents. We should stress out that MultiVedec produced only one classifier, C5 Boost produced and ensemble of classifier. So we can state that MultiVeDec was the overall best approach, but anyhow we should remember the "No free lunch theorem" [15], stating that we have to find the best method for each single problem. MultiVeDec tries to compensate that theorem by using different methods simultaneously. We can also observe interesting difference between MultiVeDec with population adaption and without. Performance improved for two databases, bat decreased for av database. We speculate, that unbalanced data was compensated in case first case, but in second case the effect of over adapting to outliers took the place.

The success of the multimethod approach can be explained with the fact that some methods converge to local optima. With the combination of multiple methodologies better local (and hopefully global) solution can be found. For example when using decision trees the genetic methodology can determine the attributes near the top of the tree and for the nodes near leafs the greedy method with different impurity measures can be used to reduce search space.

7 Conclusion

In this paper we introduced a multimethod approach used to combine different knowledge extraction methods. We also introduced an improvements of multimethod approach used to combine different induction methods. The existing hybrid systems usually work sequentially or in parallel on the fixed structure and order performing whole tasks. On the contrary multimethod approach works simultaneously with several methods on a single task, i. e. some parts are induced with different classical heuristics, some parts with hybrid methods and still another parts with evolutionary programming. We also presented some techniques used to combine different methods and introduced a technique for dynamic adaptation of the problem.

Presented multimethod approach enables quick and modular way to integrate different methods into existing system and enables the simultaneous application of several methods. It also enables partial application of method operations to improve and recombine methodologies and has no limitation to the order and number of applied methodologies It encourages standardization of knowledge representation and unifies the operational view on the methods.

References

1. S. Thrun and L. Pratt, editors. Learning to Learn. Kluwer Academic Publishers, 1998.
2. T. G. Dietterich. Ensemble Methods in Machine Learning. In J. Kittler and F. Roli (Ed.) *First International Workshop on Multiple Classifier Systems, Lecture Notes in Computer Science* (pp. 1–15) New York: Springer Verlag, 2000.
3. K. McGarry, S. Wermter, J. MacIntyre The Extraction and Comparison of Knowledge From Local Function Networks *International Journal of Computational Intelligence and Applications*, Vol. 1 Issue 4, pp: 369–382, 2001.

4. C. J. Iglesias. The Role of Hybrid Systems in Intelligent Data Management: The Case of Fuzzy/neural Hybrids, Control Engineering Practice, Vol. 4, no. 6, pp 839–845, 1996.
5. L. Todorovski, S. Dzeroski. Combining Multiple Models with Meta Decision Trees. In *Proceedings of the Fourth European Conference on Principles of Data Mining and Knowledge Discovery*. Springer, pp 54–64, 2000.
6. D.E. Goldberg, Genetic Algorithms in Search, Optimization, and Machine Learning, Addison Wesley, Reading MA, 1989.
7. J.R. Quinlan. C4.5: Programs for Machine Learning, Morgan Kaufmann publishers, San Mateo, 1993.
8. M. Šprogar, Peter Kokol, Špela Hleb Babiè, Vili Podgorelec, Milan Zorman, Vector decision trees, *Intelligent data analysis*, vol. 4, no. 3, 4, pp. 305–321, 2000.
9. M. Lenië, P. Kokol, M. Zorman, P. Povalej, B. Stiglic, Improved Knowledge Mining with the Multimethod Approach, Workshop on The Foundation of Data Mining, ICDM, 2002.
10. M. Lenië, P. Kokol: Combining classifiers with multimethod approach. In: Second international conference on Hybrid Intelligent Systems, Soft computing systems: design, management and applications, (Frontiers in artificial intelligence and applications, Vol. 87). Amsterdam: IOS Press, pp 374–383, 2002.
11. J. Zhang, Y. Liu, SVM Decision Boundary Based Discriminative Subspace Induction, *Technical report CMU-RI-TR-02-15*, Robotics Institute, Carnegie Mellon University, 2002.
12. V. Podgorelec, Peter Kokol, R. Yamamoto, G. Masuda, N. Sakamoto, Knowledge discovery with genetically induced decision trees, In *International ICSC congress on Computational intelligence: methods and applications (CIMA'2001), June 19–22, University of Wales, Bangor, Wales, United Kingdom*, ICSC, 2001.
13. V. Podgorelec, P. Kokol. Evolutionary decision forests – decision making with multiple evolutionary constructed decision trees, *Problems in applied mathematics and computational intelligence*, World Scientific and Engineering Society Press, pp. 97–103, 2001.
14. M. Zorman, P. Kokol, V. Podgorelec, Medical decision making supported by hybrid decision trees, *Proceedings of the ICSC symposia on Intelligent systems and applications (ISA'2000)*, December 11–15, 2000.
15. D. H. Wolpert and W. G. Macready, No Free Lunch Theorems for Search, Technical report, SFI-TR-95-02-010, Santa Fe, NM, 1995.
16. V. N. Vapnik, The nature of statistical learning theory, Springer Verlag, New York, 1995.
17. D. Rebernak, M. Lenië, P. Kokol, V. Žumer: Finding Boundary Subjects for Medical Decision Support with Support Vector Machines, In proceedings of Computer Based Medical Systems 2003.

Part III

General Knowledge Discovery

Posting Act Tagging Using Transformation-Based Learning

Tianhao Wu, Faisal M. Khan, Todd A. Fisher, Lori A. Shuler and William M. Pottenger

Computer Science and Engineering, Lehigh University
{tiw2, fmk2, taf2, lase, billp}@lehigh.edu

Abstract. In this article we present the application of transformation-based learning (TBL) [1] to the task of assigning tags to postings in online chat conversations. We define a list of posting tags that have proven useful in chat-conversation analysis. We describe the templates used for posting act tagging in the context of template selection. We extend traditional approaches used in part-of-speech tagging and dialogue act tagging by incorporating regular expressions into our templates. We close with a presentation of results that compare favorably with the application of TBL in dialogue act tagging.

1 Introduction

The ephemeral nature of human communication via networks today poses interesting and challenging problems for information technologists. The sheer volume of communication in venues such as email, newsgroups, and chat-rooms precludes manual techniques of information management. Currently, no systematic mechanisms exist for accumulating these artifacts of communication in a form that lends itself to the construction of models of semantics [5]. In essence, dynamic techniques of analysis are needed if textual data of this nature is to be effectively mined.

At Lehigh University we are developing a text mining tool for analysis of chat-room conversations. Project goals concentrate on the development of functionality to answer questions such as "What topics are being discussed in a chat-room?", "Who is discussing which topics?" and "Who is interacting with whom?" In order to accomplish these objectives, it is necessary to first identify threads in the conversation (i.e., topic threads). One of the first steps in our approach to thread identification is the automatic assignment of tags that characterize postings. These tags identify the type of posting; for example, Greet, Bye, etc. We term this classification task Posting Act Tagging.

Posting act tagging aids in both social and semantic analysis of chat data. For example, the question tag type, which consists of Yes-No-Question and

T. Wu et al.: *Posting Act Tagging Using Transformation-Based Learning*, Studies in Computational Intelligence (SCI) **6**, 321–331 (2005)
www.springerlink.com

Wh-Question tags, identifies postings that give clues to both the start and topic of a topic thread within a chat conversation. Other postings tagged as a Greet or Bye, for example, may not contribute significantly to the semantics of a particular topic thread. These types of postings, however, may yield information important to the social analysis of the conversation – e.g., "who is talking with whom?" Thus the tag type assigned to a posting aids the model-building process.

Posting act tagging idea is similar to dialogue act tagging. Reference [4] demonstrates that dialogue act tagging can be widely used in dialogue recognition. Chat conversations are similar to spoken dialogues in some ways. In fact, a chat conversation is a kind of dialogue in written form. Therefore, we expect that techniques applied in dialogue recognition may also be useful in chat conversation analysis.

Chat conversations differ, however, in significant ways from dialogues[1]. Chat conversations are usually informal, and multiple topics may be discussed simultaneously. In addition, multiple people are often involved. Participants do not always wait for responses before posting again. Furthermore, abbreviations and emotion icons are frequently used in chat conversations, mixed together with chat-system-generated information.

Based on these and related issues, we have extended dialog act tagging as presented in [3, 4, 6, 7] to classify postings in chat conversations.

In the following sections we present our approach to posting act tagging. We detail the posting tag types that we have used, including some new types specific to chat conversations, in Sect. 2. In Sect. 3 we briefly describe the machine-learning framework that we have employed, transformation-based learning (TBL) [1]. In Sect. 4, we present the application of TBL to posting act tagging. We discuss preliminary experimental results, including a statistical analysis comparing our results with those obtained in dialogue act tagging, in Sect. 5. Finally, we discuss conclusions and future work in Sect. 6 and acknowledge those who have contributed to this work in Sect. 7.

2 Posting Act Tags

Table 1 is our posting act tag list and includes 15 tag types. The tag types come from three different sources. The Accept, Bye, Clarify, Greet, and Reject tags are drawn from the VerbMobil project [6]. The Statement, Wh-Question, Yes-No-Question, Yes-Answer, No-Answer, Continuer and Other tag types derive from the dialogue act tagging research reported in [7]. As noted above, the final three tag types are specific to chat conversations and were included based on our research: Emotion, Emphasis, and System.

[1] E.g., the classification work reported in [3] is based on recorded conversations of phone calls to schedule appointments.

Altogether there are over 40 tags employed in dialogue act tagging [8]. In this article, we select a subset of higher frequency dialogue act tags as our posting act tags, and add three chat-specific tags as noted above.

Statement is the most often used tag in dialogue act tagging. It covers more than 36% of the utterances. The Statement tag also has a high frequency in our posting act tagging because we use Statement to cover more than one tag used in dialogue act tagging. Statement can be split into several tags if more detailed tagging information is desired. The tag Other is used for postings that do not fit readily into the other categories (i.e., are untagged).

Since [6] and [7] give clear definitions of the tags used in their work, we briefly define the System, Emotion, and Emphasis tags that we have added.

System postings are generated by chat-room software. For example, when a person joins or leaves a chat room, the chat-room software usually posts a System message.

People express strong feelings in Emotion postings. These feelings include "surprise", "laughing", "happiness", "sadness", etc. Most chat-room software supports emotion icons, and these icons give clues to participants' emotional states.

Emphasis postings are the postings in which people emphasize something. For instance, people often use "do" just before a verb to put more emphasis on the verb. Another example is the use of "really" to likewise emphasize a verb.

In this section, we introduced posting act tags for chat conversations. Some of tags are derived from dialogue act tagging; others are specific to chat conversations. Table 1 list all tags, example postings for each tag, and tag distribution in our datasets (training and testing). In the following two sections, we

Table 1. List of Tags, Examples and Frequency

Tag	Example	%
Statement	I'll check after class	42.5
Accept	I agree	10.0
System	Tom [JADV@11.22.33.44] has left #sacba1	9.8
Yes-No-Question	Are you still there?	8.0
Other	*********	6.7
Wh-Question	Where are you?	5.6
Greet	Hi, Tom	5.1
Bye	See you later	3.6
Emotion	LOL	3.3
Yes-Answer	Yes, I am.	1.7
Emphasis	I do believe he is right.	1.5
No-Answer	No, I'm not.	0.9
Reject	I don't think so.	0.6
Continuer	And ...	0.4
Clarify	Wrong spelling	0.3

describe how TBL is applied in the discovery of rules for classifying postings automatically.

3 Overview of Transformation-Based Error-Driven Learning

Transformation Based Learning (TBL) is an emerging technique with a variety of potential applications within textual data mining. TBL has been utilized for tasks such as part-of-speech tagging, dialogue act tagging, and sentence boundary disambiguation, to name a few. TBL performs admirably in these tasks since they rely on the contextual information within textual corpora.

The core functionality of TBL is a three-step process composed of an initial state annotator, templates, and a scoring function [2]. The initial state annotator begins by labeling unannotated input (e.g., postings) with tags based on simple heuristics. Using a scoring function, the annotated input is then compared to a "ground truth" consisting of the same text with the correct labels. TBL automatically generates transformation rules that rewrite labels in an attempt to reduce the error in the scoring function.

Potential rewrite rules are automatically generated from preexisting human-expert-generated templates. The input in question is then re-annotated using newly generated rules, and once again compared with the ground truth. The procedure selects the best rule (the one with minimal error) and saves it to a final rule sequence. This cycle repeats until the reduction in error reaches a predetermined minimum threshold.

At heart, TBL is a greedy learning algorithm. Within each learning iteration, a large set of different transformation rules can be generated. The rule with the best performance (least error as measured by the scoring function) is chosen. The final set of rules can be used for classification of new input.

4 Using TBL in Posting Act Tagging

In this section, we discuss the application of the three steps in the TBL learning process discussed in Sect. 3 to posting act tagging with a special emphasis on template selection. We extend traditional approaches used in template selection in part-of-speech tagging and dialogue act tagging by incorporating regular expressions into the templates.

4.1 Initial State Annotator

Since the Statement tag occurs most frequently in our data sets, we simply tag each posting as Statement in the initial annotator of TBL. If the initial state has high accuracy, the learning process will be more efficient because TBL is a greedy algorithm. Therefore, Statement is the best choice for the initial state of each posting in posting act tagging.

4.2 Template Selection

Through manual study of patterns in chat data we developed a number of rule templates. In this section we discuss the antecedents of seven such templates.

1. "A particular string 'W' appears within the current posting, where 'W' is a string with white space preceding and following." Domain-expert-identified regular expressions are used to replace "W" during learning. This template was chosen since manual inspection of chat data yielded the result that certain words are often crucial in posting tagging. For example, a posting with the word "Why" often indicates that this is a Wh-Question posting. This is also true for dialogue act tagging [3]. However, just using a single word in this template is not sufficient for posting act tagging. This is due to the fact that chat conversations are complex. One of complexities is that typos and variations of words frequently occur. For instance, "allright", "all right", "alright", "allrigggggght" all have similar meaning. It is not feasible to include all variations explicitly in a template. As a result, we employed regular expressions. For example, the regular expression "al+( )?rig+ht" covers all four variations of "all right". In Sect. 5 we present a statistical comparison between explicit representations of words vs. the use of regular expressions to confirm this intuitive result.
2. "A character 'M' appears in the current posting, where 'M' is any punctuation mark." Punctuation marks are valuable in posting tagging. For example, a question mark usually indicates that a posting is indeed a question.
3. "A word with part of speech tag 'T' appears in the current posting, where 'T' is a part of speech tag from the Brown tag set."[2] Part of speech tags often aid in identifying posting tags. For instance, the part of speech tag WRB (when, how, etc.) can be used to identify Wh-Question postings.
4. "The current posting's length (the number of words) is 'L', or the current posting's length is greater than 'L', where 'L' is a heuristically chosen constant." In this case, we observe that some postings' tags are related to their length. For example, Yes-Answer and No-Answer postings are usually shorter while Statement postings are often long.
5. "The author of the preceding or following posting is the same as the author of the current posting." Each participant in the chat environment is termed an author. We noted that authors often separate their sentences into several consecutive postings. Thus, it is likely that a posting is a Continuer if its neighbor postings have the same author. For example, the posting "<Tom> and at least try it" is a Continuer of the posting "<TOM> go to play basketball". These two postings come together and have the same author and topic.
6. "The first character of a posting is 'C'." The first character gives a crucial clue in classifying system postings. For instance, system postings do

[2] See Chap. 4 in [10] for a listing of the Brown Tags used in part-of-speech tagging.

not have (human) authors, whereas author names (i.e., screen names) are usually delimited using characters such as "(" and ")", "<" and ">", or "["and"]" in chat conversations. In IRC chat conversations, "Tom [JADV@11.22.33.44] has joined #sacbal" is an example System posting that is generated by the chat server when a user joins the current conversation. Conversely, a delimiter followed by an author name is often the start of a non-system posting. For example, "<Tom> How is everyone?" Therefore, the first character is useful in discriminating whether a posting is a System posting or not.

7. "The previous or following posting's tag is 'T' " is also helpful to determine the current posting's tag. For example, a Yes-Answer or No-Answer is normally nearby a Yes-No-Question.

We have listed seven antecedents useful as templates in TBL. It is necessary, however, that templates have consequents. In our research it sufficed to have a single consequent for all seven antecedents: "Change the current posting's tag to 'B', where 'B' is any tag". An example of a rule generated using template number one is: "if al+( )?rig+ht is in a posting, then change the posting's tag to Accept".

The learning process instantiates rules based on these templates using features present in the input postings. As a result, each template generates numerous rules. Within a given iteration during training, all of the rules so generated are applied to the postings in the training set and the rule with the best performance is chosen. We use a scoring function described in Sect. 4.3 to measure learning performance.

4.3 Scoring Function

As in [1], we use accuracy, which is defined in (1), as the scoring function for TBL. If a posting's tag assigned during learning is the same as the correct tag in the ground truth, that posting is a true positive (TP). The definition of accuracy for posting act tagging is thus:

$$Accuracy = \frac{\#\ of\ TP}{\#\ of\ Total\ Postings} \tag{1}$$

In this section, we have described the implementation of the three core steps of the TBL learning process for posting act tagging, with an emphasis on template selection. Our experimental results reported in the next section show that this approach is viable.

5 Experimental Results

In this section, we describe the datasets for training and testing. We use the widely applied technique of cross-validation to evaluate our models. We also analyze each tag's precision and recall for the test datasets. Finally, we do a statistical comparison between explicit representations of words vs. the use of regular expressions in templates.

Our datasets include nine IRC chat conversations containing 3129 postings in all. Each posting in each data set was manually tagged by a human expert, thereby creating our ground truth. The distribution of each tag over all chat conversations was depicted in Table 1 (in Sect. 2). Table 2 portrays the characteristics of the nine data sets.

Table 2. Training and Testing Data

Conversations	Postings	Authors
Conversation 1	384	9
Conversation 2	736	16
Conversation 3	97	7
Conversation 4	184	3
Conversation 5	262	6
Conversation 6	634	16
Conversation 7	368	7
Conversation 8	246	9
Conversation 9	218	3

In evaluating our approach we employed nine-fold cross-validation. Eight of the nine datasets were combined to form nine training sets, and the remaining dataset was used for testing. Table 3 presents the resulting nine test accuracies. The first column details the nature of each training set. For example, (2,3,4,5,6,7,8,9) means we used chat conversations two through nine to form the training set. The second column is the number of rules learned based on the given training set. The third column reports the dataset used for

Table 3. Cross-validation results

Training Sets	# of Rules Learned	Test Set	Test Accuracy (%)
2,3,4,5,6,7,8,9	59	1	80.46875
1,3,4,5,6,7,8,9	63	2	76.90217
1,2,4,5,6,7,8,9	59	3	77.31959
1,2,3,5,6,7,8,9	60	4	71.19565
1,2,3,4,6,7,8,9	58	5	76.33588
1,2,3,4,5,7,8,9	56	6	76.81388
1,2,3,4,5,6,8,9	61	7	78.26087
1,2,3,4,5,6,7,9	64	8	80.89431
1,2,3,4,5,6,7,8	60	9	79.81651

testing, and the last column is the accuracy that results from the application of the learned rule sequence on the given test dataset.

The best accuracy we achieved on any single test set is 80.89%, which is somewhat less than the best single-test-set accuracy reported in [4] (84.74%) for dialogue act tagging. From Table 3, our average test accuracy is 77.56% with $\sigma = 2.92\%$. Compared to the best average accuracy reported in [3] of 75.12%, our accuracy is slightly better. We conducted a statistical analysis using a one-tailed t-test to compare our results with those reported in [3]. We determined with greater than 95% confidence that our accuracy is significantly greater than that reported in [3]. As a result, we conclude that our approach to posting act tagging compares favorably with related work in the field of dialogue act tagging and is therefore viable.

The average precision and recall for each tag individually is depicted in Table 4. For example, we see that TBL succeeds in discovering System, Other, and Greet tags because all of them have both high precision and high recall. Yes-No-Question, Statement, Emotion, Bye, and Yes-Answer have more modest but still reasonable precisions and recalls. Accept has high precision with low recall. On the other hand, the classification rules generated by TBL have relatively poor performance on Emphasis, No-Answer, Reject, Continuer, and Clarify tags, with precision and recall close to zero. One reason for this relatively poor performance is the sparseness of representation – these tags are not well represented in the ground truth (i.e., 13 occurrences of Continuer, 20 of Reject, 29 of No-Answer, 8 of Clarify, and 46 of Emphasis).

Table 4. Average precision and recall for each tag

Tag	Precision	Recall
Statement	0.747266	0.925508
Accept	0.714286	0.273312
System	0.99026	0.993485
Yes-No-Question	0.687023	0.72
Other	0.900433	0.985782
Wh-Question	0.564103	0.5
Greet	0.885906	0.830189
Emotion	0.896104	0.663462
Bye	0.957447	0.79646
Yes-Answer	0.506173	0.773585
Emphasis	0.5	0.021739
No-Answer	0	0
Reject	0	0
Continuer	0	0
Clarify	0	0

As noted there are 3129 postings in all, and all postings were used in the nine-fold cross-validation evaluation. As a result, all of these tags have low occurrence frequencies in our datasets (Table 1). Figure 1 shows that the F-measure [11] (a combination of precision and recall) becomes reasonable once a tag occurs in at least 1.7% of the postings. A second reason for the poor performance of the classifier with these particular tags may have to do with the need for more specialized templates to handle these tag types.

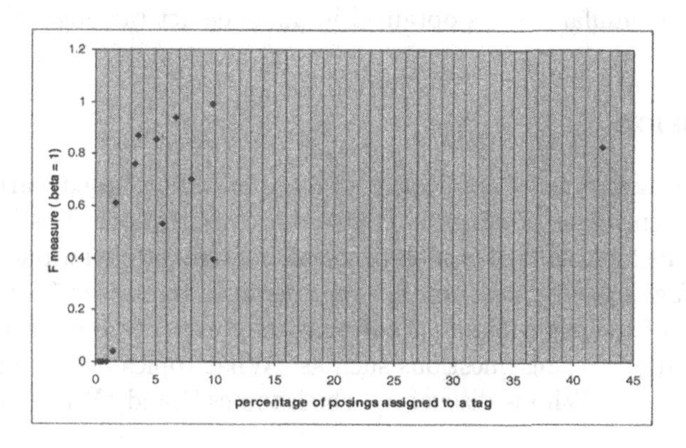

Fig. 1. Percentage of postings per tag vs. F-measure

To make the comparison between explicit representations of words vs. the use of regular expressions, we employed only template one for both training and testing. Each chat conversation was used in turn, first as a training set to generate a TBL rule sequence. In order to obtain a statistically significant sample, we chose combinations of seven out of the eight remaining datasets as test sets. In this way we generated eight test sets for each training set for a total of 9*8 = 72 test results. Table 5 depicts example test results using dataset number one as a training set. We applied a one-tailed t-test

Table 5. Explicit representation vs. regular expressions in dataset one

Test Set	Accuracy Using Explicit Representation (%)	Accuracy Using Regular Expressions (%)
1	52.1652563	53.3101045
2	53.0052265	54.0505226
3	53.2820281	54.3232232
4	51.8066635	52.9798217
5	55.5449025	58.4368737
6	52.691358	53.5802469
7	52.9576153	53.842571
8	52.8735632	54.0229885

to the two distributions, one using an explicit representation and the second using regular expressions. Based on this we determined with a confidence of over 94% that regular expressions perform significantly better than an explicit representation. Therefore, we conclude that regular expressions should be used when constructing templates such as template number one.

Our experimental results provide evidence that TBL can be usefully applied to the problem of posting act tagging. We achieved reasonable and stable test set performance for all nine of our test datasets, and our results compare favorably with similar results obtained in dialogue act tagging.

6 Conclusion

We have presented a novel application of transformation-based learning to the problem of identifying postings in chat-room conversations. Posting act tagging aids in the formation of models of social and semantic relationships within chat data. Tagging of this nature thus represents an important first step in the construction of models capable of automatically extracting information from chat data and answering questions such as "What topics are being discussed in a chat room?", "Who is discussing which topics?" and "Who is interacting with whom?".

In the work reported in this article a well known natural language processing algorithm, transformation-based learning, has been applied to posting act tagging. We developed seven templates that have proven useful in learning rules for posting act tagging. Furthermore, we have shown that the use of regular expressions in templates improves test set accuracy.

One of the tasks that lie ahead is to deal with multiple-sentence postings that call for more than one tag (on a single posting). Transformation-based learning, however, is not suited for learning problems of this nature, and as a result we are developing a new algorithm, BLogRBL [9], to handle such cases. Another task relates to the difficulty in manually creating generic regular expressions for templates. Little work has been done in the automatic generation of such regular expressions. At Lehigh University, however, we are engaged in a project with Lockheed-Martin and the Pennsylvania State Police to develop this capability. Finally, it is necessary to identify additional templates for tags that are not well represented in the training data. Our future work will focus on these problems.

Acknowledgements

This work was supported in part by NSF EIA grant number 0196374. The authors gratefully acknowledge the help of family members and friends, and co-authors William M. Pottenger and Tianhao Wu gratefully acknowledge the continuing help of their Lord and Savior, Yeshua the Messiah (Jesus Christ) in their lives.

References

1. Brill, Eric. A report of recent progress in Transformation-based Error-driven Learning. Proceedings of the ARPA Workshop on Human Language Technology. 1994.
2. Brill, Eric. Transformation-Based Error Driven Learning and Natural Language Processing: A Case Study in Part-of-Speech Tagging. Computational Linguistics 21(94): 543–566. 1995.
3. Ken Samuel, Sandra Carberry and K. Vijay-Shanker. Dialogue Act Tagging with Transformation-Based Learning. Proceedings of COLING/ACL'98, pp. 1150–1156. 1998.
4. E. Shriberg, R. Bates, P. Taylor, A. Stolcke, D. Jurafsky, K. Ries, N. Cocarro, R. Martin, M. Meteer, and C. Van Ess-Dykema. Can prosody aid the automatic classification of dialog acts in conversational speech?, Language and Speech, 41:439–487. 1998.
5. William M. Pottenger, Miranda R. Callahan, Michael A. Padgett. Distributed Information Management. Annual Review of Information Science and Technology (ARIST Volume 35). 2001.
6. Alexandersson, J., Buschbeck-Wolf, B., Fujinami, T., Maier, E., Reithinger, N., Schmitz, B. & Siegel, M. Dialogue acts in VERBMOBIL-2. Verbmobil Report 204, DFKI, University of Saarbruecken. 1997.
7. Jurafsky, D., Shriberg, E., Fox, B. & Curl, T. Lexical, Prosodic, and Syntactic Cues for Dialog Acts. Proceedings of ACL/COLING 98 Workshop on Discourse Relations and Discourse Markers, pp. 114–120, Montreal. (1998).
8. Stolcke, A., Ries, K., Coccaro, N., Shriberg, E., Bates, R., Jurafsky, D., Taylor, P., Martin, R., Meteer, M., and Van Ess-Dykema, C. Dialogue Act Modeling for Automatic Tagging and Recognition of Conversational Speech. Computational Linguistics, 26:3. 2000.
9. Tianhao Wu, and William M. Pottenger. Error-Driven Boolean-Logic-Rule-Based Learning: A Powerful Extension of Transformation-Based Learning. Lehigh CSC 2002 Technical Reports LU-CSE-02-008. October, 2002.
10. Christopher D. Manning and Hinrich Schütze. Foundations of Statistical Natural Language Processing, MIT Press, 2000.
11. Van Rijsbergen. Information Retrieval. Butterworths, London. 1979.

Identification of Critical Values in Latent Semantic Indexing

April Kontostathis[1], William M. Pottenger[2], and Brian D. Davison[2]

[1] Ursinus College, Department of Mathematics and Computer Science, P.O. Box 1000 (601 Main St.), Collegeville, PA 19426
akontostathis@ursinus.edu

[2] Lehigh University, Department of Computer Science and Engineering, 19 Memorial Drive West, Bethlehem, PA 18015
billp,davison@cse.lehigh.edu

In this chapter we analyze the values used by Latent Semantic Indexing (LSI) for information retrieval. By manipulating the values in the Singular Value Decomposition (SVD) matrices, we find that a significant fraction of the values have little effect on overall performance, and can thus be removed (changed to zero). This allows us to convert the dense term by dimension and document by dimension matrices into sparse matrices by identifying and removing those entries. We empirically show that these entries are unimportant by presenting retrieval and runtime performance results, using seven collections, which show that removal of up 70% of the values in the term by dimension matrix results in similar or improved retrieval performance (as compared to LSI). Removal of 90% of the values degrades retrieval performance slightly for smaller collections, but improves retrieval performance by 60% on the large collection we tested. Our approach additionally has the computational benefit of reducing memory requirements and query response time.

1 Introduction

The amount of textual information available digitally is overwhelming. It is impossible for a single individual to read and understand all of the literature that is available for any given topic. Researchers in information retrieval, computational linguistics and textual data mining are working on the development of methods to process this data and present it in a usable format.

Many algorithms for searching textual collections have been developed, and in this chapter we focus on one such system: Latent Semantic Indexing (LSI). LSI was developed in the early 1990s [5] and has been applied to a wide variety of tasks that involve textual data [5, 8, 9, 10, 19, 22]. LSI is based upon a linear algebraic technique for factoring matrices called Singular

A. Kontostathis et al.: *Identification of Critical Values in Latent Semantic Indexing*, Studies in Computational Intelligence (SCI) **6**, 333–346 (2005)
www.springerlink.com

Value Decomposition (SVD). In previous work [15, 16, 17] we noted a strong correlation between the distribution of term similarity values (defined as the cosine distance between the vectors in the term by dimension matrix) and the performance of LSI. In the current work, we continue our analysis of the values in the truncated term by dimension and document by dimension matrices. We describe a study to determine the most critical elements of these matrices, which are input to the query matching step in LSI. We hypothesize that identification and zeroing (removal) of the least important entries in these matrices will result in a more computationally efficient implementation, with little or no sacrifice in the retrieval effectiveness compared to a traditional LSI system.

2 Background and Related Work

Latent Semantic Indexing (LSI) [5] is a well-known technique used in information retrieval. LSI has been applied to a wide variety of tasks, such as search and retrieval [5, 8], classification [22] and filtering [9, 10]. LSI is a vector space approach for modeling documents, and many have claimed that the technique brings out the "latent" semantics in a collection of documents [5, 8].

LSI is based on a mathematical technique called Singular Value Decomposition (SVD) [11]. The SVD process decomposes a term by document matrix, A, into three matrices: a term by dimension matrix, T, a singular value matrix, S, and a document by dimension matrix, D. The number of dimensions is min (t, d) where t = number of terms and d = number of documents. The original matrix can be obtained, through matrix multiplication of TSD^T. In the LSI system, the T, S and D matrices are truncated to k dimensions. Dimensionality reduction reduces noise in the term-document matrix resulting in a richer word relationship structure that reveals latent semantics present in the collection. Queries are represented in the reduced space by $T_k^T q$, where T_k^T is the transpose of the term by dimension matrix, after truncation to k dimensions. Queries are compared to the reduced document vectors, scaled by the singular values ($S_k D_k$), by computing the cosine similarity. This process provides a mechanism to rank the document set for each query.

The algebraic foundation for Latent Semantic Indexing (LSI) was first described in [5] and has been further discussed by Berry, et al. in [1, 2]. These papers describe the SVD process and interpret the resulting matrices in a geometric context. They show that the SVD, truncated to k dimensions, gives the optimal rank-k approximation to the original matrix. Wiemer-Hastings shows that the power of LSI comes primarily from the SVD algorithm [21].

Other researchers have proposed theoretical approaches to understanding LSI. Zha and Simon describe LSI in terms of a subspace model and propose a statistical test for choosing the optimal number of dimensions for a given collection [23]. Story discusses LSI's relationship to statistical regression and Bayesian methods [20]. Ding constructs a statistical model for LSI using the cosine similarity measure, showing that the term similarity and document similarity matrices are formed during the maximum likelihood estimation, and LSI is the optimal solution to this model [6].

Although other researchers have explored the SVD algorithm to provide an understanding of SVD-based information retrieval systems, to our knowledge, only Schütze has studied the values produced by LSI [18]. We expand upon this work in [15, 16, 17], showing that the SVD exploits higher order term co-occurrence in a collection, and showing the correlation between the values produced in the term-term matrix and the performance of LSI. In the current work, we extend these results to determine the most critical values in an LSI system.

Other researchers [4, 12] have recently applied sparsification techniques to reduce the computation cost of LSI. These papers provide further empirical evidence for our claim that the retrieval performance of LSI depends on a subset of the SVD matrix elements. Gao and Zhang have simultaneously, but independently, proposed mechanisms to take the dense lower-dimensional matrices that result from SVD truncation, and make them sparse [12]. Chen et al. implicitly do this as well, by encoding values into a small set of discrete values [4]. Our approach to sparsification (described in Sect. 3) is somewhat different from the ones used by Gao and Zhang; furthermore, we present data for additional collections (they used three small collections), and also describe the run time considerations, as well as the retrieval effectiveness, of sparsification.

3 Sparsification of the LSI Input Matrices

In what follows we report the results of a study to determine the most critical elements of the T_k and $S_k D_k$ matrices, which are input to LSI. We are interested in the impact, both in terms of retrieval quality and query run time performance, of the removal (zeroing) of a large portion of the entries in these matrices.

In this section we describe the algorithm we used to remove values from the T_k and $S_k D_k$ matrices and describe the impact of this sparsification strategy on retrieval quality and query run time performance.

3.1 Methodology

Our sparsification algorithm focuses on the values with absolute value near zero, and we ask the question: "How many values can we remove without severely impacting retrieval performance?" Intuitively, the elements of the row vectors in the $T_k S_k$ matrix and the column vectors in the $S_k D_k$ matrix can be used to describe the importance of each term (document) along a given dimension.

Several patterns were identified in our preliminary work. For example, removal of all negative elements severely degrades retrieval performance, as does removal of "too many" of the S_kD_k elements. However, a large portion of the T_k values can be removed without a significant impact on retrieval performance.

We chose an approach that defines a common truncation value, based on the values in the T_kS_k matrix, which is used for both the T_k and S_kD_k matrices. Furthermore, since the negative values are important, we wanted to retain an equal number of positive and negative values in the T_k matrix. The algorithm we used is outlined in Fig. 1. We chose positive and negative threshold values that are based on the T_kS_k matrix and that result in the removal of a fixed percentage of the T_k matrix. We use these values to truncate both the T_k and S_kD_k matrices.

```
Compute TkSk and SkDk
Determine PosThres: The threshold that would result in removal of x%
        of the positive elements of TkSk
Determine NegThres: The threshold that would result in removal of x%
        of the negative elements of TkSk
For each element of Tk, change to zero, if the corresponding element
        of TkSk falls outside PosThres and NegThres
For each element of SkDk, change to zero, if it falls outside
        PosThres and NegThres
```

Fig. 1. Sparsification Algorithm

3.2 Evaluation

Retrieval quality for an information retrieval system can be expressed in a variety of ways. In the current work, we use precision and recall to express quality. Precision is defined as the percentage of retrieved documents which are relevant to the query. Recall is the percentage of all relevant documents that were retrieved.

These metrics can be applied in two ways. First, we can compute recall and precision at rank $= n$, where n is a constant. In this case, we look at the first n documents returned from the query and compute the precision and recall using the above definitions. An alternative approach involves computing precision at a given recall level. In this second case, we continue to retrieve documents until a given percentage of correct documents has been retrieved (for example, 25%), and then compute the precision. In the results that follow we apply this second approach to evaluate of our sparsification strategy.

Precision and recall require the existence of collections that contain a group of documents, a set of standard queries and a set of relevance judgments (a list of which documents are relevant to which query, and which are not relevant).

We used seven such collections during the course of our study. The collections we used are summarized in Table 1. These collections were downloaded from a variety of sources. MED, CISI, CRAN, NPL, and CACM were downloaded from the SMART web site at Cornell University. LISA was obtained from the Information Retrieval Group web site at the University of Glasgow. The OHSUMED collection was downloaded from the Text Retrieval Conference (TREC) web site at the National Institute of Standards and Technology. Not all of the documents in the OHSUMED collection have been judged for relevance for each query. In our experiments, we calculated precision and recall by assuming that all unjudged documents are not relevant. Similar studies that calculate precision using only the judged documents are left to future work.

Table 1. Collections used to compare Sparsification Strategy to Traditional LSI

Identifier	Description	Docs	Terms	Queries
MED	Medical abstracts	1033	5831	30
CISI	Information science abstracts	1450	5143	76
CACM	Communications of the ACM abstracts	3204	4863	52
CRAN	Cranfield collection	1400	3932	225
LISA	Library and Information Science Abstracts	6004	18429	35
NPL	Larger collection of very short documents	11429	6988	93
OHSUMED	Clinically-oriented MEDLINE subset	348566	170347	106

The Parallel General Text Parser (PGTP) [3] was used to preprocess the text data, including creation and decomposition of the term document matrix. For our experiments, we applied the log entropy weighting option and normalized the document vectors for all collections except OHSUMED. The sparsification algorithm was applied to each of our collections, using truncation percentages of 10% to 90%. Retrieval quality and query runtime performance measurements were taken at multiple values of k. The values of k for the smaller collections ranged from 25 to 200; k values from 50 to 500 were used for testing the larger collections.

3.3 Impact on Retrieval Quality

The retrieval quality results for three different truncation values for the collections studied are shown in Figs. 2–8. In these experiments, we computed the precision at three different recall levels (.25, .50, and .75) and calculated the mean precision for each query. The average precision shown in Figs. 2–8 is the average of these means for all queries in the collection.

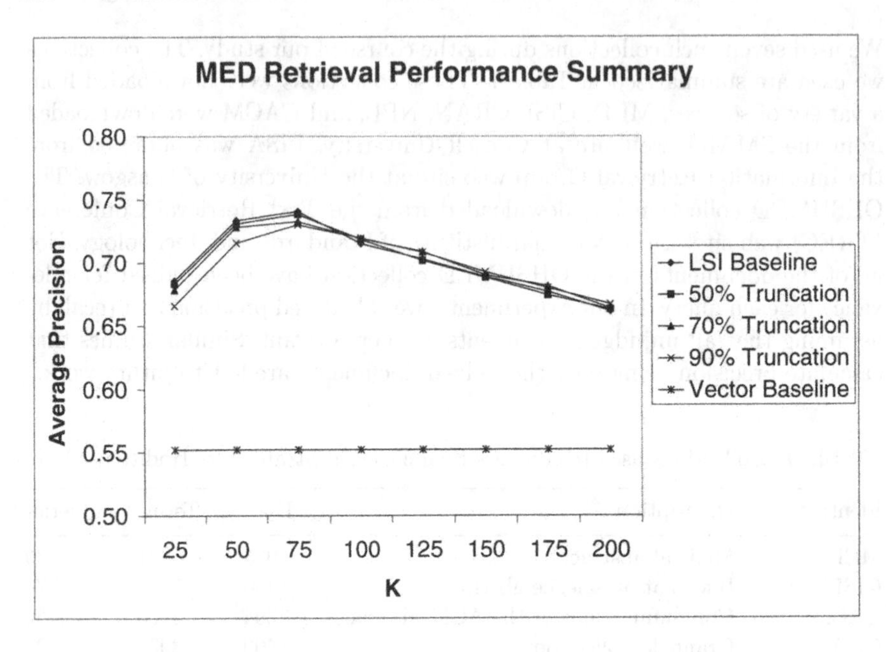

Fig. 2. Sparsification Performance Summary for MED

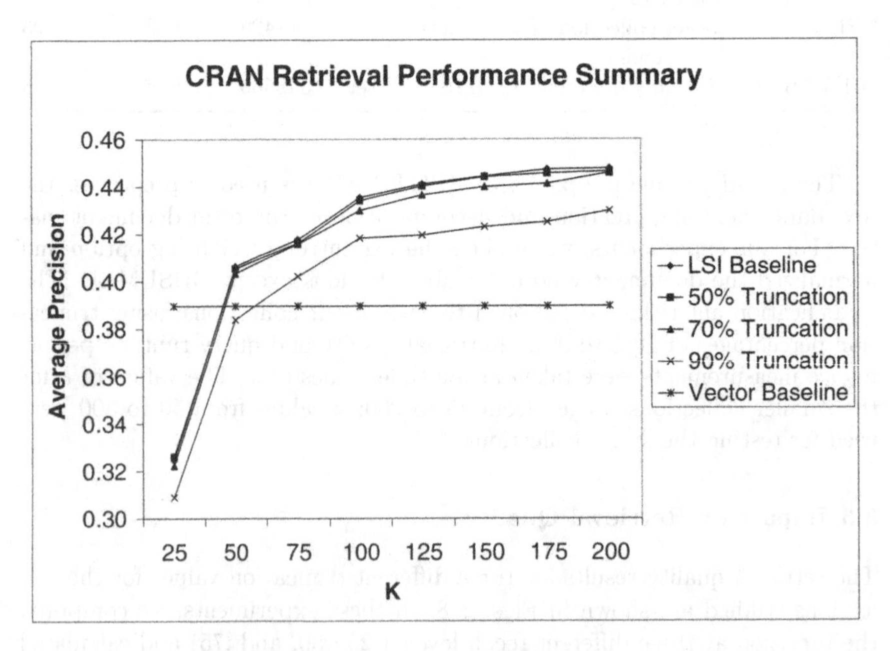

Fig. 3. Sparsification Performance Summary for CRAN

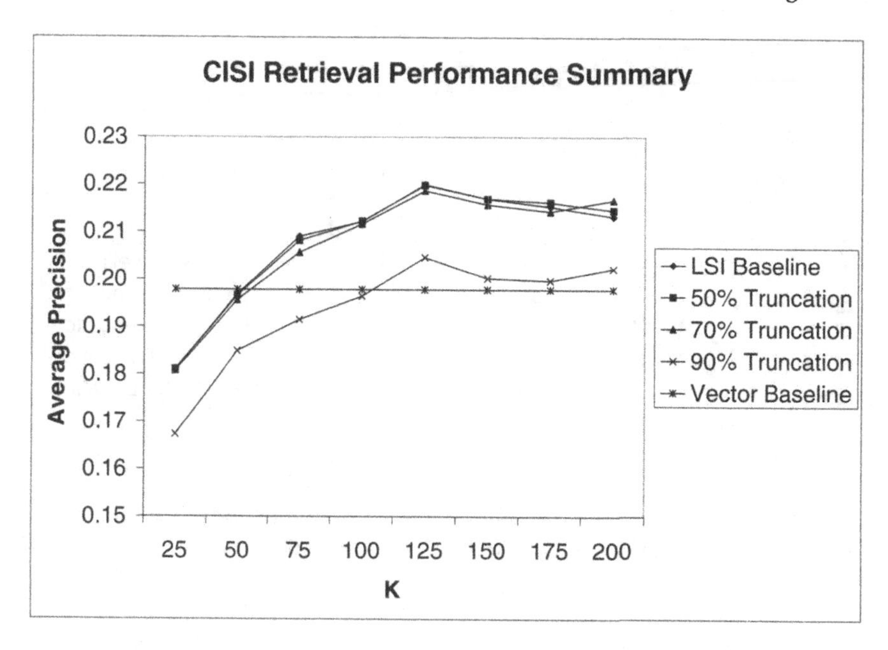

Fig. 4. Sparsification Performance Summary for CISI

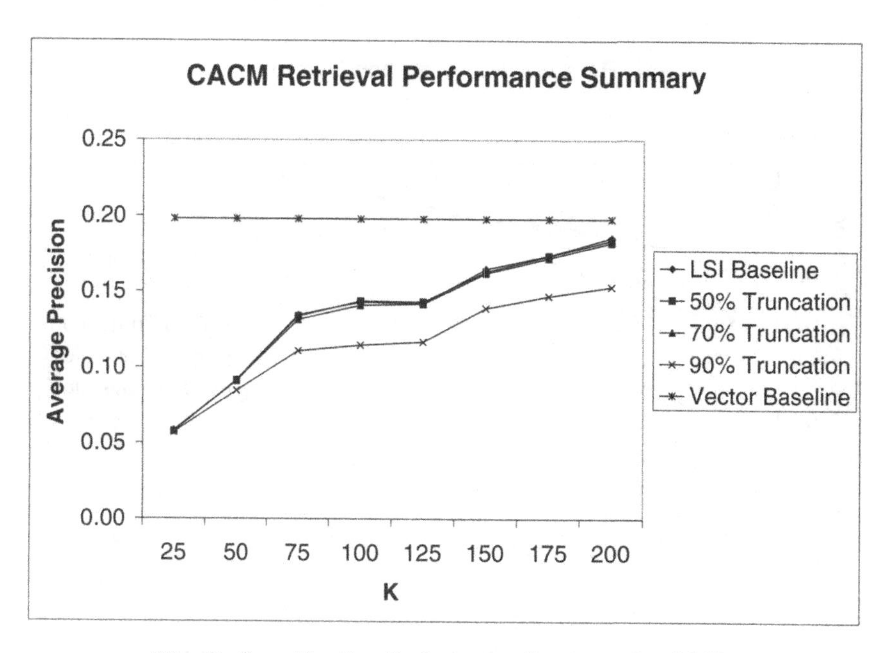

Fig. 5. Sparsification Performance Summary for CACM

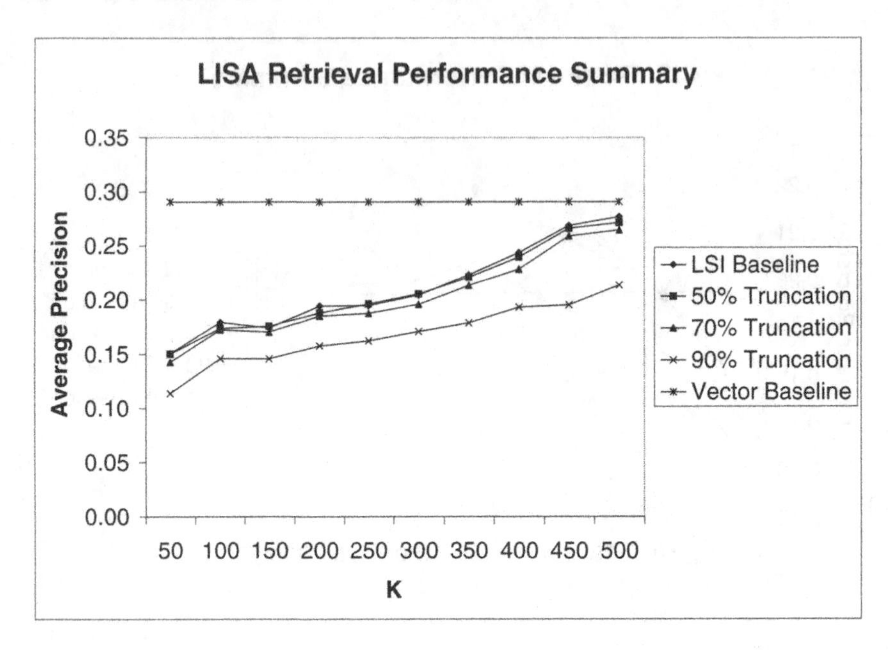

Fig. 6. Sparsification Performance Summary for LISA

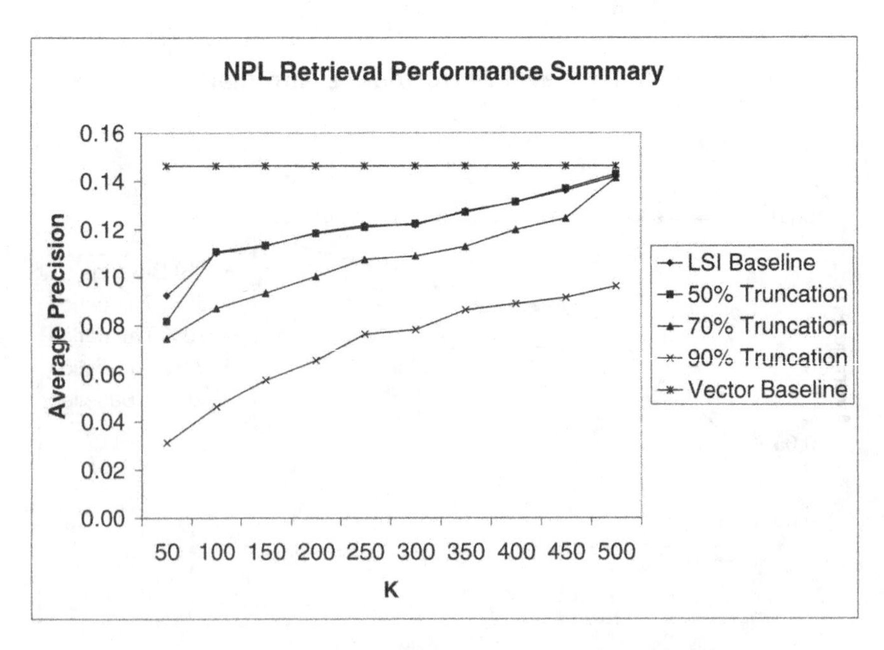

Fig. 7. Sparsification Performance Summary for NPL

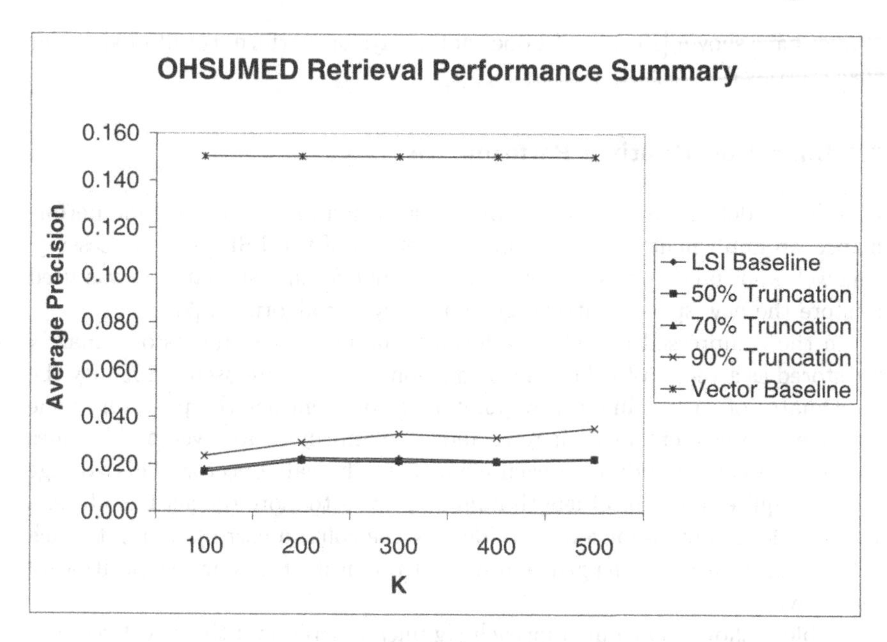

Fig. 8. Sparsification Performance Summary for OHSUMED

Two baselines were used to measure retrieval quality. Retrieval quality for our sparsified LSI was compared to a standard LSI system, as well as to a traditional vector space retrieval system.

Comparison to Standard LSI Baseline

Removal of 50% of the T_k matrix values resulted in retrieval quality that is indistinguishable from the LSI baseline for the seven collections we tested. In most cases, sparsification up to 70% can be achieved, particularly at better performing values of k, without a significant impact on retrieval quality. For example, $k = 500$ for NPL and $k = 200$ for CACM have performance near or greater than the LSI baseline when 70% of the values are removed. The data for OHSUMED appears in Fig. 8. Notice that sparsification at the 90% level actually improves LSI average precision by 60% at $k = 500$.

Comparison to Traditional Vector Space Retrieval Baseline

Figures 2–8 show that LSI outperforms traditional vector space retrieval for CRAN, MED and CISI even at very small values of k. However, traditional vector space is clearly better for OHSUMED, CACM, LISA and NPL at small k values. LSI continues to improve as k increases for these collections. As expected, we found no obvious relationship between the performance of the sparsified LSI system and traditional vector space retrieval. As other research studies have shown [13, 14], LSI does not always outperform traditional vector space retrieval.

3.4 Impact on Runtime Performance

In order to determine the impact of sparsification on query run time performance we implemented a sparse matrix version of the LSI query processing. The well-known compressed row storage format for sparse matrices was used to store the new sparse matrices generated by our algorithm [7].

In the compressed row storage format, the nonzero elements of a matrix are stored as a vector of values. Two additional vectors are used to identify the coordinates of each value: a row pointer vector identifies the position of the first nonzero element in each row, and a column indicator vector identifies the column corresponding to each element in the value vector. This storage format requires a vector of length num-nonzeroes to store the actual values, a vector of length num-nonzeroes to identify the column corresponding to each value, and a vector of length num-rows to identify the starting position of each row.

Table 2 shows that our approach significantly reduces the RAM requirements of LSI. This reduction is due to our sparsification strategy, which produce sparse matrices. Implementation of the compressed row storage format for a non-sparsified LSI system would result in an increase in the memory requirements.

Table 2. RAM Savings for T and D matrices

Collection	k	Sparsification Level (%)	Sparsified RAM (MB)	LSI RAM (MB)	Improvement (%)
MED	75	70	3.7	7.9	53
CISI	125	70	6.3	12.6	50
CRAN	200	70	8.3	16.3	49
CACM	200	70	14.4	24.6	42
NPL	500	70	93.1	140.5	34
LISA	500	70	66.8	115.7	42
OHSUMED	500	70	3217	3959	19
MED	75	90	1.6	7.9	79
CISI	125	90	2.9	12.6	77
CRAN	200	90	3.6	16.3	78
CACM	200	90	6.3	24.6	75
NPL	500	90	29.0	140.5	79
LISA	500	90	28.3	115.7	76
OHSUMED	500	90	2051	3959	48

When comparing the runtime considerations of our approach to LSI, we acknowledge that our approach requires additional preprocessing, as we implement two additional steps, determining the threshold value and applying the threshold to the T_k and S_kD_k matrices. These steps are applied once per collection, however, and multiple queries can then be run against the collection.

Table 3 indicates that the S_kD_k sparsification ranges from 18% to 33%, when 70% of the T_k values are removed. A much larger S_kD_k sparsification range of 43%–80% is achieved at a 90% reduction in the T_k matrix.

Table 3. Percentage of Document Vector Entries Removed

Collection	k	Term Spars (%)	Doc Spars (%)	Run Time Improvement (%)
MED	75	70	23	–1
CISI	125	70	26	1
CRAN	200	70	29	6
CACM	200	70	28	3
NPL	500	70	33	–3
LISA	500	70	29	10
OHSUMED	500	70	18	3
MED	75	90	47	16
CISI	125	90	53	27
CRAN	200	90	61	37
CACM	200	90	64	40
NPL	500	90	80	66
LISA	500	90	66	54
OHSUMED	500	90	43	30

The query processing for LSI is comprised of two primary tasks: development of the pseudo query, which relies on the T_k matrix, and the comparison of the pseudo query to all documents in the collection, which uses the S_kD_k matrix. The performance figures shown in Table 3 are determined based on the aggregate CPU time for the distance computation between the pseudo query vector and all the document vectors. After the distance computation for each document is derived, the scores are stored in a C++ Standard Template Library (STL) map structure. This structure allowed us to sequentially retrieve a number of documents in relevance score order until the required percentage of relevant documents had been retrieved. This structure was used for both our sparsified LSI tests and our traditional LSI tests.

The number of cpu cycles required to run all queries in each collection was collected using the clock() function available in C++. Measurements were taken for both the baseline LSI code and the sparsified code. Each collection was tested twice, and the results in Table 3 represent the average of the two runs for selected levels of sparsification.

Sparsification of the matrix elements results in an improvement in query runtime performance for all collections, with the exception of MED and NPL, at 70% sparsification. The 90% sparsification data shown in Table 3 reveals significant improvements in query run time performance for all collections tested.

4 Conclusions

Our analysis has identified a large number of term and document vector values that are unimportant. This is a significant component in the development of a theoretical understanding of LSI. As researchers continue working to develop a thorough understanding of the LSI system, they can restrict their focus to the most important term and document vector entries.

Furthermore, we have shown that query run time improvements in LSI can be achieved using our sparsification strategy for many collections. Our approach zeroes a fixed percentage of both positive and negative values of the term and document vectors produced by the SVD process. Our data shows that, for small collections, we can successfully reduce the RAM requirements by 45% (on average), and the query response time an average of 3%, without sacrificing retrieval quality. If a slight degradation in retrieval quality is acceptable, the RAM requirements can be reduced by 77%, and query run time can be reduced by 40% for smaller collections using our approach.

On the larger TREC collection (OHSUMED), we can reduce the runtime by 30%, reduce the memory required by 48% and improve retrieval quality by 60% by implementing our sparsification algorithm at 90%.

Acknowledgements

This work was supported in part by National Science Foundation Grant Number EIA-0087977 and the National Computational Science Alliance under IRI030006 (we utilized the IBM pSeries 690 cluster). The authors appreciate the assistance provided by their colleagues at Lehigh University and Ursinus College. Co-author William M. Pottenger also gratefully acknowledges his Lord and Savior, Yeshua (Jesus) the Messiah, for His continuing guidance in and salvation of his life.

References

1. Michael W. Berry, Zlatko Drmac, and Elizabeth R. Jessup. Matrices, vector spaces, and information retrieval. *SIAM Review*, 41(2):335–362, 1999.

2. Michael W. Berry, Susan T. Dumais, and Gavin W. O'Brien. Using linear algebra for intelligent information retrieval. *SIAM Review*, 37(4):575–595, 1995.
3. Michael W. Berry and Dian I. Martin. Principal component analysis for information retrieval. In E. J. Kontoghiorghes, editor, *Handbook of Parallel Computing and Statistics*. Marcel Dekker, New York, 2004. In press.
4. Chung-Min Chen, Ned Stoffel, Mike Post, Chumki Basu, Devasis Bassu, and Clifford Behrens. Telcordia LSI engine: Implementation and scalability issues. In *Proceedings of the Eleventh International Workshop on Research Issues in Data Engineering (RIDE 2001)*, Heidelberg, Germany, April 2001.
5. Scott C. Deerwester, Susan T. Dumais, Thomas K. Landauer, George W. Furnas, and Richard A. Harshman. Indexing by latent semantic analysis. *Journal of the American Society of Information Science*, 41(6):391–407, 1990.
6. Chris H. Q. Ding. A similarity-based probability model for latent semantic indexing. In *Proceedings of the Twenty-second Annual International ACM/SIGIR Conference on Research and Development in Information Retrieval*, pp. 59–65, 1999.
7. Jack Dongarra. Sparse matrix storage formats. In Z. Bai, J. Demmel, J. Dongarra, A. Ruhe, and H. van der Vorst, editors, *Templates for the Solution of Algebraic Eigenvalue Problems: A Practical Guide*, pp. 372–378. SIAM, Philadelphia, 2000.
8. Susan T. Dumais. LSI meets TREC: A status report. In D. Harman, editor, *The First Text REtrieval Conference (TREC-1), National Institute of Standards and Technology Special Publication 500-207*, pp. 137–152, 1992.
9. Susan T. Dumais. Latent semantic indexing (LSI) and TREC-2. In D. Harman, editor, *The Second Text REtrieval Conference (TREC-2), National Institute of Standards and Technology Special Publication 500-215*, pp. 105–116, 1994.
10. Susan T. Dumais. Using LSI for information filtering: TREC-3 experiments. In D. Harman, editor, *The Third Text REtrieval Conference (TREC-3), National Institute of Standards and Technology Special Publication 500-225*, pp. 219–230, 1995.
11. George E. Forsythe, Michael A. Malcolm, and Cleve B. Moler. *Computer Methods for Mathematical Computations*. Prentice Hall, 1977.
12. Jing Gao and Jun Zhang. Sparsification strategies in latent semantic indexing. In M. W. Berry and W. M. Pottenger, editors, *Proceedings of the 2003 Text Mining Workshop*, May 2003.
13. Thomas Hofmann. Probabilistic Latent Semantic Indexing. In *Proceedings of the 22nd Annual ACM Conference on Research and Development in Information Retrieval*, pp. 50–57, Berkeley, California, August 1999.
14. Parry Husbands, Horst Simon, and Chris Ding. On the use of singular value decomposition for text retrieval. In M. Berry, editor, *Proc. of SIAM Comp. Info. Retrieval Workshop*, October 2000.
15. April Kontostathis and William M. Pottenger. A framework for understanding Latent Semantic Indexing (LSI) performance. *Information Processing and Management*. To appear.
16. April Kontostathis and William M. Pottenger. Detecting patterns in the LSI term-term matrix. In *IEEE ICDM02 Workshop Proceedings, The Foundation of Data Mining and Knowledge Discovery (FDM02)*, December 2002.
17. April Kontostathis and William M. Pottenger. A framework for understanding LSI performance. In S. Dominich, editor, *Proceedings of SIGIR Workshop on Mathematical/Formal Methods in Information Retrieval*, July 2003.

18. Hinrich Schütze. Dimensions of meaning. In *Proceedings of Supercomputing*, pp. 787–796, 1992.
19. Hinrich Schütze. Automatic word sense discrimination. *Computational Linguistics*, 24(1):97–124, 1998.
20. Roger E. Story. An explanation of the effectiveness of Latent Semantic Indexing by means of a Bayesian regression model. *Information Processing and Management*, 32(3):329–344, 1996.
21. Peter Wiemer-Hastings. How latent is latent semantic analysis? In *Proceedings of the Sixteenth International Joint Conference on Artificial Intelligence*, pp. 932–937, 1999.
22. Sarah Zelikovitz and Haym Hirsh. Using LSI for text classification in the presence of background text. In H. Paques, L. Liu, and D. Grossman, editors, *Proceedings of CIKM-01, tenth ACM International Conference on Information and Knowledge Management*, pp. 113–118, Atlanta, GA, 2001. ACM Press, New York.
23. Hongyuan Zha and Horst Simon. A subspace-based model for Latent Semantic Indexing in information retrieval. In *Proceedings of the Thirteenth Symposium on the Interface*, pp. 315–320, 1998.

Reporting Data Mining Results in a Natural Language

Petr Strossa[1], Zdeněk Černý[2], and Jan Rauch[1,2]

[1] Department of Information and Knowledge Engineering,
[2] EuroMISE centrum – Cardio, University of Economics, Prague,
nám. W. Churchilla 4, 130 67 Praha 3, Czech Republic
kizips@vse.cz, cernyz@vse.cz, rauch@vse.cz

Summary. An attempt to report results of data mining in automatically generated natural language sentences is described. Several types of association rules are introduced. The presented attempt concerns implicational rules – one of the presented types. Formulation patterns that serve as a generative language model for formulating implicational rules in a natural language are described. An experimental software system AR2NL that can convert implicational rules both into English and Czech is presented. Possibilities of application of the presented principles to other types of association rules are also mentioned.

Key words: Data mining, association rules, natural language processing, formulation patterns

1 Introduction

The goal of this paper is to draw attention to a possibility of presentation of the data mining results in a **natural language (NL).** We are specifically interested in **association rules (ARs)** of the form $\varphi \approx \psi$ where φ and ψ are Boolean attributes derived from the columns of the analysed data matrix $\mathcal{M}$.

The AR $\varphi \approx \psi$ says that φ and ψ are associated in the way corresponding to the symbol $\approx$. The data mining procedure *4ft-Miner* [6] mines for ARs of this form – see Sect. 2. The 4ft-Miner procedure is a part of the *LISp-Miner* academic system [6]. The "classical" ARs with confidence and support [1] can be also expressed this way.

Let us start with an example of an AR:

$$\mathrm{ED}(univ) \wedge \mathrm{RS}(mng) \Rightarrow_{0.9,45} \mathrm{AJ}(sits)$$

This example is inspired by the medical project STULONG briefly described in Sect. 3. The following attributes are used in the rule:

P. Strossa et al.: *Reporting Data Mining Results in a Natural Language*, Studies in Computational Intelligence (SCI) **6**, 347–361 (2005)
www.springerlink.com

ED – *Education*, with possible values *univ* (university), *basic* (basic school), *appr* (apprentice school), etc.;

RS – *Responsibility In a Job*, with possible values *mng* (managerial worker), *prt* (partly independent worker), etc.;

AJ – *Activity In a Job*, with possible values *sits* (he mainly sits), *mdrt* (moderate activity), etc.

The rule is related to a four-fold contingency table (4ft table for short) an example of which is shown as Table 1.

Table 1. 4ft table

	$\mathrm{AJ}(sits)$	$\neg\,\mathrm{AJ}(sits)$
$(\mathrm{ED}(univ) \wedge \mathrm{RS}(mng))$	45	5
$\neg\,(\mathrm{ED}(univ) \wedge \mathrm{RS}(mng))$	60	30

The table says that there are 45 patients satisfying both $(\mathrm{ED}(univ) \wedge \mathrm{RS}(mng))$ and $\mathrm{AJ}(sits)$, there are further 5 patients satisfying $(\mathrm{ED}(univ) \wedge \mathrm{RS}(mng))$ and not satisfying $\mathrm{AJ}(sits)$, etc.

The goal of data mining is to find unsuspected relationships and to summarize the data in novel ways that are both understandable and useful to the data owner [4]. A problem can arise when the data owner (a physician in our example) gets a set of ARs true in the analysed data – it can be simply too much of formalised information. The association rules can be more intelligible when they are presented in a natural language.

The symbol $\Rightarrow_{0.9,45}$ represents a *founded implication.* It is defined in the next section. The association rule $\mathrm{ED}(univ) \wedge \mathrm{RS}(mng) \Rightarrow_{0.9,45} \mathrm{AJ}(sits)$ can be reformulated in a natural language in various ways. Here are two examples:

- *45 (i.e. 90%) of the observed patients confirm this dependence: if the patient has university education and responsibility of a manager, then he mainly sits in his job.*
- *90% of the observed patients that have reached university education and work in a managerial position also mainly sit in their job.*

In our experience, it can be very useful when the output rules are presented both in a formal way and in a natural language. It can reasonably help the user – non specialist in data mining to understand the results.

We aim to show that there is quite a simple NL model that can be used to automatically transform ARs generated from various data tables into various NLs, e.g., Czech or English. This model is presented in Sect. 4; the main features of the experimental system implementing the model, called AR2NL, are described in Sect. 5.

The idea of automated conversion of ARs of various types has been first formulated in [5]. A part of the related theory has been presented in [9]. The core of the experimental system AR2NL has been implemented as a subject of a master thesis [2].

The work presented here is a first step in a supposed, rather long-time, research activity. The main contribution of this paper is to present results of a research in a possibly new branch in presentation of data mining results. Despite a large effort we were not able to find a paper dealing with similar topic.

The current experience with the AR2NL system can be summarized in the following way:

- The NL presentation of ARs increases interest of field specialists in data mining results. It concerns namely the field specialists that have education only in humanities. The NL sentences like *"90% of the observed patients that have reached university education and work in a managerial position also mainly sit in their job"* attracts these specialists much more than even very sophisticated measures of intensity of the ARs.
- It has been shown that this remarkable improvement can be achieved by quite simple NL tools (formulation patterns, see Sect. 4, "elementary" expressions and morphological tables, see 5.2). In our opinion the complexity of this linguistic task roughly corresponds to secondary education in one's native language combined with standard education in informatics.
- The use of AR2NL does not, of course, decrease the number of output rules. However we use it together with the 4ft-Miner mining procedure that has very rich tools both for tuning the definition of the set of rules to be generated and verified and for filtering and sorting the output rules, see [6]. This way, a relatively small set of interesting ARs can be focused on, *and* they can be presented in a NL.
- The data structures and algorithms developed for the AR2NL system can be easy modified and enhanced to meet further goals concerning the presentation of the data mining results in NLs.

We suppose to continue in the activity concerning presentation of data mining results in natural language. An overview of supposed further work is in Sect. 6.

2 Association Rules and the 4ft-Miner Procedure

The 4ft-Miner procedure mines for ARs of the form $\varphi \approx \psi$ where φ and ψ are Boolean attributes automatically generated by the procedure from the columns of the analysed data matrix $\mathcal{M}$.

The symbol $\approx$ stands for a particular *4ft-quantifier*. It denotes a condition concerning a 4ft table. The association rule $\varphi \approx \psi$ is true in the analysed data matrix $\mathcal{M}$ if the condition related to the 4ft-quantifier $\approx$ is satisfied in the 4ft table of φ and ψ in $\mathcal{M}$, see Table 2.

Table 2. 4ft table of φ and ψ in $\mathcal{M}$

$\mathcal{M}$	ψ	$\neg\psi$	
φ	a	b	r
$\neg\varphi$	c	d	
	k		n

Here a is the number of rows in $\mathcal{M}$ satisfying both φ and ψ, b is the number of the rows satisfying φ and not satisfying ψ, c is the number of rows not satisfying φ and satisfying ψ, and d is the number of rows satisfying neither φ nor ψ. Two examples of the 4ft-quantifiers follow:

The 4ft-quantifier $\Rightarrow_{p,Base}$ of *founded implication* for $0 < p \leq 1$ and $Base > 0$ [3] is defined by the condition

$$\frac{a}{a+b} \geq p \wedge a \geq Base \; .$$

It means that at least 100p per cent of objects satisfying φ satisfy also ψ and that there are at least *Base* objects of $\mathcal{M}$ satisfying both φ and ψ. The fraction $\frac{a}{a+b}$ is the *confidence* of the AR. The frequency a corresponds to the *support* of the AR, which is equal to $\frac{a}{a+b+c+d}$. This means that the parameter p defines the minimal level of *confidence* and the parameter *Base* corresponds to the minimal level of *support*.

The 4ft-quantifier $\sim^{+}_{p,Base}$ of *above average* for $0 < p$ and $Base > 0$ is defined by the condition

$$\frac{a}{a+b} \geq (1+p)\frac{a+c}{a+b+c+d} \wedge a \geq Base \; .$$

It means that among the objects satisfying φ there is at least 100p per cent more objects satisfying ψ than among all the objects and that there are at least *Base* objects satisfying both φ and ψ.

There are 16 types of 4ft-quantifiers implemented in the 4ft-Miner procedure. Examples of further 4ft-quantifiers are in [6].

The left part of an association rule $\varphi \approx \psi$ is called *antecedent*, the right part (i.e. ψ) is called *succedent*. The input of the 4ft-Miner procedure consists of

- the analysed data matrix;
- several parameters defining the set of relevant antecedents;
- several parameters defining the set of relevant succedents;
- the 4ft quantifier.

Both the antecedent and the succedent are conjunctions of literals. A literal is an expression of the form $A(\alpha)$ where A is an attribute and α is a proper subset of all possible values of the attribute A. The literal $A(\alpha)$ is true in a row of the analysed data matrix if the value of A in this row belongs to α. In this paper we only deal with literals of the form $A(a)$ where a is a single value of A

(pedantically we should write $A(\{a\})$). The conjunction $\mathrm{ED}(univ) \wedge \mathrm{RS}(mng)$ is an example of an antecedent (see Introduction). For more details concerning the 4ft-Miner procedure see [6].

3 The STULONG Project

The STULONG project concerns a detailed study of the risk factors of atherosclerosis in the population of middle aged men. The risk factors of atherosclerosis were observed for 1,417 men in the years 1976–1999 according to a detailed project methodology. The primary data cover both entry and control examinations. 244 attributes have been surveyed with each patient at the entry examinations. 10,610 control examinations were further made, each examination concerns 66 attributes.

For more details about the STULONG project see [10] and `http://euromise.vse.cz/challenge2003/project/`.

Our example used in Sect. 5 deals with the following attributes:

- *Activity In a Job* (abbreviated AJ) with values *he mainly sits, he mainly stands, he mainly walks, he carries heavy loads*;
- *Activity After a Job* (abbr. AA) with values *he mainly sits, moderate activity, great activity*;
- *Education* (abbr. ED) with values *basic school, apprentice school, secondary school, university*;
- *Responsibility in a job* (abbr. RS) with values *manager, partly independent worker, other, pensioner because of ischemic heart diseases, pensioner for other reasons.*

4 Formulation Patterns for Implicational Rules

Let us now look a bit more formally at the mutual relation between the formal logical expression like

$$\mathrm{ED}(univ) \wedge \mathrm{RS}(mng) \Rightarrow_{0.9,45} \mathrm{AJ}(sits)$$

and its possible natural language (NL) formulations, e.g.:

(1) *45 (i.e. 90%) of the observed patients confirm this dependence: if the patient has university education and responsibility of a manager, then he mainly sits in his job.*

(2) *A combination of university education and a managerial job implies a sedentary job. This fact is confirmed by 45 (i.e. 90%) of the observed patients.*

(3) *90% of the observed patients that have reached university education and work in a managerial position also mainly sit in their job.*

If we consider the formal expression to be an instance of a general "logical pattern"

$$A \Rightarrow_{p,Base} S$$

then we certainly can also think of the given NL formulations as of (respective) instances of **NL formulation patterns** that could be presented in a form near to the following:

(1) ***Base*** *(i.e.* ***100p%*** *) of the patients confirm this dependence: if the patient has* ***NLF(A)****, then he* ***NLF(S)****.*
(2) *A combination of* ***NLF(A₁)*** *and* ***NLF(A₂)*** *implies* ***NLF(S)****. This fact is confirmed by* ***Base*** *(i.e.* ***100p%****) of the patients.*
(3) ***100p%*** *of the patients that* ***NLF(A)*** *also* ***NLF(Z)****.*

In this form of the NL formulation patterns, the symbols ***NLF(A)*** and ***NLF(S)*** stand for *suitable partial NL formulations* corresponding to the antecedent and to the succedent, respectively.[1] There is only one little problem concealed in it: the formal expression like ***NLF(X)*** does not represent an invariant linguistic form in all the formulation patterns. Some of the patterns require a literal expressed by **a noun phrase** ("university education", "a managerial position", "a sedentary job"), while others may require **a verb phrase** ("has reached university education", "works as a manager", "mainly sits in his job"), or (although it cannot be seen in our examples) **a prepositional phrase** ("with university education", "in a managerial position"), **an adjectival phrase** ("university-educated") or **a participial phrase** ("having reached university education", "working as a manager", "mainly sitting in his job").

Some of these phrase types can be grammatically derived from other ones that are once given: e.g., if one knows the verb phrase "works as a manager", it is not at all complicated to derive the participial phrase "work*ing* as a manager"; similarly the prepositional phrase (if it exists for a given literal) must be somehow derived from a noun phrase ("a managerial position" → "*in* a managerial position"). But this does not hold universally. It is sufficient to look at the verb phrase "works as a manager" and the noun phrase "a managerial position": there is no simple *grammatical* relation between them, they rather show a more-or-less *idiomatic* choice of words.

It turns out that these **types of "elementary" linguistic information** should be assigned to every possible literal, i.e. to every pair ⟨attribute, value⟩:

[1] As it can be seen in pattern (2), some of the patterns suppose (at least) two literals in the antecedent. Symbols like ***NLF(A₁)*** etc. are required in such patterns.

In fact, any *greater* number of antecedents than explicitly given in the pattern is easily convertible by means of the same pattern: any sequence of the type *NLF*(A_1)*, NLF*(A_2)*, ... and NLF*(A_n) (including the special cases when $n = 1$ or $n = 2$) is simply transformed into the sequence *NLF*(A_1)*, NLF*(A_2)*, ... and NLF*(A_r) where $r > n$. The only "general linguistic knowledge" the transformation procedure must have is that the last two antecedents are always separated with "and", the others (if there are more of them) with commas.

- a **noun phrase** representing the literal (e.g., "a managerial position");
- the specific way **how to transform the noun phrase into a prepositional phrase** (e.g., use the preposition "with" followed by the noun phrase, or use the preposition "in" followed by the noun phrase, or, for languages like Czech, use the preposition x followed by the noun phrase in case y);
- a **verb phrase** representing the literal (e.g., "works as a manager");
- an **adjectival phrase** representing the literal (e.g., "university-educated");
- a **word-order position** relative to the word "patient" (or other name of the observed entity type) in which **the given adjectival phrase** should be placed (e.g., in English, "university-educated" has to be placed to the left from "patient", while "born in 1950" has to be placed to the right, and it may be found interesting that the same is true about the Czech equivalents of these two adjectival phrases).

As for the *verb phrases*, we have discovered that there are some of them really *unique in shape* for a particular value of an attribute (like "has a university degree" for ED(*univ*)), but on the other hand they often can be simply and efficiently composed from a *universal part* corresponding to the attribute (e.g., "has finished" for ED) and a *specific part* corresponding to a value of the attribute (e.g., "university" for ED(*univ*)). Sometimes a similar mechanism can be of use for noun phrases and adjectival phrases, too.

Of course, any of the linguistic information items described above may be empty in a particular situation. This generally means that a NL formulation pattern requiring a specific type of phrase cannot be used for the given literals.

The final form of the NL formulation patterns is approximately this one:

(1) ***Base*** *(i.e.* ***100p%****) of the patients confirm this dependence: if the patient has* ***NP(A)****, then he* ***VP(S)****.*
(2) *A combination of* ***NP(A₁)*** *and* ***NP(A₂)*** *implies* ***NP(S)****. This fact is confirmed by* ***Base*** *(i.e.* ***100p%****) of the patients.*
(3) ***100p%*** *of the patients that* ***plural(VP(A))*** *also* ***plural(VP(S))****.*

Here ***NP(X)*** means a noun phrase representing X, ***VP(X)*** means a verb phrase representing X, etc.; ***plural(...)*** is an example of a **morphological directive** that must be fulfilled by a corresponding module. (See [9] for more details concerning morphology in our context.)

The morphological module need not "know" all the morphology (and vocabulary) of the language in consideration. It does not need any special dictionary. There is only one additional type of information necessary from the morphological point of view in the linguistic data structures described in this section: in all the "elementary" noun phrases, verb phrases and adjectival phrases, an **inflection pattern code** must be assigned **to every word that may sometimes require a morphological change** according to the directives in the formulation patterns. Thus, for instance, a particular verb phrase like "mainly sits in his job" must in fact be given (somewhere – see the next section for more details) in such a way:

mainly ***IF(sits,v1t)*** *in* ***IF(his,pron1)*** *job*

Here, the expression ***IF(x,y)*** represents an *inflected form* of the word x corresponding to the inflection pattern y and the *morphological directive* contained in the formulation pattern (if there is no explicit morphological directive, no transformation takes place and the word is used in its "basic" form x[2]); *v1t* is an English verb inflection code denoting the following paradigm: singular form ending with *-ts*, plural form ending with *-t*,[3] and participle ending with *-tting*. Similarly, *pron1* is an English pronoun inflection code denoting this paradigm: singular form *his*, plural form *their*.[4]

5 The AR2NL System

In this section we briefly describe the current version of the system named *AR2NL* (for *Association Rules to Natural Language*), which converts ARs into a NL.[5]

5.1 Main Features

The main features of the system can be summarized in the following points:

- The AR2NL system is a web application. It means that an end user can use its services through any web browser with the support of (X)HTML and CSS standards.
- The whole system has been written in the PHP scripting language and it works with any database system that is accessible via the ODBC interface.
- Theoretically it is possible to convert the ARs into any language[6] "written from the left to the right". (We have successfully accomplished to convert ARs into English and Czech. A Finnish version is in preparation.)

The AR2NL system consists of two parts:

[2] However, the form we technically call "basic" need not always be the lemma generally given in dictionaries: e.g., for verbs our "basic" form is 3rd person singular, not the infinitive. This reflects a simplification of the language for our purpose and an optimization of the linguistic data structures needed.

[3] Only 3rd person finite forms are needed in our application, and the tense of a verb representing a particular literal never changes.

[4] The morphological tables contain *end segments* of the words corresponding to the specified inflection codes and morphological directives. If an "inflection pattern" is in fact a complete irregularity, the *end segments* may expand as far as to the whole word forms.

[5] A new version of the system is now being tested, which makes more use of XML database schemes and deals with some details described here in a more general way. Some of the improvements are listed in Sect. 6.

[6] Some languages can require minor extensions of the system as described here for English.

- The first part enables comfortable filling of tables that contain linguistic data necessary for the conversion. (Particular tables are described in the Subsect. 5.2.) Another function of this part is to maintain the consistency of the linguistic database.
- The second part of the AR2NL system constitutes a "public" part of the system. This part executes the conversion of ARs into a NL.

The AR2NL system further requires an access to the metabase of the LISp-Miner system, in which all data about the analyzed data, tasks definition, generated association rules, etc. are saved. (For more details see `http://lispminer.vse.cz` and [8].) The system reads ARs from the metabase and subsequently converts them into NL sentences.

5.2 Data Model

All the data necessary for the conversion of the rules are held in several tables. In the following text the particular tables are briefly described.

The **FPFI** (Formulation Patterns for Founded Implication) **table** contains formulation patterns – see Sect. 4.[7]

Every single formulation pattern constitutes a skeleton of a sentence that is gradually rewritten in the process of conversion. Here is an example of a formulation pattern in the real inner form:

> &&X patients show this relation: if the patient has &A1^N and &A2^N, then he [also] &S^V.

This formulation pattern is equivalent to the formulation pattern (1) introduced in Sect. 4:

> ***Base*** *(i.e.* ***100p%****) of the patients confirm this dependence: if the patient has* ***NP(A)****, then he* ***VP(S)****.*

The text string "Base (i.e. 100p%) of the" is replaced by the symbol "X"; the "&&" string before "X" means that this symbol is to be rewritten according to the FPA table (see below). The symbols "A1", "A2" and "S" denote, respectively, the first antecedent literal, the second antecedent literal and the succedent literal of a rule.[8] Figure 1 shows the content of the FPFI table for English.

[7] The current version of the system is only able to convert ARs of the *founded implication* type, but formulation patterns for other types of ARs are in preparation.

[8] The two antecedents in the general form of the pattern represent purely a convention. A special procedure incorporated in the system can adapt the pattern for any number of antecedents that is admissible for the type of formulation. (Which numbers are admissible is coded in the NumOfAnt column of the table.) See also footnote 1 on page 352. In the newest version of the system the formulation patterns usually contain only one antecedent, which does not affect the funcionality.

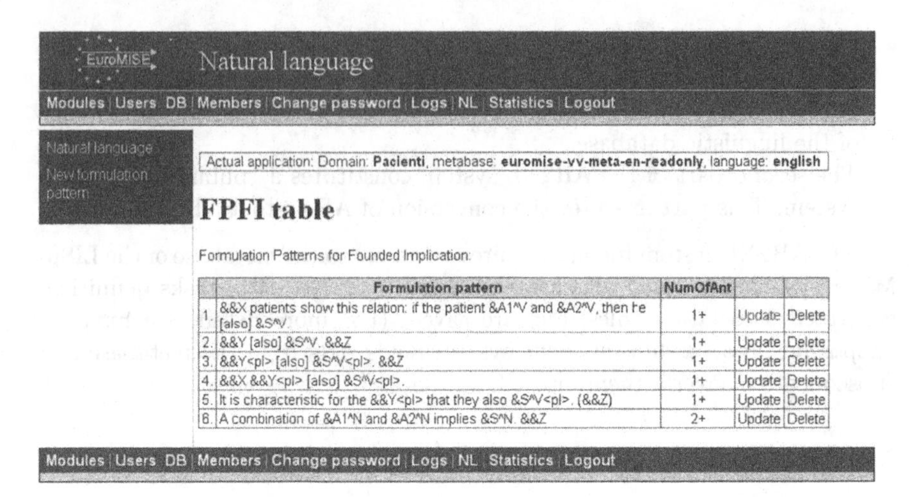

Fig. 1. Web interface

The **FPA** (Formulation Patterns – Auxiliary) **table** holds substitution for all "higher-order non-terminal symbols" that are used in formulation patterns. This approach efficiently allows for a greater variability of the resulting NL sentences.

The symbol "X" in the previous example of a formulation pattern is just such a "higher-order non-terminal symbol". The FPA table determines what to substitute for this symbol. The symbol "X" can be replaced, e.g., with the string "&B, (i.e. &P%)" – where "B" and "P" represent *input values* (they correspond to *Base* and p introduced in Sect. 2 – exactly B = *Base*, P = $100p$).

The contents of the tables FPFI and FPA is domain-independent except for the names of observed entities and some pronouns related to them (like "patient" and "his" for STULONG applications, where only male patients are concerned).[9] The linguistic data specific for a particular application of AR2NL are recorded in the tables E1 and E2. These tables contain "elementary" [10] NL expressions. The types of these expressions are introduced in Sect. 4.

The **E1 table** contains NL expressions associated to the attributes. The **E2 table** holds NL expressions associated to the pairs ⟨attribute, value⟩. E.g., we can express the literal ED(*univ*) in English, with the help of the tables E1 and E2: "has finished university" or "has university education". For other examples see 5.3, item 5.

The data necessary for the morphological changes of words in the expressions given above are saved in tables MN and MV.

[9] In the newest version of the system, these words are concentrated in a special database file used in the process of formulation pattern rewriting.

[10] The term "elementary" is used to express that we deal with an elementary item that is saved in a database table.

The **MN** (Morphology – Nouns) **table** contains patterns defining the declension of nouns, adjectives and pronouns (i.e., parts of speech that distinguish singular and plural forms universally, and variable number of cases in some languages). It is supposed that every language will have its own MN table, so the number of columns of this table may be variable according to the number of cases in the given language.

The **MV** (Morphology – Verbs) **table** contains patterns defining the necessary morphological transformations of verbs.[11]

An example of using the patterns from tables MN and MV is given in Subsect. 5.3, item 6.

The morphological tables could be prepared as universal for every particular language included in the capabilities of the system – although our present instances of the tables for English and Czech are not so universal (they basically contain only the inflection patterns necessary for the words actually used in all the other tables).

5.3 Conversion of Association Rules

The whole process of the conversion is described in this subsection.

Let us suppose that we have got an AR in this form (the rule being produced by the LISp-Miner system):

Activity In a Job (he mainly sits) ∧ Activity After a Job (he mainly sits) $\Rightarrow_{0.62,21}$ Education (university)

The process of AR conversion is done in the following steps:

1. A formulation pattern is selected from the table FPFI. The selected pattern may have, e.g., this form:
   &&X &&Y⟨pl⟩ [also] &S^V⟨pl⟩.
   (This formulation pattern corresponds to the pattern no. 4 in Fig. 1.)
2. The substitution for the symbols X and Y is performed – they are replaced according to the FPA table (the symbol "&" generally means "replace the next symbol with something other"; "&&", as already mentioned in the previous subsection, means more specifically "replace the next symbol according to the FPA table"):
   "X" → "&B (i.e. &P%) of the"
   "Y" → "_n1:patient that &A1^V and &A2^V"
   But the grammatical directive "⟨pl⟩" present at the symbol Y in the original formulation pattern must be propagated in this way:
   "Y⟨pl⟩" → "_n1:patient⟨pl⟩ that &A1^V⟨pl⟩ and &A2^V⟨pl⟩"
   Thus the generated sentence has got this form:

[11] Generally, 3rd person singular and plural forms should be defined here if the language distinguishes between them, as well as the active participle formation. Again, some languages may require more different forms.

&B (i.e. &P%) of the _n1:patient⟨pl⟩ that &A1^V⟨pl⟩ and &A2^V⟨pl⟩ [also] &S^V⟨pl⟩.[12]

3. The symbols B and P are replaced with concrete values:
   "B" → "21"
   "P" → "62"
   The generated sentence has got this form:
   21 (i.e. 62%) of the _n1:patient⟨pl⟩ that &A1^V⟨pl⟩ and &A2^V⟨pl⟩ [also] &S^V⟨pl⟩.
4. A selection of optional sequences from the formulation pattern is performed. The optional sequences are written in square brackets; they are used to accomplish greater variability of a resulting sentence. In our case the program decides whether to retain the word "also":
   "[also]" → "also"
   The generated sentence has got this form:
   21 (i.e. 62%) of the _n1:patient⟨pl⟩ that &A1^V⟨pl⟩ and &A2^V⟨pl⟩ also &S^V⟨pl⟩.
5. The symbols A1, A2 and S are replaced with the appropriate data taken from the tables E1 and E2. The symbol A1 stands for the first antecedent literal – in our case it is *Activity After a Job (he mainly sits)*; the auxiliary symbol "V" determines what kind of phrase to create ("V" stands for verb phrase), and the expression in the angle brackets ("⟨pl⟩") contains grammatical directives ("pl" stands for *plural form*). By and large, the expression "&A1^V⟨pl⟩" says: "create a verb phrase from the first antecedent literal and change the words that may require[13] it to plural forms". All the modifications made in this step are following:
   "&A1^V⟨pl⟩" → "mainly _v1t:sits⟨pl⟩ in _pron1:his⟨pl⟩ job"
   "&A2^V⟨pl⟩" → "mainly _v1t:sits⟨pl⟩ after the job"
   "&S^V⟨pl⟩" → "_v4:has⟨pl⟩ reached university education"
   The generated sentence gets this form:
   21 (i.e. 62%) of the _n1:patient⟨pl⟩ that mainly _v1t:sits⟨pl⟩ in _pron1:his⟨pl⟩ job and mainly _v1t:sits⟨pl⟩ after the job also _v4:has⟨pl⟩ reached university education.
6. In this final phase the morphological changes of specific words are performed. These words have got a form "*_PatternName:word⟨Directive⟩*", where *PatternName* denotes a row and *Directive* denotes a column in one of the morphological tables – MN or MV – according to which the change

[12] Now we have got a formulation pattern for a rule with exactly two antecedents. For the case described it is all right because a rule with two antecedents has been given as the input. If there were more than two antecedents (or only one of them) in the input, an adaptation procedure would be called, the principles of which have been described in footnote 1 (see p. 352).

[13] Such words are recognised by the inflection pattern code assigned to them in the expression being processed.

is to be made.[14] In our particular case these changes are performed:
"_n1:patient⟨pl⟩" → "patients"
"_v1t:sits⟨pl⟩" → "sit"
"_pron1:his⟨pl⟩" → "their"
"_v4:has⟨pl⟩" → "have"
The final result of the whole conversion is a sentence in English:
21 (i.e. 62%) of the patients that mainly sit in their job and mainly sit after the job also have reached university education.

An example of AR2NL output can be seen in Fig. 2.

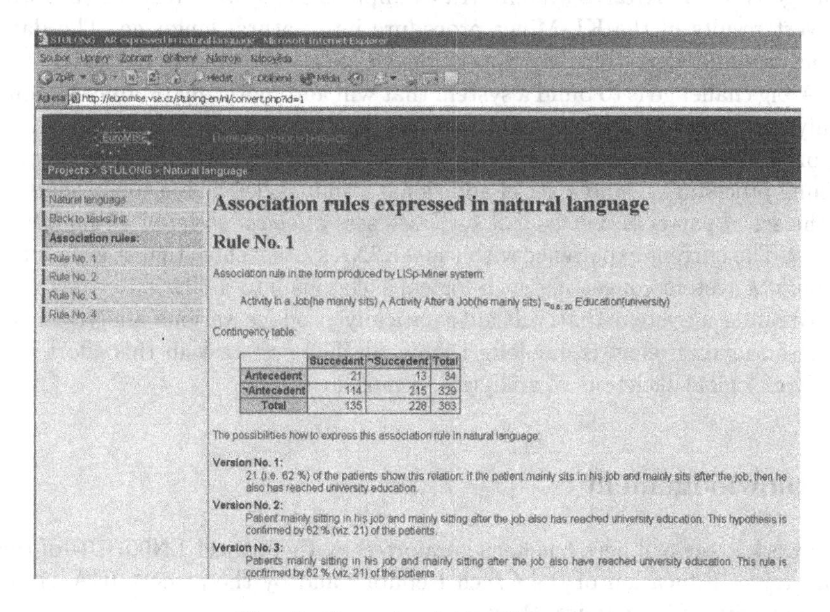

Fig. 2. Example of AR2NL output

The output contains three variants of the NL form of the rule:

Activity In a Job (he mainly sits) $\wedge$ Activity After a Job (he mainly sits) $\Rightarrow_{62\,\%,21}$ Education (university)

6 Further Work

The presented experimental system AR2NL is a first step in the supposed long-time research activity concerning the presentation of data mining results in a natural language. The next step in the further work will be improving

[14] The appropriate morphological table can be recognised by the first letter of the inflection pattern code: v → MV, other letters → MN.

the AR2NL system (namely the conversion of ARs with other types of 4ft-quantifiers – e.g. "above average", – but also other details are being improved, e.g., a variable entity name is introduced into the formulation patterns instead of the word "patient", ...). Further we also suppose to test the developed approach on Finnish, i.e., the AR2NL system will be adapted to convert ARs into Finnish.

The AR2NL system co-operates with the 4ft-Miner procedure, which is a part of the academic LISp-Miner system [6] for KDD research and teaching. There are several additional data mining procedures involved in the LISp-Miner system, e.g. *KL-Miner* [7]. We suppose to build the additional systems analogous to the AR2NL system. An example is the system *KL2NL* that will convert results of the KL-Miner procedure into natural language. The data structures and algorithms developed for AR2NL will be multiply used.

A big challenge is to build a system that will automatically produce various analytical reports on mining effort in a NL [5]. The core of such an analytical report will be both a somehow structured set of patterns – results of particular mining procedures – and a set of additional formulas describing the properties of the set of patterns. Let us call such two sets a *logical skeleton of analytical report*. The current experience with the AR2NL system show that it is possible to build a system converting such logical skeletons into a NL.

Building a system that will automatically produce various analytical reports on mining effort is our long time goal. The first task in this effort is a study of logical skeletons of analytical reports.

Acknowledgement

The work described here has been supported by the project LN00B107 of the Ministry of Education of the Czech Republic and by the project IGA 17/04 of University of Economics, Prague.

References

1. Aggraval R et al (1996) Fast Discovery of Association Rules. In: Fayyad UM et al (eds) Advances in Knowledge Discovery and Data Mining. AAAI Press, Menlo Park (CA)
2. Černý Z (2003) WWW support for applications of the LISp-Miner system. MA Thesis, (in Czech). University of Economics, Prague
3. Hájek P, Havránek T (1978) Mechanising Hypothesis Formation – Mathematical Foundations for a General Theory. Springer, Berlin Heidelberg New York
4. Hand D, Manilla H, Smyth P (2001) Principles of Data Mining. MIT
5. Rauch J (1997) Logical Calculi for Knowledge Discovery in Databases. In: Zytkow J, Komorowski J (eds) Principles of Data Mining and Knowledge Discovery. Springer, Berlin Heidelberg New York

6. Rauch J, Šimůnek M (2002) An Alternative Approach to Mining Association Rules. In: This book
7. Rauch J, Šimůnek M, Lín V (2002) KL-Miner Rauch J, Šimůnek M (2002) Alternative Approach to Mining Association Rules. In: This book
8. Šimůnek M (2003) Academic KDD Project LISp-Miner. In: Abraham A et al (eds) Advances in Soft Computing – Intelligent Systems Design and Applications. Springer, Tulsa (Oklahoma)
9. Strossa P, Rauch J (2003) Converting Association Rules into Natural Language – an Attempt. In: Kłopotek MA, Wierzchoń ST, Trojanowski K (eds) Intelligent Information Processing and Web Mining. Springer, Berlin Heidelberg New York
10. Tomečková M, Rauch J, Berka P (2002) STULONG – Data from Longitudinal Study of Atherosclerosis Risk Factors. In: Berka P (ed) ECML/PKDD-2002 Discovery Challenge Workshop Notes. Universitas Helsingiensis, Helsinki

An Algorithm to Calculate the Expected Value of an Ongoing User Session

S. Millán[2], E. Menasalvas[1], M. Hadjimichael[4], and E. Hochsztain[3]

[1] Departamento de Lenguajes y Sistemas Informaticos Facultad de Informatica, U.P.M, Madrid, Spain
ernes@fi.upm.es

[2] Universidad del Valle, Cali Colombia
millan@eisc.univalle.edu.co

[3] Facultad de Ingeniería Universidad ORT Uruguay
esthoc@adinet.com.uy

[4] Naval Research Laboratory, Monterey, CA, USA
hadjimic@nrlmry.navy.mil

Summary. The fiercely competitive web-based electronic commerce environment has made necessary the application of intelligent methods to gather and analyze information collected from consumer web sessions. Knowledge about user behavior and session goals can be discovered from the information gathered about user activities, as tracked by web clicks. Most current approaches to customer behavior analysis study the user session by examining each web page access. Knowledge of web navigator behavior is crucial for web site sponsors to evaluate the performance of their sites. Nevertheless, knowing the behavior is not always enough. Very often it is also necessary to measure sessions value according to business goals perspectives. In this paper an algorithm is given that makes it possible to calculate at each point of an ongoing navigation not only the possible paths a viewer may follow but also calculates the potential value of each possible navigation.

1 Introduction

The continuous growth of the World Wide Web together with the competitive business environment in which organizations are moving has made it necessary to know how users use web sites in order to decide the design and content of the web site. Nevertheless, knowing most frequent user paths is not enough, it is necessary to integrate web mining with the organization site goals in order to make sites more competitive. The electronic nature of customer interaction negates many of the features that enable small business to develop a close human relationship with customers. For example, when purchasing from the web, a customer will not accept an unreasonable wait for web pages to be delivered to the browser.

S. Millán et al.: *An Algorithm to Calculate the Expected Value of an Ongoing User Session*, Studies in Computational Intelligence (SCI) **6**, 363–375 (2005)
www.springerlink.com

One of the reasons of failure in web mining is that most of the web mining firms concentrate on the analysis exclusively of clickstream data. Clickstream data contains information about users, their page views, and the timing of their page views. Intelligent Web mining can harness the huge potential of clickstream data and supply business critical decisions together with personalized Web interactions [6].

Web mining is a broad term that has been used to refer to the process of information discovery from sources in the Web (Web content), discovery of the structure of the Web servers (Web structure) and mining for user browsing and access patterns through logs analysis (Web usage) [8].

In particular, research in Web usage has focused on discovering access patterns from log files. A Web access pattern is a recurring sequential pattern within Web logs. Commercial software packages for Web log analysis, such as WUSAGE [1], Analog [3], and Count Your Blessings [2] have been applied to many Web servers. Common reports are lists of the most requested URLs, a summary report, and a list of the browsers used. These tools, either commercial or research-based, in most cases offer only summary statistics and frequency counts which are associated with page visits. New models and techniques are consequently under research. At present, one of the main problems on Web usage mining has to do with the pre-processing stage of the data before application of any data mining technique. Web servers commonly record an entry in a Web log file for every access. Common components of a log file include: IP address, access time, request method, URL of the page accessed, data transmission protocol, return code and number of bytes transmitted. The server log files contain many entries that are irrelevant or redundant for the data mining tasks and need to be cleaned before processing. After cleaning, data transactions have to be identified and grouped into meaningful sessions. As the HTTP protocol is stateless, is impossible to know when a user leaves the server. Consequently, some assumptions are made in order to identify sessions. Once logs have been preprocessed and sessions have been obtained there are several kinds of access pattern mining that can be performed depending on the needs of the analyst (i.e. path analysis, discovery of association rules, sequential patterns, clustering and classification) [4, 7, 9, 12, 14].

Nevertheless, this data has to be enhanced with domain knowledge about the business, if useful patterns are to be extracted which provide organizations with knowledge about their users' activity. According to [15] without demonstrated profits, a business is unlikely to survive.

An algorithm that takes into account both the information of the server logs and the business goals to improve traditional web analysis was proposed in [10]. In this proposal the focus is on the business and its goals, and this is reflected by the computation of web link values. The authors integrated the server logs analysis with both the background that comes from the business goals and the available knowledge about the business area.

Although this approach makes it possible to know the session value it cannot be used to predict the future value of the complete session in a given page during a navigation. It can only compute the value of the traversed path up to a given page.

On the other hand, an algorithm to study the session (sequences of clicks) of the user in order to find subsessions or subsequences of clicks that are semantically related and that reflect a particular behavior of the user even within the same session was proposed in [11]. This algorithm enables the calculation of rules that, given a certain path, can predict with a certain level of confidence the future set of pages (subsession) that the user will visit. Nevertheless, rules alone are not enough because we can estimate the future pages the user will visit but not how valuable this session will be according to the site goals. In this sense it is important to see that not all the subsessions will be equally desirable.

In this paper we integrate both approaches. We present an algorithm to be run on-line each time a user visits a page. Based on the behavior rules obtained by the subsession algorithm presented in [11], our algorithm calculates the possible paths that the user can take as well as the probability of each possible path. Once these possible paths are calculated, and based on the algorithm to calculate the value of a session, the expected value of each possible navigation is computed. We will need to obtain these values to be able to predict the most valuable (in terms of the site goal) navigation the user can follow in an ongoing session. The different paths themselves are not of interest – we are concerned with discovering, given a visited sequence of pages, the value of the page sequences. This is the aim of the proposed algorithm. With these values we provide the site administrator with enhanced information to know with action to perform next in order to provide the best service to navigators and to be more competitive.

The remainder of the paper is organized as follows: in Sect. 2, the algorithms to compute subsession and to calculate the value of a session are presented. In Sect. 3, we introduce the new approach to integrate the previous algorithms. In Sect. 4, an example of the application of the algorithm as well as the advantages and disadvantages of the work are explained. Section 5, presents the main conclusion and the future works.

2 Preliminaries

In this section we first present some definitions that are needed in order to understand the proposed algorithm and next we briefly describe the algorithms in which our approach is based.

2.1 Definitions

Web-site: As in [5] we define a web site as a finite set of web pages. Let $\mathcal{W}$ be a web-site and let Ω be a finite set representing the set of pages contained in $\mathcal{W}$. We assigned a unique identifier α_i to each page so that a site containing m

pages will be represented as $\Omega = \alpha_1, \ldots, \alpha_m$. $\Omega(i)$ represents the *ith* element or page of Ω, $1 \leq i \leq m$. Two special pages denoted by α_0 and α_∞ are defined to refer to the page from which the user enters in the web site and the page that the user visits after he ends the session respectively [13].

Web-site representation: We consider a **Web-site** as a directed graph. A directed graph is defined by (N, E), where N is the set of nodes and E is the set of edges. A node corresponds to the concept of web page and an edge to the concept of hyperlink.

Link: A link is an edge with origin in page α_i and endpoint in page α_j. It is represented by the ordered pair (α_i, α_j).

Link value: The main user action is to select a link to get to the next page (or finish the session). This action takes different values depending on the nearness or distance of the target page or set of target pages. The value of the link (α_i, α_j) is represented by the real number v_{ij}, $(v_{ij} \in \Re)$ for $0 \leq i, j \leq n$):

- If $v_{ij} > 0$ we consider that the user navigating from node i to node j is getting closer to the target pages. The larger the link value the greater the links effect is in bringing the navigator to the target. (If $v_{ij} > 0, v_{il} > 0, v_{ij} > v_{il}$: then we consider that it is better to go from page α_i to page α_j than going from page α_i to α_l).
- If $v_{ij} < 0$ we consider that the navigator, as he goes from page αi to page αj, is moving away from the target pages. (If $v_{ij} < 0$, $v_{il} < 0$, $v_{ij} < v_{il}$: then it is worse to go from page α_i to page α_j than to go from page α_i to page α_k).
- If $v_{ij} = 0$ we consider that the link represents neither an advantage nor a disadvantage in the objective's search.

A **path** $\alpha_{p(0)}, \alpha_{p(1)}, \ldots, \alpha_{p(k)}$ is a nonempty sequence of visited pages that occurs in one or more sessions. We can write *Path[i]* $= \alpha_{path(i)}$. A path *Path* $= (\alpha_{path(0)}, \alpha_{path(1)}, \ldots, \alpha_{path(n-1)})$ is said to be *frequent* if *sup(Path)* $> \varepsilon$. A path *Path* $= (\alpha_{path(0)}, \alpha_{path(1)}, \ldots, \alpha_{path(n-1)})$ is said to be *behavior-frequent* if the probability to reach the page $\alpha_{path(n-1)}$ having visited $\alpha_{path(0)}$, $\alpha_{path(1)}, \ldots, \alpha_{path(n-2)}$ is higher than a established threshold. This means that $\forall i,\ 0 \leq i < n\ /\ \mathrm{P}(\alpha_{path(i)} | \alpha_{path(0)}, \ldots, \alpha_{path(i-1)}) > \delta$.

A **sequence** of pages visited by the user will be denoted by S,with $|S|$ the length of the sequence (number of pages visited). Sequences will be represented as vectors so that S[i]$\in \Omega$ (1$\leq i \leq$n) will represent the *ith* page visited. In this paper, path and sequence will be interchangeable terms.

The **Added Value of a *k*-length Sequence** *S[1],...,S[k]*: It is computed as the sum of the link values up to page $S[k]$ to which the user arrives traversing links *(S[1],S[2]), (S[2],S[3]),... (S[k-1],S[k])*. It is denoted by *AV(k)* and computed as $AV(k) = v_{S[1],S[2]} + v_{S[2],S[3]} + \cdots + v_{S[k-1],S[k]}$ $2 \leq k \leq n$.

Sequence value: It is the sum of the traversed link values during the sequence (visiting n pages). Thus it is the added value of the links visited by a user until he reaches the last page in the sequence $S[n]$. It is denoted by *AV(n)*.

The **Average Added Value of a k-length Sequence** *S[1],S[2],. . ., S[k]*: It represents the added accumulated value for each traversed link up to page k, for $2 \leq k \leq n$. It is denoted by AAV(k) and is computed as the accumulated value up to page k divided by the number of traversed links up to page k ($k-1$). AAV(k) = $AV\ (k)/(k-1)$.

Session. We define session as the complete sequence of pages from the first site page viewed by the user until the last.

2.2 Session Value Computation Algorithm

In [10], an algorithm to compute the value of a session according both to user navigation and web site goals was presented. The algorithm makes it possible to calculate how close a navigator of a site is from the targets of the organization. This can be translated to how the behavior of a navigator in a web site corresponds to the business goals. The distance from the goals is measured using the value of the traversed paths.

The input of the algorithm is a values matrix $V[m, m]$ that represents the value of each link that a user can traverse in the web site. These values are defined according to the web site organization business processes and goals. The organization business processes give a conceptual frame to compute links values, and makes it possible to know a user is approaching or moving away from the pages considered as goals of the site. Thus, these values included in value matrix V must be assigned by business managers. Different matrices can be defined for each user profile, and make it possible to adapt the business goals according to the user behavior.

The original algorithm outputs are the added accumulated value and average accumulated value evolution during the session. We will only consider the added accumulated value.

2.3 Pseudocode of the Sequence Value Algorithm

Input: Value links matrix *V[m,m]*

Initialization:

AV = 0 //Added Value=0

AAV = 0 //Average added value=0

k = 1 //number of nodes=1

read *S[k]* //read the first traversed page in the sequence

Output: Sequence Accumulated Value

Pseudocode

While new pages are traversed

$k = k + 1$ //compute the traversed page sequential number

read $S[k]$// read the next traversed page

/* the selected link is $(S[k-1], S[k])$ $1 \leq S[k-1] \leq m-1$ $1 \leq S[k] \leq m$ $2 \leq k \leq n$ */

AV = AV + *V(S[k–1],S[k])*

// Add link traversed value to accumulated value

2.4 Subsession Calculation

An approach to studying the session (sequences of clicks) of the user is proposed in [11]. The purpose is to find subsessions or subsequences of clicks that are semantically related and that reflect a particular behavior of the user even within the same session. This enables examination of the session data using different levels of granularity. In this work the authors propose to compute frequent paths that will be used to calculate subsessions within a session.

The algorithm is based on a structure that has been called FBP-tree (Frequent Behavior Paths tree). The FBP-Tree represents paths through the web site. After building this tree, frequent behavior rules are obtained that will be used to analyze subsessions within a user session. The discovery of these subsessions will make it possible to analyse, with different granularity levels, the behavior of the user based on the pages visited and the subsessions or subpaths traversed. Thus, upon arriving at an identifiable subsession, it can be stated with a degree of certainty the path the user will take to arrive at the current page.

In order to find frequent paths that reveal a change in the behavior of the user within an ongoing session are calculated. The first step in obtaining frequent paths is discovering *frequent behavior paths* – paths which indicate frequent user behavior.

Given two paths, $Path_{IND}$ and $Path_{DEP}$, a *frequent-behavior rule* is a rule of the form:

$$Path_{IND} \rightarrow Path_{DEP}$$

where $Path_{IND}$ is a frequent path, called the independent part of the rule and $Path_{DEP}$ is a behavior-frequent path, called the dependent part. Frequent-behavior rules must have the following property:

$$P(Path_{DEP} | Path_{IND}) > \delta$$

The rule indicates that if a user traverses path $Path_{IND}$, then with a certain probability the user will continue visiting the set of pages in $Path_{DEP}$. The confidence of a frequent-behavior rule is denoted as *conf* $(Path_{IND} \rightarrow Path_{DEP})$ and it is defined as the probability to reach $Path_{DEP}$ once $Path_{IND}$ has been visited.

Pseudocode of the FBP-tree Algorithm

Input: FTM : frequent transition matrix *(N x N)*
L : list of sessions
Output: FBP-Tree : frequent behavior path tree (each FBP-tree node or leaf has a hit counter).
Pseudocode:
For each *s* in *L*
{

```
for i in N
{
j = i + 1
while j<N and FTM[s(j-1),s(j)]≠0
{
sub-s={α_s(i),...,gα_s(j)}
                    if(∀k/i≤k<j: FTM[s(k),s(k+1)]≠0)
{
if (∃ sub-s in FBP-Tree)
FBP-Tree.increment_hit(sub-s)
else
FBP-Tree.insert_path(sub-s)
                                }}}}
```

Once the tree has been calculated and frequent paths are obtained, it is scanned from the leaves to the root node. Then, taking into account the support of each path, the rules are calculated as follows:

Pseudocode of the frequent behavior rule algorithm

Input: FBP-Tree : frequent behavior path tree
σ : minimum support of the rule
κ : minimum confidence of the rule
Output: FBP-Rules : frequent behavior path rules.
Pseudocode:

```
For each l in FBP-Tree.leaves
{
while l ≠ FBP-Tree.root
{
if (l.hit< σ)
FBP-Tree.prune(l)
else
{
q=queue{}
while(l.hit/l.parent.hit> κ)
{
q.append(l.page)
l=l.parent()
}
if(¬q.empty())
{
r=rule{}
r.P_IND({FBP-Tree.root.page(),..,l.page})
```

```
r.P_DEP(q)
FBP-Rules.add(r)
}
FBP-Tree.prune(l)}}
```

3 Expected Path Value

In this section we propose an algorithm in order to calculate the path expected value. The algorithm makes it possible to calculate the average sequence value of pages likely to be visited by a user in an ongoing session. In order to do so, the algorithm uses the frequent-behavior rules obtained by the algorithm described in Sect. 2.4 to later obtain the value of each possible path to follow. These values are computed by using the algorithm presented in Sect. 2.4. In some pages a user visits it is possible to predict the possible paths he can follow. These paths are computed based on the rules obtained by the frequent-behavior-rules algorithm (see Sect. 2.4). Some preliminary concepts are needed in order to understand the proposed approach.

3.1 Preliminary Concepts

- **Frequent Rules Set (FRS)**
  $FRS = \{r_i | r_i : Path_{IND} \rightarrow Path_{DEP}$ and $P(Path_{DEP} | Path_{IND}) > \delta\}$
  where
  $r_i = (Path_{IND}, Path_{DEP}, P(Path_{DEP} | Path_{IND}))$
- **Equivalent Sequences (*Q* ≈ *P*)** Let P, Q be two paths, $Q = (\alpha_{q(0)}, \alpha_{q(1)}, \ldots, \alpha_{q(m)})$ and $P = (\alpha_{p(0)}, \alpha_{p(1)}, \ldots, \alpha_{p(m)})$. $Q \approx P$ if and only if $\forall\, j\ (0 \leq j < m)\ Q[j] = P[i]$.
- **Decision Page** (α_D). Let $\alpha_D \in \Omega$ be a page and $Path = (\alpha_{p(0)}, \alpha_{p(1)}, \ldots, \alpha_{p(k)})$ be a path, α_D is a decision page in *Path* if:
  $\alpha_D = \alpha_{p(k)}$ and $\exists r_i \in FRS$ such that $Path \approx r_i{\cdot}Path_{IND}$
  In other case α_D is called a non-decision page.
- **Predicted Paths** Predicted paths are all possible dependent subsequences with their associated probabilities:
  $(Path_{DEP}, P(Path_{DEP} | Path_{IND}))$
  When the user has arrived at a decision page there may be several different inferred paths which follow, or none. The paths that can follow have different consequences for the web site's organization. These consequences can be measured using each path's value. A path value shows the degree to which the user's behavior is bringing him or her closer to the site's goals (target pages).
- **Predicted Subsession Value**
  It is the value of each of a possible dependent path. It is computed using the function Sequence_Value that returns the value of the sequence using the sequence value algorithm (see Sect. 2.2). Subsequence value measures

the global accordance between the subsequence and the site goals according to the given links value matrix.

- **Possible_Dependent_Sequences (IND, FRS) function.**
  Based on Frequent Behavior rule Algorithm in our approach we have defined the set Frequent Rule Set (FRS) that represents the set of behavioral rules obtained by the algorithm in a certain moment. We also define a function Possible_Dependent_Subsequences (IND, FRS) that given a sequence of pages IND and a set of behavioral rules will give all the possible sequences that can be visited after visiting IND and the associated probability. This function will simply scan the set of FRS and will look for the rules that match IND in order to obtain the mentioned result.
- **Sequence_Value (seq) function** Based on the algorithm seen in Sect. 2.2 a function is defined that given a sequence computes its value.
- **Subsession Expected Value E(V)** Let V be a random variable representing the value of each possible sequence.

Let P(Path $\text{DEP}_i|\text{Path}_{IND}$) be the conditional probability that a user will begin visiting DEP_i after having visited the sequence of pages represented by IND. This probability satisfies the condition: $P(Path_{DEP_i}|Path_{IND}) = 1$ The conditional subsession expected value given a sequence of pages already visited IND (E(V|IND)) is defined as follows: $IND(E(V|IND)) = (V_i * (P(Path_{DEP_i}|Path_{IND}))$ where Path_{DEPi} represents the path that has the value V_i.

3.2 Expected Path Value Algorithm

For each page in a sequence the algorithm checks whether it is a decision page, with respect to the current path traversed.

If it is a non-decision page, it means that we have no knowledge related to the future behavior (no rule can be activated from the FRS) and no action is taken. If it is a decision page then at least one behavioral rule applies so that we can calculate the possible values for each possible path the user can follow from this point. These possible values can be considered as a random variable. Consequently we can calculate the expected value of this variable. This gives us an objective measure of the behavior of the user up to a certain moment. This way if the expected value is positive then it will mean that no matter which possible path the user would take, in average the resulting navigation will be profitable for the web site. Nevertheless, if the expected value happens to be negative then it would mean that in the long run the navigation if the user will end in a undesirable effect for the web site. The algorithm provides the site with added value information which can be used, for example, to dynamically modify the web site page content to meet the projected user needs.

Pseudocode of the expected path value algorithm

```
Input:
Seq //Sequence of pages visited by a user
Output: value of each possible path and expected added value of all
possibilities
Pseudocode
V = 0; // value of the sequence
FRS = Frequent_Behavior_Rules();
If the last page in Seq is a decision page
pathDEP = Possible_Dependent_Subsequences (Seq,FRS);
V = 0;
For each path Pa in PathDEP compute:
Vi = Sequence_Value(Pa);
V = V +Vi*P(Pa|Seq);
Else;
```

Note that if we have additional information related to the effect that certain web site actions would have on the probabilities of the activated rules, we could calculate both the expected value before and after a certain action.

4 Example

Let us suppose that during a user session in the web site (see Table 1), there are no rules to activate until the sixth page encountered in the navigation. This means that there are no rules that satisfy the property $P(P_{DEP}|P_{IND}) > \delta$. Let's assume that $\delta = 0.24$.

When the user arrives to the 6th traversed page three frequent-behavior rules are activated so that 3 sequences (α_7), $(\alpha_9\alpha_8)$, $(\alpha_{10}\alpha_{11})$ are frequently followed from that point (see Table 2)

Table 1. Sequence visited by a user

Sequence	Decision Page?
α_5	No
$\alpha_5\alpha_2$	No
$\alpha_5\alpha_2\alpha_1$	No
$\alpha_5\alpha_2\alpha_1\alpha_6$	No
$\alpha_5\alpha_2\alpha_1\alpha_6\alpha_3$	No
$\alpha_5\alpha_2\alpha_1\alpha_6\alpha_3\alpha_4$	Yes

The probabilities and values associated with each activated rule are presented in Table 2.

Table 2. Activated FRS for $\alpha_3\alpha_4$, conditional probabilities and value of the consequent paths

Activated Rule	Prob(Dep|Ind)	Value
$\alpha_3\alpha_4 \rightarrow \alpha_7$	0,25	–15
$\alpha_3\alpha_4 \rightarrow \alpha_9\alpha_8$	0,3	16
$\alpha_3\alpha_4 \rightarrow \alpha_{10}\alpha_{11}$	0,35	-3
Dummy rules	0,1	0
Total	1	

Once this information is available, the algorithm computes the subsession expected value. In our example, this subsession expected value is 0. This means that up to this moment, on average, further navigation by the user along any frequent path will neither bring the user closer to, nor farther from, the target pages. E(V|IND) = 0.25*(–15)+0.3*16+0.35*(–1.05)+0.1*0 = 0 (see Table 3).

Table 3. Example 1: Subsession Expected value calculation

Sequence	Prob(Dep_i|Ind)	Value(V_i)	(Prob(Dep_i|Ind) $* V_i$
Dep1α_7	0,25	–15	–3,75
Dep2$\alpha_9\alpha_8$	0,3	16	4,80
Dep1$\alpha_{10}\alpha_{11}$	0,35	-3	–1,05
Dummy rules	0,1	0	0
Total	1		

Another example is presented in Table 4. In this example, before knowing which PATH_{DEP} the user will follow, the subsession expected value can be calculated. In this example, the expected subsession value at this point in the navigation is 6.75. Thus, as the expected value is positive, the average final result at this point is profitable for the site. So, in this case we can estimate that the user will act according to the web site goals.

Table 4. Example 2: Expected subsession value calculation

Sequence	Prob(Dep_i|Ind)	Value(V_i)	(Prob(Dep_i|Ind) $* V_i$
Dep_i	0,75	10	7,50
Dep_j	0,25	-3	–0,75
Total	1		6,75

Figure 1 illustrates the advantages of the algorithm. In the example that is represented, one can see (given that the y-axis illustrates the value of the ongoing session), that up to the decision page the value of the session is positive. At this point, there are three possibilities (according to the FRS).

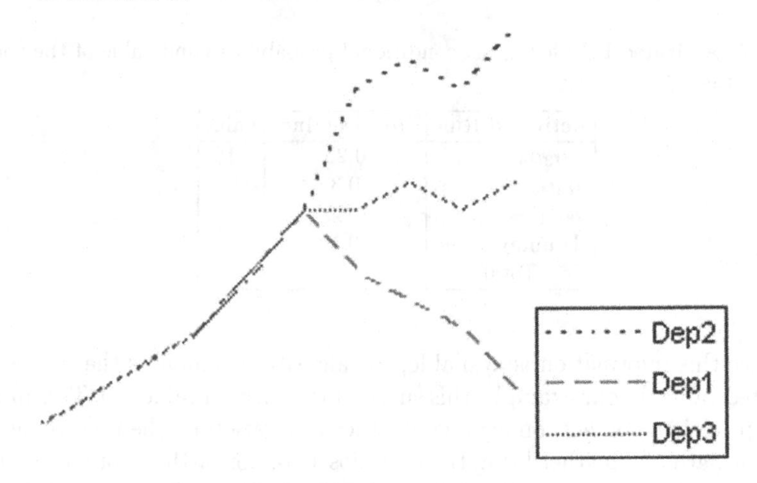

Fig. 1. Expected value of sessions

If the user follows paths Dep2 or Dep3 the result would be positive, while following path Dep1 leads to decreased session value. Nevertheless, we cannot say for sure which will be followed, but we have an algorithm that tells us the expected value (in this case positive) of the future behavior. This means that in some cases (if Dep1 is followed) the behavior will not be positive for the web site but on average we can say that the result will be positive, and we can act taking advantage of this knowledge.

Figure 1: The change in session value (y-axis) with path progression (x-axis), depending on the various dependent paths (Dep1, Dep2, Dep3) followed.

Note that additional information about past actions could yield more information indicating which paths might be followed.

5 Conclusions and Future Work

An integrated approach to calculate the expected value of a user session has been presented. The algorithm makes it possible to know at any point if the ongoing navigation is likely to lead the user to the desired target pages. This knowledge can be used to dynamically modify the site according to the user's actions. The main contribution of the paper is that we can quantify the value of a user session while he is navigating. In a sense this makes the relationship of the user with the site closer to real life relationships.

It is important to note that the algorithm can be applied recursively to all the possible branches in a subsession in order to refine the calculation of the expected value.

We are currently working on an extension of the algorithm in which the impact of actions performed by the site in the past are evaluated in order to include this knowledge in the algorithm.

Acknowledgments

The research has been partially supported by Universidad Politécnica de Madrid under Project WEB-RT and Programa de Desarrollo Tecnológico (Uruguay).

References

1. http://www.boutell.com/wusage.
2. http://www.internetworld.com/print/monthly/1997/06/iwlabs.htm l.
3. http://www.statlab.cam.ac.uk/ sret1/analalog.
4. E. Han B. Mobasher, N. Jain and J. Srivastava. Web mining: Pattern discovery from www transaction. In *Int. Conference on Tools with Artificial Intelligence*, pp. 558–567, 1997.
5. J. Adibi C. Shahabi, A.M. Zarkesh and V. Shah. Knowledge discovery from user's web-page navigation. In *Proceedings of the Seventh International Workshop on Research Issues in Data Engineering High Performance Database Management for Large-Scale Applications (RIDE'97)*, pp. 20–31, 1997.
6. Oren Etzioni. The world-wide web: Quagmire or gold mine? *Communications of the ACM*, 39(11):65–77, November 1996.
7. Daniela Florescu, Alon Y. Levy, and Alberto O. Mendelzon. Database techniques for the world-wide web: A survey. *SIGMOD Record*, 27(3):59–74, 1998.
8. Jiawei Han and Micheline Kamber. *Data Mining:Concepts and Techniques.* Morgan Kaufmann publishers, 2001.
9. Mukund Deshpande Jaideep Srivastava, Robert Cooley and Pa ng Ning Tan. Web usage mining: Discovery and applications of usage patter ns from web data. *SIGKDD Explorations.*, 1:12–23, 2000.
10. Hoszchtain E. Menasalvas E. Sessions value as measure of web site goal achievement. In *SNPD'2002*, 2002.
11. Pea J. M Hadjimichael M. Marbn Menasalvas E., Milln S. Subsessions: a granular approach to click path analysis. In *WCCI'2002*, 2002.
12. Carsten Pohle Myra Spiliopoulou and Lukas Faulstich. Improving the effectiveness of a web site with web usage mining. In *Web Usage Analysis nad User Profiling, Masand and Spiliopoulou (Eds.), Spriger Verlag, Berlin*, pp. 142–162, 1999.
13. R. Krisnapuram O. Nasraoiu and A. Joshi. Mining web access logs using a fuzzy relational clustering algorithm based on a robust estimator. In *8th International World Wide Web Conference, Toronto Cana da*, pp. 40–41, May 1999.
14. M. Perkowitz and O. Etzioni. Adaptive web sites: Automatically synthesizing web pages. In *Fifteenth National Conference on Artificial Intelligence (AAAI/IAAI'98)Madison, Wisconsin*, pp. 727–732, July 1998.
15. From: Gregory Piatetsky-Shapiro. Subject: Interview with jesus mena, ceo of webminer, author of data mining your website. page http://www.kdnuggets.com/news/2001/n13/13i.html, 2001.

Signals and Communication Technology

(continued from page ii)

Chaos-Based Digital Communication Systems
Operating Principles, Analysis Methods, and Performance Evalutation
F.C.M. Lau and C.K. Tse ISBN 3-540-00602-8

Adaptive Signal Processing
Application to Real-World Problems
J. Benesty and Y. Huang (Eds.)
ISBN 3-540-00051-8

Multimedia Information Retrieval and Management
Technological Fundamentals and Applications
D. Feng, W.C. Siu, and H.J. Zhang (Eds.)
ISBN 3-540-00244-8

Structured Cable Systems
A.B. Semenov, S.K. Strizhakov, and I.R. Suncheley ISBN 3-540-43000-8

UMTS
The Physical Layer of the Universal Mobile Telecommunications System
A. Springer and R. Weigel
ISBN 3-540-42162-9

Advanced Theory of Signal Detection
Weak Signal Detection in Generalized Obeservations
I. Song, J. Bae, and S.Y. Kim
ISBN 3-540-43064-4

Wireless Internet Access over GSM and UMTS
M. Taferner and E. Bonek
ISBN 3-540-42551-9